Nonlinear Systems and Complexity

Series Editor
Albert C. J. Luo
Southern Illinois University Edwardsville
Edwardsville, IL, USA

More information about this series at http://www.springer.com/series/11433

Albert C.J. Luo • Hüseyin Merdan
Editors

Mathematical Modeling and Applications in Nonlinear Dynamics

Editors
Albert C.J. Luo
Department of Mechanical and Industrial Engineering
Southern Illinois University Edwardsville
Edwardsville, IL, USA

Hüseyin Merdan
Department of Mathematics
TOBB University of Economics and Technology
Ankara, TURKEY

ISSN 2195-9994 ISSN 2196-0003 (electronic)
Nonlinear Systems and Complexity
ISBN 978-3-319-26628-2 ISBN 978-3-319-26630-5 (eBook)
DOI 10.1007/978-3-319-26630-5

Library of Congress Control Number: 2015960740

Springer Cham Heidelberg New York Dordrecht London

Printed on acid-free paper

Springer International Publishing AG Switzerland is part of Springer Science+Business Media (www.springer.com)

Preface

This edited book collects seven chapters on mathematical modeling and applications in nonlinear dynamics for a deeper understanding of complex phenomena in nonlinear systems. The chapters of this edited book are selected from the 3rd International Conference on Complex Dynamical Systems: New Mathematical Concepts and Applications in Life Sciences (CDSC 2014), held at Ankara, Turkey, 24–26 November 2014. The aim of this conference was to promote research on differential equations and discrete and hybrid equations, especially in life sciences and chemistry. This conference was for the 60th birthday celebration of Professor Marat Akhmet, who is a faculty member of the Mathematics Department at Middle East Technical University, Turkey. After peer review, 54 papers were accepted for presentation from 17 countries. The chapters of this edited book are based on the invited lectures with extended results in nonlinear dynamical systems, and the edited book is dedicated to Prof. Akhmet's 60th birthday. The edited chapters include the following topics:

- Integrate-and-fire biological models with continuous/discontinuous couplings
- Analytical periodic solutions in nonlinear dynamical systems
- Dynamics of hematopoietic stem cells
- Dynamics of periodic evolution processes in pharmacotherapy
- Ultimate solution boundedness for differential equations with several delays
- Delay effects on the dynamics of the Lengyel–Epstein reaction-diffusion model
- Semilinear impulsive differential equation in an abstract Banach space

During this conference, comprehensive discussions on the above topics were made, led by invited recognized scientists. From such discussions, young scientists and students learned new methods, ideas, and results.

The editors would like to thank TÜBİTAK (The Scientific and Technological Research Council of Turkey), TOBB University of Economics and Technology, Ankara, Turkey, and the Institute of Informatics and Control Problems, Almaty,

Kazakhstan, for all financial support, and the authors and reviewers for supporting the conference and collection. We hope the results presented in this edited book will be useful for other specialists in complex dynamical systems.

Ankara, TURKEY
Edwardsville, IL, USA

Hüseyin Merdan
Albert C.J. Luo

Contents

Chapter 1
The Solution of the Second Peskin Conjecture and Developments

M.U. Akhmet

Abstract The integrate-and-fire cardiac pacemaker model of pulse-coupled oscillators was introduced by C. Peskin. Because of the pacemaker's function, two famous synchronization conjectures for identical and nonidentical oscillators were formulated. The first of Peskin's conjectures was solved in the paper (J. Phys. A 21:L699–L705, 1988) by S. Strogatz and R. Mirollo. The second conjecture was solved in the paper by Akhmet (Nonlinear Stud. 18:313–327, 2011). There are still many issues related to the nature and types of couplings. The couplings may be impulsive, continuous, delayed, or advanced, and oscillators may be locally or globally connected. Consequently, it is reasonable to consider various ways of synchronization if one wants the biological and mathematical analyses to interact productively. We investigate the integrate-and-fire model in both cases—one with identical and another with not-quite-identical oscillators. A combination of continuous and pulse couplings that sustain the firing in unison is carefully constructed. Moreover, we obtain conditions on the parameters of continuous couplings that make possible a rigorous mathematical investigation of the problem. The technique developed for differential equations with discontinuities at nonfixed moments (Akhmet, *Principles of Discontinuous Dynamical Systems*, Springer, New York, 2010) and a special continuous map form the basis of the analysis. We consider Peskin's model of the cardiac pacemaker with delayed pulse couplings as well as with continuous couplings. Sufficient conditions for the synchronization of identical and nonidentical oscillators are obtained. The bifurcation of periodic motion is observed. The results are demonstrated with numerical simulations.

1.1 Introduction and Preliminaries

In the paper [50], C. Peskin developed the integrate-and-fire model of the cardiac pacemaker [32] to a population of identical pulse-coupled oscillators. Thus, a cardiac pacemaker model was proposed where the signal to fire arises not from an

M.U. Akhmet (✉)
Department of Mathematics, Middle East Technical University, 06800 Ankara, Turkey
e-mail: marat@metu.edu.tr

A.C.J. Luo, H. Merdan (eds.), *Mathematical Modeling and Applications in Nonlinear Dynamics*, Nonlinear Systems and Complexity 14,
DOI 10.1007/978-3-319-26630-5_1

outside stimuli, but in the population of cells itself. Well-known conjectures of self-synchronization were formulated and solutions of these conjectures for *identical oscillators* [45, 50] stimulated mathematicians as well as biologists for the intensive investigations in the field [7, 16, 19, 25, 33, 36, 44, 47, 52, 58, 60, 62].

A specialized bundle of about 10,000 neurons located in the upper part of the right atrium of the heart is known as the sinoatrial node. It fires at regular intervals to cause the heart to beat, with a rhythm of about 60 to 70 beats per minute for a healthy, resting heart. The electrical impulse from the pacemaker triggers a sequence of electrical events in the heart to control the orderly sequence of muscle contractions that pump the blood out of the heart. That is why it is called the *cardiac pacemaker* in the literature. The cells of the sinoatrial node are able to depolarize spontaneously toward the threshold firing and then recover [9]. The electrical activity of the cardiac pacemaker produces a strong pattern of voltage change. While the nerve cells require a stimulus to fire, cells of the cardiac pacemaker can be considered to be "self-firing." They repetitively go through a depolarizing discharge and then recover to fire again. This action is analogous to a relaxation oscillator in electronics. The circuit involves a capacitor, which is charged by the energy of a battery (the membranes of the sinoatrial node and the ion transport processes play the role), and a resistor, which controls the flashing rate of the light. In the case of the sinoatrial node, there is input from the physiology of the body related to oxygen demand and other factors that control the rate of firing of the sinoatrial node and hence the heart rate. The question naturally arises of how the neurons organize their *firing in unison*. The simplest explanation is that the fastest neuron drives all the others, bringing them to the threshold. If that were the case, then the injury of a single cell could significantly change the frequency of the heartbeat. To avoid this important shortcoming, in the paper [50], Peskin proposed a cardiac pacemaker model where signals to fire do not arise from an outside stimuli but instead originate in the population of cells itself. Moreover, the paper proposed that a cardiac pacemaker is a population of neurons with weak couplings such that synchrony emerges as a result of the interaction of all cells, rather than a single cell dominating.

In the papers [3–6], we introduced a new method for the investigation of biological oscillators. The method seems to be universal to analyze integrate-and-fire oscillators. In particular, we solved the second Peskin conjecture in [3, 5]. It was proved that an ensemble of an arbitrary number of oscillators synchronizes even if they are *not quite identical*.

In this chapter we extend the approach to the model with delayed pulse coupling. Conditions that guarantee the synchronization of the model are found. Our system is different than that in [16] since we suppose that the pulse coupling is instantaneous if oscillators are close to each other and are near threshold. In upcoming papers, we plan to consider other models, varying types of the delay involvement, as well as inhibitory models such that analogs of results in [16] and [62] can be obtained. Moreover, we plan to develop for these systems the theory of the bifurcation of periodic solutions. Some open problems are discussed in Sect. 1.5. The method of the analysis of nonidentical oscillators is based on results of the theory of differential equations with discontinuities at nonfixed moments [2].

The cells that create rhythmical impulses for contraction of the cardiac muscle, and control the heart rate, are called pacemaker cells. Peskin developed a model of an encoding neuron [32] for a population of identical pulse-coupled oscillators [50]. The synchronization of the system, viewed as firing in unison, was proved for two [50] and more than two [45] identical oscillators. In fact, Peskin proposed a model, which is a hybrid of continuous and discrete equations, that admits synchrony. The suggestion was so attractive that it has been used not only for cardiac models, but also, for example, for coupled neurons [8]. The paper [45] has been the most stimulating and intensive analysis of the problem [8, 9, 20, 21, 53, 62–64].

The mathematical problems connected to synchrony emerge in numerous applications—not only in a model of a heartbeat [32, 50], but also in models of firefly flashing [11, 24], insulin-secreting cells of the pancreas [53], neural networks [20–28, 37, 51], and so on. There is still much uncertainty with respect to the types of coupling in a population (these may be impulsive, continuous, delayed, advanced, regular, or random) [10, 11, 13, 16, 19, 24, 30, 35, 42, 45, 51, 63], and with respect to the structural complexity of networks, the connection may be local or global, with various quantitative characteristics and geometrical configurations [13, 57, 59]. It is clear that the larger the diversity of mathematical models, the more opportunities to tackle the biological issues.

It is natural that the problem has been considered in the more general form. In [45] the method of phase diagrams is effectively used to discuss the models. In the paper [3] we suggested a special map that helped us to solve the synchronization problem for nonidentical oscillators. A version of the model is considered such that perturbations can be evaluated to save the synchronization. Other problems of the theory are considered, in particular the relationship between synchronization and spatial structure. Nevertheless, an analysis of models with a general form of dynamics remains unconsidered. In this chapter we extend our proposals of [3] to a case suitable for various applications. They can easily be developed further such that the results have an important meaning for the theory of integrate-and-fire models of biological oscillators in both exhibitory and inhibitory cases as well as for different types of couplings: continuous, delayed, and so forth. Moreover, we suppose that the approach can be utilized for various types of motions of the systems (e.g., periodic, almost periodic, chaotic) since results of the discrete equations are now available for applications. To prove assertions of this chapter concerning multidimensional systems of nonidentical oscillators, we need an advanced understanding of the theory of dynamical systems with discontinuities at variable moments of time [2]. This is one more reason why we decided to write this chapter separately from [3], where only Peskin's model was considered with the simple theoretical methods. Oscillators considered in this chapter are connected with each other not only at the firing moments, but permanently. That is, the differential equations are not separated as they are, for example, in the papers [45, 50]. One can admit that this fact provides more biological sense to investigations. The chapter consists of the main results, simulations, and a discussion of the possible generalization.

The main object of our investigation in the next section is an integrate-and-fire model that consists of n nonidentical pulse-coupled oscillators, $x_i, i = 1, 2, \ldots, n$.

Set $x = (x_1, x_2, \ldots, x_n)$. If the system does not fire, the oscillators satisfy the following equations:

$$x_i' = f(x_i) + \phi_i(x), \tag{1.1}$$

where $0 \leq x_i \leq 1 + \zeta_i(x), i = 1, 2, \ldots, n$. When the oscillator $x_j(t), j = 1, \ldots, n$, increases its value from zero and meets the surface $x_j = 1 + \zeta_i(x)$ the first time, such that $x_j(t) = 1 + \zeta_i(x)$, then the oscillator fires, $x_j(t+) = 0$. Firing changes the values of all oscillators with $i \neq j$ such that at the same moment t,

$$x_i(t+) = \begin{cases} 0, & \text{if } x_i(t) + \epsilon + \epsilon_i \geq 1 + \zeta_i(x), \\ x_i(t) + \epsilon + \epsilon_i, & \text{otherwise.} \end{cases} \tag{1.2}$$

It is assumed also that there exist positive constants μ_i and ξ_i such that $|\phi_i(x)| < \mu_i$ and $|\zeta_i(x)| < \xi_i$, for all x and $i = 1, 2, \ldots, n$. In what follows, we call real numbers $\epsilon, \mu_i, \xi_i, \epsilon_i$ *parameters*, assuming the first one is positive. Moreover, the constants ξ_i, ϵ_i, μ_i are called *parameters of perturbation*. If all of them are zeros, one obtains the model of identical oscillators. We assume that $\epsilon + \epsilon_i - \xi_i > 0$ and $\epsilon + \epsilon_i + \xi_i < 1$, for all i, and that the function f is positive-valued and lipschitzian. Moreover, assume that all functions involved in the discussion are continuous, that the system (1.1) satisfies conditions of a theorem of existence and uniqueness, and that each solution of the system is continuable to the threshold's value.

We have chosen the all-to-all coupling such that each firing elicits jumps in all nonfiring oscillators. If several oscillators fire simultaneously, then other oscillators react as if just one oscillator has fired. In other words, any firing acts only as a signal that abruptly provokes a state change; the intensity of the signal is not important, and pulse strengths are not additive. The opposite case also is discussed in this chapter.

Two oscillators are synchronized if they fire in unison. A system of oscillators is synchronized if all of them fire in unison.

Next, in Sect. 1.3 we extend the method and these results to the model with continuous couplings. Sufficient conditions for synchronization are found. The research utilizes results and proposals from [10, 12–63]. We investigate the integrate-and-fire model for both cases—with identical and not-quite-identical oscillators. A combination of continuous and pulse couplings that sustain the firing in unison is carefully constructed. Moreover, we find conditions on the parameters of continuous couplings that make possible a rigorous mathematical investigation of the problem.

Let us first consider n identical oscillators, which are characterized by voltage state variables $x_1, x_2, \ldots, x_n$ with values in $[0, 1]$. The following assumptions describe the model and its coupling style.

$(A1)$. If $x_j(t) = 1$, then the oscillator fires, and there exists a positive number ϵ such that

$$x_i(t+) = 0, \text{ if } x_i(t) \geq 1 - \epsilon \qquad (1.3)$$

for all $i \neq j$.

Fix a positive τ. If $t = s$ is a firing moment of x_j, then the interval $[s, s + \tau]$ is said to be the e^j-interval or e-interval for all $x_i, i \neq j$. We say that an oscillator $x_i(t)$ *is continuously excited* if t is in an e-interval, and $x_i(t) < 1$.

$(A2)$. When $x_i(t)$ is not continuously excited, then

$$x_i' = S - \gamma x_i. \qquad (1.4)$$

Otherwise, there exists a positive real number η such that

$$x_i' = (S + \eta) - \gamma x_i. \qquad (1.5)$$

$(A3)$. Positive constants S, γ, η, and ϵ satisfy the following inequalities:

(i) $\gamma < S$;
(ii) $\eta \leq \epsilon$;
(iii) $e^{\gamma\tau} - 1 < \min\{1, \frac{\epsilon}{S-\gamma+\eta}\}$.

We call the collection of n oscillators $x_1, x_2, \ldots, x_n$, *the integrate-and-fire model of continuously coupled identical biological oscillators* if conditions $(A1)$–$(A3)$ hold.

One should emphasize that the coupling is all-to-all, and exciting strengths are not additive. The model of the present chapter admits two types of coupling: the *continuous* one, which is described by $(A2)$; and the *impulsive* coupling given by $(A1)$. In the first case the motion of oscillators remains continuous if they are not near the threshold. Nevertheless, the rate of oscillators jumps to response. Otherwise, by assumption, $(A1)$ oscillators are coupled impulsively.

This assumption is natural since firing provokes other oscillators instantaneously, if they are near thresholds, and are therefore in a state ready to fire. From the proofs of this chapter it will be seen that the constant η in $(A2)$ can be replaced with a function defined on the real axis, continuous and nonzero on e-intervals. That is why it is reasonable to say that oscillators are *continuously* coupled.

To illustrate the last remark, let us provide the following simulation. Consider three oscillators: x_1, x_2, and x_3 with initial values $0.2, 0.5$, and 0.9, respectively. They satisfy (1.4) and (1.5) with $S = 2, b = 2, \eta = 2.1, \tau = 0.05, \epsilon = 0.15$. The motion of these oscillators is seen in Fig. 1.1.

Couplings $\alpha^2(t-t_0)e^{\alpha(t-t_0)}$, where t_0 is the firing moment, were used in the paper [62] to find that with "fast enough excitatory coupling both the fully synchronized and the asynchronous state are unstable. In this case individual units fire quasi-periodically even though the network as a whole shows a periodic firing pattern."

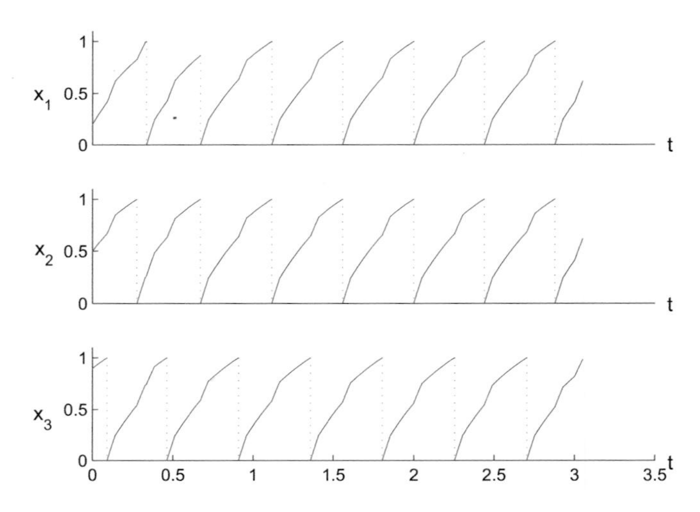

Fig. 1.1 The intervals where oscillators satisfy Eqs. (1.4) and (1.5) are depicted

The results of our chapter are different from those of [62]. Firing in unison is achieved, and this synchrony is stable. The difference can be explained with the smallness of α-functions near the firing moments. We plan to discuss the clustering phenomenon of integrate-and-fire models in an upcoming paper.

Since the dynamics of systems considered in the present chapter are discontinuous, we strongly believe that they can be investigated with the methods developed for differential equations with variable moments of discontinuity [2] in the future. Controllability, phase locking, frequency locking, synchrony, almost periodic solutions, and even chaos can be considered in this theory.

Delays arise naturally in many biological models [46]. In particular, they were considered in firefly models [10] as delay between stimulus and response, and in continuously coupled neuronal oscillators [33]. The authors of [16] considered the phenomenon for the analysis of Mirollo and Strogatz in such a way that identical oscillators were investigated. The dynamics of two oscillators were discussed mathematically, and a multioscillatory system was analyzed by using computer simulations. It was found that the excitatory model of two units "can get only out-of-phase synchronization since in-phase synchronization proved to be not stable." In the paper [19] a model without a leakage was discussed; that is, oscillators increase at a constant rate between moments of firing. It was found that a periodic solution is reached after a finite time. Consequently, research into integrate-and-fire models, which admit delays and fire in unison, is still on the agenda. Section 1.4 investigates synchrony of delayed integrate-and-fire oscillators.

1.2 The Solution of the Peskin Second Conjecture

1.2.1 Construction of the Prototype Map

In this subsection we shall define the map, which is the basic instrument of our investigation. The map will be considered in a general form such that it can be the basis for new investigations in the future. Let the model of two identical oscillators, $x_1(t), x_2(t), t \geq 0$, be given such that

$$x_1' = g(x_1, x_2),$$
$$x_2' = g(x_2, x_1), \tag{1.6}$$

where $0 \leq x_i \leq 1, i = 1, 2, \ldots, n$, and the function g is positive and lipschitzian in both arguments. When the oscillator x_j fires at the moment t such that $x_j(t) = 1$, then the oscillator fires, $x_j(t+) = 0$. Firing changes the value of another oscillator with $i \neq j$, such that

$$x_i(t+) = \begin{cases} 0, & \text{if } x_i(t) + \epsilon \geq 1, \\ x_i(t) + \epsilon, & \text{otherwise.} \end{cases} \tag{1.7}$$

Denote by $u(t, t_0, u_0) = (u_1, u_2)$, the solution of (1.6) such that $u(t_0, t_0, u_0) = u_0$. Conditions on g imply that the solution exists, is unique, and is continuable to the threshold for all t_0 and u_0. Consider the solution $u(t) = u(t, 0, (0, v + \epsilon))$. Find the moment $t = s$ such that $u_2(s) = 1$, and define $\bar{L}(v) = u_1(s)$ on $(0, 1 - \epsilon)$. From the conditions on g it implies that s is a strictly decreasing continuous function of v, and then that $\bar{L}$ is a strictly decreasing continuous function of v. It is clearly seen that $\lim_{v \to 1-\epsilon} \bar{L}(v) = 0$, and there is a unique fixed point, v^*, of the function, $\bar{L}(v^*) = v^*$. Let $\eta = \lim_{v \to 0+} \bar{L}(v)$.

Now, let us define a map $L : [0, 1] \to [0, 1]$, such that

$$L(v) = \begin{cases} \bar{L}(v), & \text{if } v \in (0, 1 - \epsilon), \\ \eta, & \text{if } v = 0, \\ 0, & \text{if } v \in [1 - \epsilon, 1]. \end{cases} \tag{1.8}$$

In what follows we need the following conditions:

(A1) $\eta > 1 - \epsilon$;
(A2) The map L^2 admits a unique fixed point v^* in $(0, 1 - \epsilon)$.

This newly defined function is continuous on $[0, 1]$. The sketch of the graph of the function L is seen in Fig. 1.2.

To make the following discussion constructive, consider the sequence of maps $L^k(v), k = 1, 2, \ldots,$ where $L^k(v) = L(L^{k-1}(v))$. The graphs of these maps with $k = 1, 2, 3$ is shown in the figure.

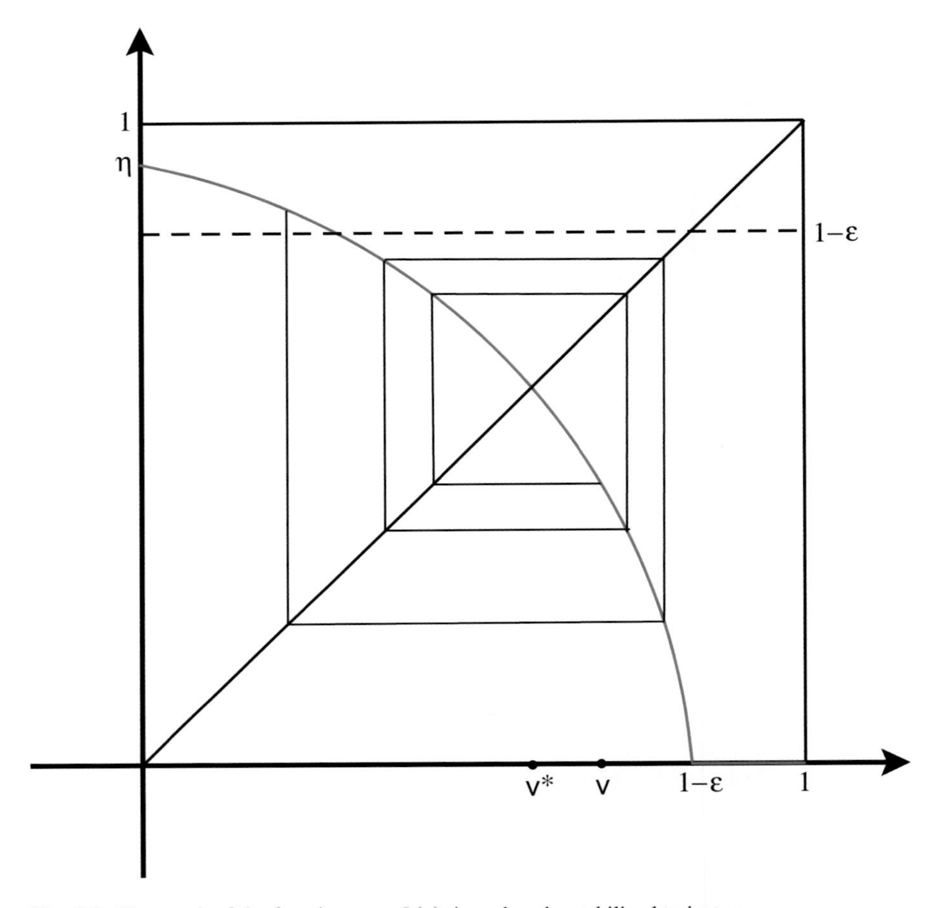

Fig. 1.2 The graph of the function $w = L(v)$, in *red*, and a stabilized trajectory

Denote $a_0 = 0, a_1 = 1 - \epsilon, a_2 = L^{-1}(1-\epsilon), a_3 = (L^2)^{-1}(1-\epsilon), \ldots$. The sequences can also be obtained through the iterations $a_0 = L^{-1}(\eta) = 0, a_{k+1} = L^{-1}(a_k), k = 0, 1, 2, \ldots$, which are seen in Fig. 1.3. It is clear that the sequences a_{2i} and a_{2i+1} are monotonic, decreasing, and increasing, respectively. Otherwise, by utilizing the intermediate-value theorem, one can show that there exists a period-2 motion of the discrete dynamics. This contradicts condition $(A2)$. Thus, both sequences converge. These limits equal to v^*. Indeed, if they are different, then there exists a period-2 motion of the dynamics, and that has been excluded earlier.

Let us show the role of the map L for our research. Suppose that t_1, t_2, t_3, are three successive firing moments of the system such that x_1 fires at t_1 and t_3, the oscillator x_2 fires at t_2, and the oscillators are not synchronized until t_3. We have that $x_1(t_1+) = 0, 0 < x_2(t_1) < 1 - \epsilon$. One can see that $x_2(t) = u(t, t_1, x_2(t_1) + \epsilon)$

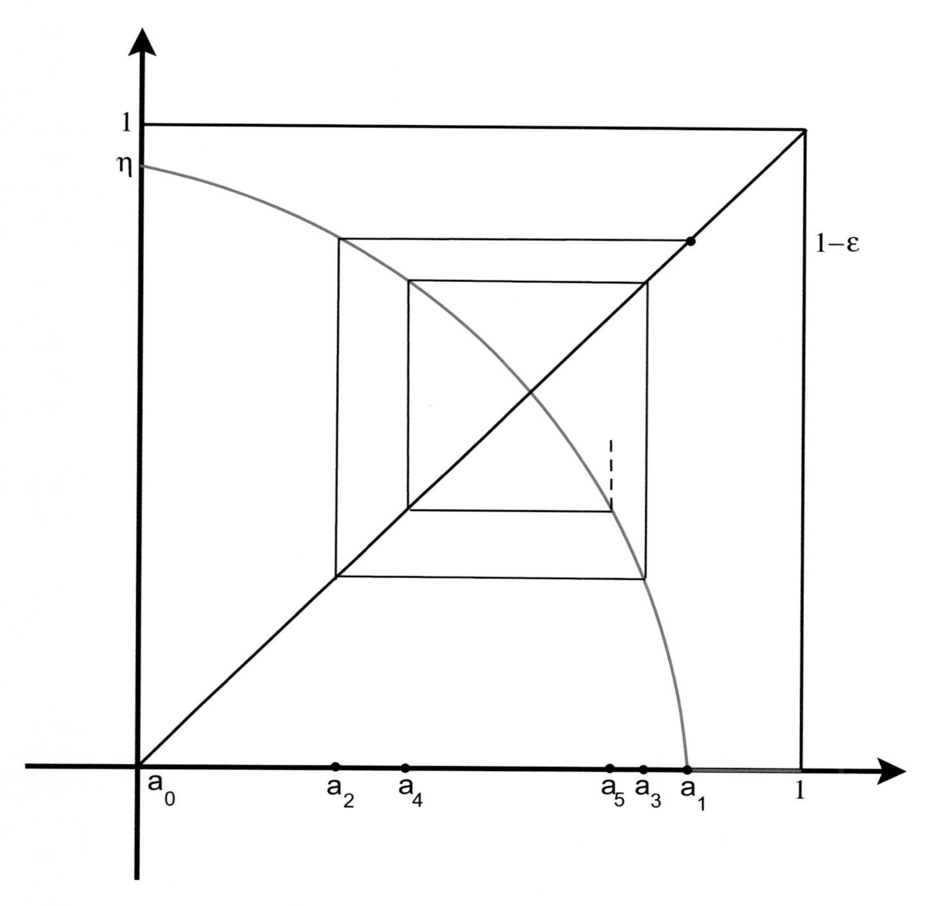

Fig. 1.3 The points $a_0 = L^{-1}(\eta) = 0, a_{k+1} = L^{-1}(a_k), k = 0, 1, 2, \ldots$

for $t_1 < t \leq t_2$, and $u(t_2, t_1, x_2(t_1) + \epsilon) = 1$. That is, $x_1(t_2) = L(x_2(t_1))$. Similarly, one can show that $x_2(t_3) = L(x_1(t_2))$. This demonstrates how the map L can be used for the analysis of the synchronization problem.

Next, we shall prove the synchronization and evaluate the rate of synchronization simultaneously. The rate evaluation will be done in a specific way: We shall indicate the set of initial points which synchronize after precisely k iterations of the map for each nonnegative integer k. Another valuable set in this sense is the collection of all points which synchronize after no more than k iterations is shaped. In the sequel, denote by B_k the region in [0, 1], where points v are synchronized after k iterations of map L. Consider Fig. 1.3 again.

One can see that $B_0 = [1-\epsilon, 1]$, and, consequently, $B_k = [a_{k-1}, a_{k+1}]$ if k is an odd positive integer; $B_k = [a_{k+1}, a_{k-1}]$ if k is an even positive integer. We have that $a_k \to v^*$ as $k \to \infty$.

Denote by C_k the region of all points which synchronize in no more than k iterations of L. One can see that $C_k = [0,1]\backslash(a_k, a_{k+1})$ if k is an even number, and it is the set $C_k = [0,1]\backslash(a_{k+1}, a_k)$ if k is an odd number. In the next section we shall use the set $C_k^c = [0,1]\backslash C_k$, that includes all points which synchronize in no less than $k+1$ iterations of the map. From our earlier discussion it follows that there is no finite time in which *all points* of the unit square synchronize. The closer v is to the equilibrium v^*, the later is the moment of synchronization.

Denote by T the natural period of oscillators, that is, the period of each of the identical units, when there are no couplings. Since each oscillator necessarily fires within an interval of length T, on the basis of the above discussion, the following assertion is valid:

Theorem 1. *Assume that* (A1) *and* (A2) *are valid. If* $(x_1(0), x_2(0)) \in C_k \times C_m$, k *and* m *are natural numbers, then the couple* x_l, x_r *synchronizes within the time interval* $[0, T(\max(k,m)+1)]$.

Example 1. Consider the model of two integrate-and-fire identical oscillators, which is pulse-coupled, of the form

$$\begin{aligned} x_1' &= S + \gamma x_1 + \beta x_2, \\ x_2' &= S + \gamma x_2 + \beta x_1, \end{aligned} \tag{1.9}$$

where the constants S and β are positive numbers, and γ -1. One can easily see that the system developed is Peskin's model [50]. The terms with coefficient β are additional in the system. They reflect the permanent influences of the partners during the process. Eigenvalues of the corresponding linear system to (1.9) are $\lambda_1 = \gamma + \beta$ and $\lambda_2 = \gamma - \beta$. We suppose that β is small such that both eigenvalues are negative. Moreover, it is assumed that $\kappa = S/\lambda_1 < -1$. The solution of the system (1.9) with value $(0, v+\epsilon)$ at $t = 0$ is equal to

$$u_1(t) = \frac{1}{2}[e^{\lambda_1 t} - e^{\lambda_2 t}](v+\epsilon) + \kappa(e^{\lambda_1 t} - 1),$$

$$u_1(t) = \frac{1}{2}[e^{\lambda_1 t} + e^{\lambda_2 t}](v+\epsilon) + \kappa(e^{\lambda_1 t} - 1).$$

That is why the necessary equations have the following forms:

$$\frac{1}{2}[e^{\lambda_1 t_2} + e^{\lambda_2 t_2}](v+\epsilon) + \kappa(e^{\lambda_1 t_2} - 1) = 1 \tag{1.10}$$

and

$$L(v) = \frac{1}{2}[e^{\lambda_1 t_2} - e^{\lambda_2 t_2}](v + \epsilon) + \kappa(e^{\lambda_1 t_2} - 1). \tag{1.11}$$

The last two formulas imply that

$$L(v) = \frac{e^{\lambda_1 t_2} - e^{\lambda_2 t_2} + 2\kappa e^{\lambda_2 t_2}(e^{\lambda_1 t_2} - 1)}{e^{\lambda_1 t_2} + e^{\lambda_2 t_2}}.$$

Differentiating (1.10), one can find that

$$\frac{\partial t_2}{\partial v} = -\frac{1}{2}\frac{[e^{\lambda_1 t_2} + e^{\lambda_2 t_2}]^2}{(1+\kappa)(\lambda_1 e^{\lambda_1 t_2} + \lambda_2 e^{\lambda_2 t_2}) + 2\beta\kappa e^{(\lambda_1+\lambda_2)t_2}}$$

and

$$\frac{\partial^2 t_2}{\partial v^2} = -\frac{1}{2}\frac{\partial t_2}{\partial v}\frac{(e^{\lambda_1 t_2} + e^{\lambda_2 t_2})(1+\kappa)[(\lambda_1 e^{\lambda_1 t_2} + \lambda_2 e^{\lambda_2 t_2})^2 - 4\beta^2 e^{(\lambda_1+\lambda_2)t_2}]}{[(1+\kappa)(\lambda_1 e^{\lambda_1 t_2} + \lambda_2 e^{\lambda_2 t_2}) + 2\beta\kappa e^{(\lambda_1+\lambda_2)t_2}]^2} +$$

$$\frac{2\kappa\beta e^{(\lambda_1+\lambda_2)t_2}(e^{\lambda_1 t_2} + e^{\lambda_2 t_2})(2\lambda_1 - 1)(e^{\lambda_1 t_2} + e^{\lambda_2 t_2})]}{}.$$

We deliberately have written the last two formulas in a form such that one can easily see that both derivatives are negative if S is sufficiently large and β is sufficiently small. Next, we evaluate the derivatives of L:

$$L'(v) = \frac{\partial t_2}{\partial v}\frac{2\kappa e^{\lambda_2 t_2}(e^{\lambda_1 t_2} - 1)(e^{\lambda_1 t_2} + e^{\lambda_2 t_2}) + 4\beta e^{(\lambda_1+\lambda_2)t_2}[1 - \kappa(e^{\lambda_1 t_2} - 1)]}{[e^{\lambda_1 t_2} + e^{\lambda_2 t_2}]^2},$$

$$L''(v) = \frac{\partial^2 t_2}{\partial v^2}\frac{2\kappa e^{\lambda_2 t_2}(e^{\lambda_1 t_2} - 1)(e^{\lambda_1 t_2} + e^{\lambda_2 t_2}) + 4\beta e^{(\lambda_1+\lambda_2)t_2}[1 - \kappa(e^{\lambda_1 t_2} - 1)]}{[e^{\lambda_1 t_2} + e^{\lambda_2 t_2}]^2} +$$

$$(\frac{\partial t_2}{\partial v})^2\frac{\kappa e^{\lambda_2 t_2}(e^{\lambda_1 t_2}+e^{\lambda_2 t_2})[(\lambda_1^2-\lambda_1)e^{\lambda_1 t_2}-1]+}{[e^{\lambda_1 t_2}+e^{\lambda_2 t_2}]^4}$$

$$\frac{4\beta e^{(\lambda_1+\lambda_2)t_2}[1-\kappa(e^{\lambda_1 t_2}-1)]\{\lambda_1+\lambda_2-2(e^{\lambda_1 t_2}+e^{\lambda_2 t_2})(\lambda_1 e^{\lambda_1 t_2}+\lambda_2 e^{\lambda_2 t_2})]-\kappa\lambda_1 e^{\lambda_1 t_2}\}}{}.$$

Again, one can find that the last derivatives are both negative if S and β are sufficiently large and small, respectively. That is, the function L is convex. Now, one can easily show that condition ($A2$) is fulfilled.

Consider formulas (1.10) and (1.11) with $\beta = 0$ and $v = 0$ to obtain $L(0) = \kappa \frac{1-v-\epsilon}{\kappa+v+\epsilon} > 1-\epsilon$. That is, if β is sufficiently small, then condition $(A1)$ is valid, and the pair synchronizes. This result of the synchronization of two identical oscillators with the right-hand side depending on both variables is new. In previous papers, the differential equations were separated.

Example 2. Consider the system of two identical oscillators with the differential equations

$$\begin{aligned} x_1' &= x_1^2 + c, \\ x_2' &= x_2^2 + c, \end{aligned} \tag{1.12}$$

where c is a positive constant. It is known that the canonical type I phase model [14] can be reduced by a transformation [20] to the form

$$u' = u^2 + c, \tag{1.13}$$

that is, to the quadratic integrate-and-fire model. This time we have added to the model the pulse coupling as described in the beginning of the section, and we investigate the synchronization problem by using the last result. We can assume, without loss of generality, that $t_1 = 0$. Since the two equations are identical, we consider a solution $u(t)$ of Eq. (1.13) for the construction of the map L. We have that $u(t, 0, v+\epsilon) = \sqrt{c}\tan(\sqrt{c}t + \arctan(\frac{v+\epsilon}{\sqrt{c}}))$ and

$$\sqrt{c}\tan(\sqrt{c}t_2 + \arctan(\frac{v+\epsilon}{\sqrt{c}})) = 1. \tag{1.14}$$

Next, $u(t_2, 0, 0) = \sqrt{c}\tan(\sqrt{c}t_2)$, and by applying (1.13) and (1.14) we find that

$$L(v) = \sqrt{c}\frac{1-(v+\epsilon)}{\sqrt{c}+(v+\epsilon)}$$

if $v \in (0, 1-\epsilon)$, and the fixed point is $v^* = \sqrt{c + \sqrt{c} + \epsilon^2/4} - (\sqrt{c} + \epsilon/2)$. Evaluate

$$L(0) = \sqrt{c}\frac{1-\epsilon}{\sqrt{c}+\epsilon},$$

to see that $\eta = L(0) > 1-\epsilon$, if $c < 1$ and ϵ is sufficiently small such that

$$\sqrt{c} + \epsilon < 1. \tag{1.15}$$

Moreover, one can verify that v^* is a unique fixed point of L^2. Thus, we obtain that if (1.15) is valid, then all conditions of Theorem 1 are fulfilled, and, consequently, the couple synchronizes if only $v \neq v^*$.

Example 3. Consider the pair of identical oscillators when $f(u) = S - \gamma u, \kappa = \frac{S}{\gamma} > 1$, that is, Peskin's model [50]. Assume again that $t_1 = 0$. One can find that $u(t, 0, v + \epsilon) = (v + \epsilon)e^{-\gamma t} + \kappa(1 - e^{-\gamma t})$ and

$$(v + \epsilon)e^{-\gamma t_2} + \kappa(1 - e^{-\gamma t_2}) = 1.$$

We have that

$$e^{-\gamma t_2} = \frac{\kappa - 1}{\kappa - (v + \epsilon)}. \tag{1.16}$$

Substituting the last expression in $u(t_2, 0, 0) = \kappa(1 - e^{-\gamma t_2})$, one obtains that

$$L(v) = \kappa \frac{1 - (v + \epsilon)}{\kappa - (v + \epsilon)}, \tag{1.17}$$

where $0 < v < 1 - \epsilon$.

There is a unique fixed point of L and L^2, and it is equal to

$$v^* = (\kappa - \frac{\epsilon}{2}) - \sqrt{\kappa^2 - \kappa + \frac{\epsilon^2}{4}}. \tag{1.18}$$

Finally, $L(0) = \kappa \frac{1-\epsilon}{\kappa-\epsilon} > 1 - \epsilon$. That is, all conditions of the last theorem are valid. Thus, we have proved the assertion in [50].

1.2.2 The General Case: The Multidimensional System of Nonidentical Oscillators

In this section we shall discuss the main object of investigation. First, we will apply the result of the last section and analyze the motion of a pair of oscillators in the multioscillatory ensemble; we'll find that the couple synchronizes if the parameters are close to zero. Next, we will prove the main theorem.

Consider a model of n nonidentical oscillators given by relations (1.1) and (1.2). Fix two of the oscillators, let's say x_l and x_r. Denote $\mathscr{C}_j^k = [0, 1 + \xi_j] \backslash C_k^c, j = 1, 2, \ldots, n$, where $C_k^c = [0, 1] \backslash C_k$, as defined in the last section, is the set which consists of all points of the unit section that synchronize after no less than $k + 1$ iterations of L.

Lemma 1. *Assume that conditions* (A1) *and* (A2) *are valid. If* $(x_l(0), x_r(0)) \in \mathscr{C}_l^k \times \mathscr{C}_r^m$, *k and m are natural numbers, then the couple* x_l, x_r *synchronizes within the time interval* $[0, T(\max(k, m) + 3)]$ *if the parameters are sufficiently close to zero, and absolute values of the parameters of perturbation are sufficiently small with respect to* ϵ.

Proof. Denote by $x(t) = (x_1(t), x_2(t), \ldots, x_n(t))$ a motion of the oscillator, and let $u(t) = (u_1, u_2, \ldots, u_n)$ be the solution to Eq. (1.1) with $u_i(t_0, t_0, u_0) = u_0^i, u_0 = (u_0^1, u_0^2, \ldots, u_0^n)$. Suppose, without loss of generality, that $k \geq m$ and $t = 0$ is a moment of firing such that $x_l(0) = 1 + \zeta_l(x(0)), x_l(0+) = 0$. We will show that the couple x_l, x_r synchronizes at some moment $0 \leq t < (k+2)T$, if the parameters are close to zero. If $1 + \zeta_r(x(0)) - \varepsilon - \varepsilon_r \leq x_r(0) \leq 1 + \zeta_r(x(0))$, then these two oscillators fire simultaneously, and we only need to prove the persistence of synchrony, which will be done later. Thus, fix another oscillator $x_r(t)$ such that $0 \leq x_r(0) < 1 + \zeta_r(x(0)) - \varepsilon - \varepsilon_r$.

We shall divide the proof into two parts. First, we will show that the couple synchronizes eventually and then keeps the synchrony state permanently. In the second part, we will evaluate the time of synchronization.

Assume that the pair does not synchronize. Then there is a sequence of firing moments, t_i, such that $0 = t_0 < t_1 < t_2 < \ldots$, the oscillator x_l fires at t_i with even i, and x_r fires at t_i with odd i. For the sake of brevity, let $u_i = x_r(t_i), i = 2j, j \geq 0, u_i = x_l(t_i), i = 2j+1, j \geq 0$.

Let's fix an even i. If the parameters are sufficiently small, then there are $m \leq n-2$ distinct firing moments of the motion $x(t)$ on the interval (t_i, t_{i+1}). Denote by $t_i < \theta_1 < \theta_2 < \ldots < \theta_m < t_{i+1}$ the moments of firing, when at least one of the coordinates of $x(t)$ fires. We have that

$$x_r(\theta_1) = x_r(t_i) + \epsilon + \int_{t_i}^{\theta_1} f(x_r(s))ds + \int_{t_i}^{\theta_1} \phi_r(s)ds, \tag{1.19}$$

where $x(t) = u(t, t_i, x(t_i+))$, is the solution of (1.1),

$$x_r(\theta_2) = x_r(\theta_1) + \epsilon + \int_{\theta_1}^{\theta_2} f(x_r(s))ds + \int_{\theta_1}^{\theta_2} \phi_r(s)ds, \tag{1.20}$$

where $x(t) = u(t, \theta_1, x(\theta_1+))$,

. . . .

$$x_r(t_{i+1}) = x_r(\theta_m) + \epsilon + \int_{\theta_m}^{t_{i+1}} f(x_r(s))ds + \int_{\theta_m}^{t_{i+1}} \phi_r(s)ds. \tag{1.21}$$

The moment t_{i+1} satisfies

$$1 + \zeta_r(x(t_{i+1})) - \epsilon - \epsilon_r \leq x_r(t_{i+1}) \leq 1 + \zeta_r(x(t_{i+1})). \tag{1.22}$$

Similar to the expressions for x_r, we have

$$x_l(\theta_1) = \int_{t_i}^{\theta_1} f(x_l(s))ds + \int_{t_i}^{\theta_1} \phi_l(s)ds,$$

$$x_l(\theta_2) = x_l(\theta_1) + \epsilon + \int_{\theta_1}^{\theta_2} f(x_l(s))ds + \int_{\theta_1}^{\theta_2} \phi_l(s)ds,$$

$$\ldots$$

$$x_l(t_{i+1}) = x_l(\theta_m) + \epsilon + \int_{\theta_m}^{t_{i+1}} f(x_l(s))ds + \int_{\theta_m}^{t_{i+1}} \phi_l(s)ds. \quad (1.23)$$

One can see that formulas (1.19) to (1.23) completely define a relation $u_{i+1} = K_i(u_i) = x_l(t_{i+1})$. A similar one can be found for odd i.

Let us construct the value of $L(u_i, \epsilon)$ now. With this aim, evaluate

$$\phi(\bar{t}_{i+1}) = x_r(t_i) + \epsilon + \int_{t_i}^{\bar{t}_{i+1}} f(\phi(s))ds, \quad (1.24)$$

where $\bar{t}_{i+1}$ satisfies

$$\phi(\bar{t}_{i+1}) = 1, \quad (1.25)$$

and

$$\psi(\bar{t}_{i+1}) = \int_{t_i}^{\bar{t}_{i+1}} f(\psi(s))ds, \quad (1.26)$$

to find that $L(u_i, \epsilon) = \psi(\bar{t}_{i+1})$. Next, we will show that the difference $K_i(u_i) - L(u_i, \epsilon)$ is small if the parameters are small.

First, we have that for $t \in [t_i, \theta_1]$, it is true that

$$\phi(t) - x_r(t) = \int_{t_i}^{t} [f(\phi(s)) - f(x_r(s))]ds - \int_{t_i}^{t} \phi_r(s)ds. \quad (1.27)$$

Then, by applying Gronwall–Bellman's lemma, one can easily find that

$$|\phi(\theta_1) - x_r(\theta_1)| \leq \mu_r(\theta_1 - t_i)e^{\ell(\theta_1 - t_i)}, \quad (1.28)$$

where T is the natural period defined in the first section for the identical oscillators. Next, similarly, we have that if $t \in [\theta_1, \theta_2]$, then

$$|\phi(\theta_2) - x_r(\theta_2)| \leq [\mu_r(\theta_1 - t_i)e^{\ell(\theta_1 - t_i)} + \epsilon]e^{\ell(\theta_2 - \theta_1)}. \quad (1.29)$$

Without loss of generality, assume that $t_{i+1} > \bar{t}_{i+1}$. Continuing evaluations made above, we can obtain $|1 - x_r(t_{i+1})| = |\phi(t_{i+1}) - x_r(t_{i+1})| = \Phi(\epsilon, \mu_r)$, where Φ is of the order $O(\epsilon, \mu_r, \xi_r)$. There are two positive numbers m, M such that $m \le f(s) \le M$ if $0 \le s \le 1 + \max_i \xi_i$. We have that $|x_r(t_{i+1}) - x_r(\bar{t}_{i+1})| \le |1 - x_r(t_{i+1})| + |1 - x_r(\bar{t}_{i+1})| \le \Phi(\epsilon, \mu_r) + \xi_r$. Consequently,

$$|t_{i+1} - \bar{t}_{i+1}| < \frac{\Phi(\epsilon, \mu_r) + \xi_r}{m - \mu_r}.$$

By applying the last inequality, (1.23) and (1.26) to evaluate the difference $|\psi(\bar{t}_{i+1}) - x_l(t_{i+1})|$, one can find that $K_i(u_i) - L(u_i, \epsilon)$ can be made arbitrarily small if the parameters are sufficiently close to zero; the parameters of perturbation are small, in absolute values, with respect to ϵ. This convergence is uniform with respect to $u_0 \in \mathscr{C}_r^k$. We can also vary the number of points θ_i and their location in the intervals (t_j, t_{j+1}) between 0 and $n - 1$. The convergence is indifferent with respect to these variations, too.

Consider now the sequence $L^i(u_0)$. It is true that $L^k(u_0) \in [1 - \epsilon, 1]$ for some $k \ge 0$. Assume, without loss of generality, that k is an even number. Since L is a continuous function, we can conclude that either $1 + \xi_r - \epsilon - \epsilon_r \le u_k < 1 + \xi_r$ or $1 + \xi_l - \epsilon - \epsilon_l \le u_{k+1} < 1 + \xi_l$ if the parameters are sufficiently close to zero, and absolute values of the parameters of perturbation are sufficiently small with respect to ϵ.

Both of these inequalities bring the system to synchronization.

In Fig. 1.4 one can see the sequence of maps K_i, and the synchronizing sequence u_i is constructed. In the figure we show not only u_i, but also the graphs of functions $w = K_i(u), u_{i+1} = K_i(u_i)$, in the neighborhood of u_i, to give a better geometrical visualization of the convergence.

To evaluate the time of synchronization, we should consider the general case, when $t = 0$ is not necessarily the firing moment. Then either x_l or x_r fires within the interval $I = [0, \bar{T}]$, where $\bar{T}$ is close to T, the natural period of the identical oscillators, as ξ_i, ζ_i, and μ_i are close to zero. Since each of the iterations of map L happens within an interval with the length of no more than T, we now obtain that the couple x_l, x_r is synchronized no later than $t = (k + 3)T$.

If two oscillators x_l and x_r are nonidentical and fire simultaneously at a moment $t = \theta$, how will they retain the state of firing in unison despite being different? To find the required conditions, let us denote by $\tau > \theta$ a moment when one of the two oscillators, let's say x_r, fires. We have that $x_l(\theta+) = x_r(\theta+) = 0$. Then $x_l(t) = x_r(t), \theta \le t \le \tau$. It is clear that to satisfy $x_l(\tau+) = x_r(\tau+) = 0$, we need $1 + \xi_l - \epsilon - \epsilon_l \le x_l(\tau)$. By applying formula (1.23) again, this time with

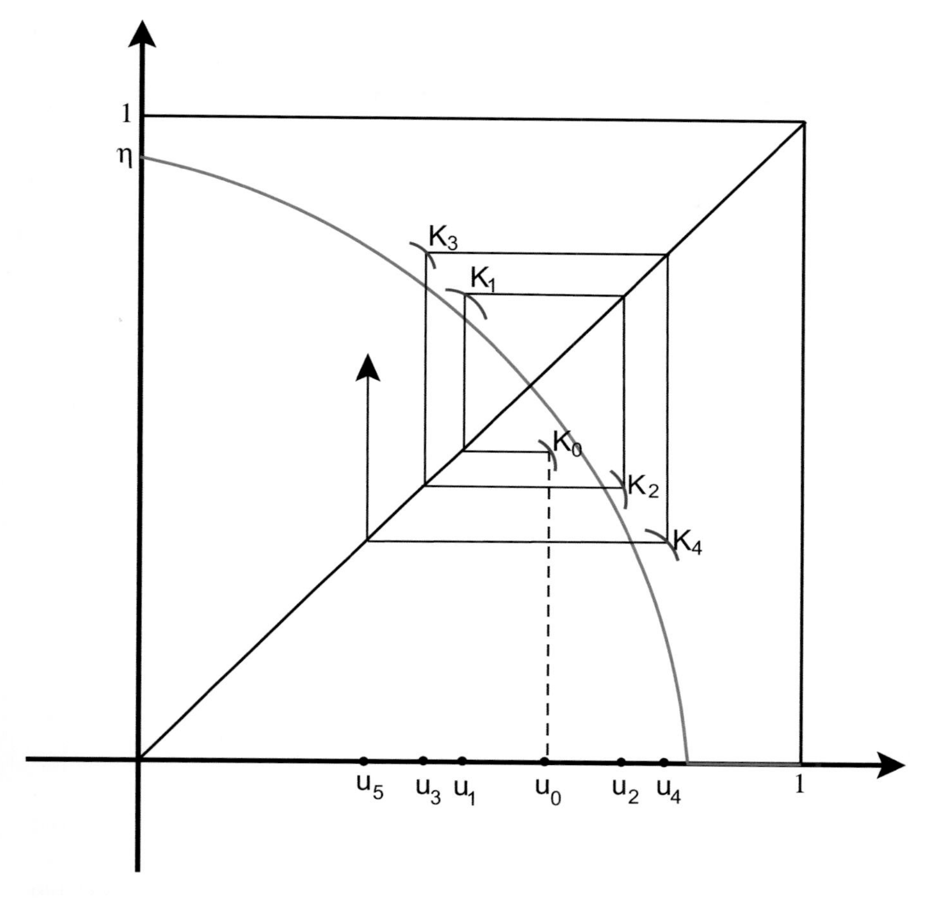

Fig. 1.4 The dynamics of a pair of oscillators from the system. The graphs of functions $w = K_i(u)$ are shown

$t_i = \theta, t_{i+1} = \tau$, one can easily obtain that the inequality is correct if parameters are close to zero. Thus, one can conclude that if a couple of oscillators is synchronized at some moment of time, then it will continue to fire in unison. The lemma is proved.

Remark 1. The last lemma not only plays an auxiliary role for the next main theorem, but it can also be considered a synchronization result for the model of two nonidentical oscillators.

Let us extend the result of the lemma for the whole ensemble.

Theorem 2. *Assume that a motion $x(t)$ of the system satisfies $x(0) \in \Pi_{i=1}^{n} \mathscr{C}_i^{k_i}$. If parameters are sufficiently close to zero, and absolute values of the parameters of perturbation are sufficiently small with respect to ϵ, then the motion synchronizes within the time interval $[0, (\max_i k_i + 3)T]$.*

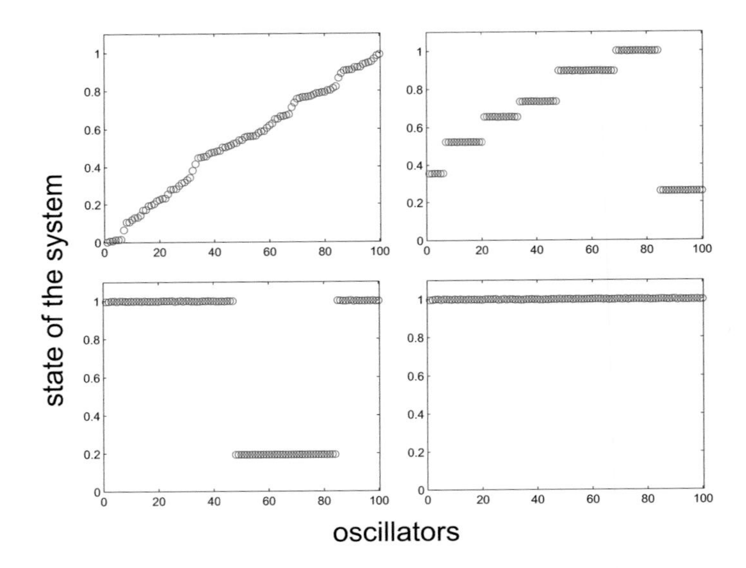

Fig. 1.5 The state of the model before the 1st, 21st, 42nd, and 63rd firings of the system. The flat sections of the graph are groups of synchronized oscillators

Proof. Fix one of the oscillators, let's say x_1, and consider the collection of couples $(x_1, x_j), j = 2, \ldots, n$. Applying the last lemma, we can say that each pair of oscillators synchronizes within $[0, (\max_i k_i + 3)T]$. The theorem is proved.

To illustrate the last theorem, we consider a group of oscillators, $x_i, i = 1, 2, \ldots, 100$, with random uniform distributed start values in $[0, 1]$. It is supposed that they satisfy the equations $x_i' = (3+0.01\bar{\mu}_i)-(2+0.01\bar{\zeta}_i)x_i$. The constants $\bar{\mu}_i, \bar{\zeta}_i$, as well as $\bar{\xi}_i$ in the thresholds $1 + 0.005\bar{\xi}_i, i = 1, 2, \ldots, 100$, are uniform random distributed numbers from $[0, 1]$. In Fig. 1.5 one can see the result of simulation with $\epsilon = 0.08$, where the state of the system is shown before the 1st, 21st, 42nd, and 63rd firings of the system. It is obvious that eventually the state of the model is in synchrony.

Let us describe a more general system of oscillators such that Theorem 2 is still true. A system of n oscillators is given such that if the ith oscillator does not fire or jump up, it satisfies the ith equation of system (1.1). If several oscillators $x_{i_s}, s = 1, 2, \ldots, k$, fire so that $x_{i_s}(t) = 1 + \xi_{i_s}$ and $x_{i_s}(t+) = 0$, then all other oscillators $x_{i_p}, p = k+1, k+1, \ldots, n$, change their coordinates by the law

$$x_{i_p}(t+) = \begin{cases} 0, & \text{if } x_{i_p}(t) + \epsilon + \sum_{s=1}^{k} \epsilon_{i_p i_s} \geq 1 + \xi_{i_p}, \\ x_{i_p}(t) + \epsilon + \sum_{s=1}^{k} \epsilon_{i_p i_s}, & \text{otherwise}. \end{cases} \tag{1.30}$$

One can easily see that the last theorem is correct for the model just described if $\epsilon + \sum_{s=1}^{k} \epsilon_{i_p i_s} > 0$, for all possible k, i_p and i_s.

Remark 2. Since the length of C_k^c diminishes to zero as $k \to \infty$, one can say that our results are consistent with conjecture $(C2)$. Indeed, it was said in the beginning that all initial points must synchronize. Then, the fixed point was excluded [50]. In [45] the condition is weakened to the exception of a set with the Lebesgue measure null. In the present chapter we just analyze another kind of smallness of the set.

Remark 3. Our preliminary analysis shows that the dynamics in the neighborhood of v^* can be very complex. We would not exclude the possibility of chaos appearance and the belongings of trajectories to a fractal if parameters are not small [61]. It does not contradict the zero Lebesgue measure of a set of nonsynchronized points. The analysis of nonidentical oscillators with nonsmall parameters may shed light on the investigation of arrhythmias, chaotic flashing of fireflies, and so on.

Remark 4. The time of synchronization for a given initial point does not change much as the number of oscillators increases (but the parameters need to be closer to zero!). This property can be viewed as a small-world phenomenon.

1.2.3 Possible Generalization

Next, it is natural to extend the result for a system of the following form:

$$
\begin{aligned}
x_1' &= g(x_1, x_2, \ldots, x_n),\\
x_2' &= g(x_2, x_1, \ldots, x_n),\\
&\ldots,\\
x_n' &= g(x_1, x_2, \ldots, x_{n-1}),
\end{aligned}
\tag{1.31}
$$

considered with condition (1.2). One can say that the coupling is now not only pulse, but is continuous also. We assume that the function g is continuous and positive again. If one supposes that the scalar-valued function g is indifferent with respect to permutations of the next-to-last variables, then we can use the map L constructed in this section in the same way as is done for the system (1.1). It is natural to accept some conditions of smallness with respect to the next-to-last variables, to obtain synchronization of the system. Nevertheless, it seems that conditions can be found except for the smallness to have the system still synchronized.

1.3 Integrate-and-Fire Models with Continuous Couplings

1.3.1 The Model of Two Identical Oscillators

Start the investigation with the simplest model of two identical oscillators. That is, assume that $n = 2$ in the description of the last section.

Let $[s, s + \tau]$ be the e-interval for $x_i(s)$. Then one can easily find that

$$x_i(t) = x_i(s)e^{-\gamma(t-s)} + \int_s^t e^{-\gamma(t-u)}(S + \eta)du, \tag{1.32}$$

for $t > s$.

Set $\kappa = \frac{S}{\gamma} > 1$. By integrating in (1.32), we have that

$$x_i(s + \tau) = x_i(s)e^{-\gamma\tau} + (\kappa + \frac{\eta}{\gamma})(1 - e^{-\gamma\tau}).$$

From $(A3)$, (iii), it follows that $x_i(t) < 1$ for all $t \in [s, s + \tau]$. That is, x_i does not fire in the e-interval if $x_i(s) < 1 - \epsilon$. Consequently, the domain of any oscillator contains only disjoint e-intervals.

Denote by t_1, t_2, t_3 three successive firing moments of the system such that x_1 fires at t_1 and t_3, the oscillator x_2 fires at t_2, and the oscillators are not synchronized until t_3. We have that $0 < x_2(t_1) < 1 - \epsilon$, and $x_2(t_1 + \tau) < 1$. From $x_2(t_2) = 1$ or $[x_2(t_1)e^{-\gamma\tau} + (\kappa + \frac{\eta}{\gamma})(1 - e^{-\gamma\tau})]e^{-\gamma(t_2-t_1-\tau)} + \kappa[1 - e^{-\gamma(t_2-t_1-\tau)}] = 1$, it follows that

$$e^{-\gamma(t_2-t_1)} = \frac{\kappa - 1}{\kappa - x_2(t_1) - \eta_1}, \tag{1.33}$$

where $\eta_1 = \frac{\eta}{\gamma}(e^{\gamma\tau} - 1) < \eta$.

Apply (1.33) in

$$x_1(t_2) = \int_{t_1}^{t_2} e^{-\gamma(t-u)}Sdu = \kappa[1 - e^{-\gamma(t_2-t_1)}]$$

to obtain $x_1(t_2) = L_C(x_2(t_1))$, where

$$L_C(v) = \kappa\frac{1 - v - \eta_1}{\kappa - v - \eta_1} \tag{1.34}$$

is a map defined for $0 < v < 1 - \eta_1$. Similarly, by using the identity of oscillators, one can find that $x_2(t_3) = L_C(x_1(t_2))$. That is, the map L_C evaluates the sequence of coordinates of the model interchanging at firing moments. Its derivatives satisfy

$$L'_C(v) = \kappa \frac{1-\kappa}{(\kappa - (v+\eta_1))^2} < 0$$

and

$$L''_C(v) = 2\kappa \frac{1-\kappa}{(\kappa - (v+\eta_1))^3} < 0$$

in $(0, 1-\eta_1)$. There is a fixed point of L_C, and it is equal to $v^* = (\kappa - \frac{\eta_1}{2}) - \sqrt{\kappa^2 - \kappa + \frac{\eta_1^2}{4}}$.

Moreover, we have

$$L'_C(v^*) = \kappa \frac{1-\kappa}{(\sqrt{\kappa^2 - \kappa + \frac{\eta_1^2}{4}} - \frac{\eta_1}{2})^2} < -1. \tag{1.35}$$

That is, v^* is a repeller.

Next, we extend the map on $[0, 1]$ in the following way. Set $L_C(0) = \omega$, where $\omega = \kappa \frac{1-\eta_1}{\kappa-\eta_1}$. It is easy to check that $1 - \epsilon < \omega < 1$. Moreover, we define $L_C(v) = 0$ if $1 - \eta_1 \leq v \leq 1$. On the basis of the analysis above, one finds that this newly introduced map is continuous and monotonic, and $[0, 1]$ is an invariant set. Hence, $L_C(v)$ is very appropriate for iteration analysis. The graph of the function $w = L_C(v)$ is seen in Fig. 1.6.

Let us show how synchronization can be investigated by analyzing iterations of L_C. Fix $t_1 \geq 0$, a firing moment, $x_1(t_1) = 1, x_1(t_1+) = 0$. While the couple x_1, x_2 does not synchronize, there exists a sequence of moments $t_1 < t_2 < t_3 < \ldots$ such that x_1 fires at t_i with odd i and x_2 fires at t_i with even i. Set $u_i = x_1(t_i)$ if i is even, and $u_i = x_2(t_i)$ if i is odd. Then $u_{i+1} = L_C(u_i), i \geq 1$. The pair synchronizes if and only if there exists $j \geq 1$ such that $x_1(t) \neq x_2(t)$, if $t \leq t_j$, and $x_1(t) = x_2(t)$ for $t > t_j$. In particular, both oscillators have to fire at t_j. That is, inequalities $1 - \epsilon \leq u_j < 1$ hold, which is possible if $0 \leq u_{j-1} \leq L^{-1}(1-\epsilon)$. We have that $L_C(0) = \omega$ satisfies this condition.

If $1 - \epsilon \leq x_2(t_1) \leq 1$, then we have that t_1 is a common firing moment of both x_1 and x_2, and it is the synchronization moment. Moreover, $1-\epsilon < L_C^2(x_2(t_1)) = \eta < 1$. That is, the map L_C brings us to synchrony, with two steps of delay. Summarizing, if there exists an integer $k \geq 0$ such that $1-\epsilon \leq L_C^k(v) \leq 1$, then a motion $(x_1(t), x_2(t))$ with $x_1(t_1+) = v, x_2(t_1+) = 0$ synchronizes. Conversely, if a motion $(x_1(t), x_2(t))$ synchronizes, then one can find a firing moment, t_1, such that $x_1(t_1+) = v, v \in [0, 1], x_2(t_1+) = 0$, and a number k such that $1 - \epsilon \leq L_C^k(v) \leq 1$. Thus, we have verified that the constructed map is in full correspondence with the synchronization goal.

Analyzing maps $L_C^k, k \geq 0$, one can easily obtain that they all have only one nonzero fixed point v^*, and $|[L_C^k(v^*)]'| > 1$. Consequently, there is no k-periodic motion, $k > 1$, of the map, and a motion stabilizes if its initial point $v \neq v^*$ (see Fig. 1.6).

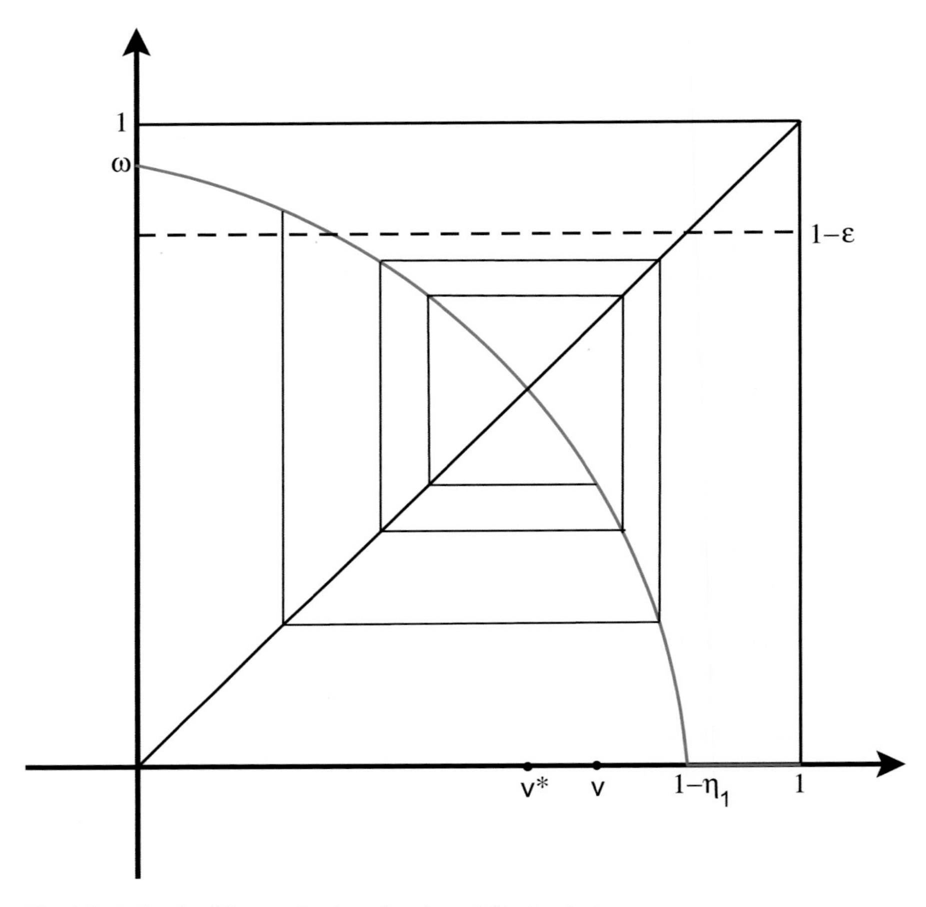

Fig. 1.6 A sketch of the map L_C, in *red*, and a stabilized trajectory

Our next goal is to locate, for each nonnegative integer k, the set of initial points such that their motions synchronize in precisely k iterations of the map. In the sequel, denote by S_k the region in $[0, 1]$ where points v are synchronized after precisely k iterations of L_C. Let $a_0 = L_C^{-1}(\eta) = 0, a_{k+1} = L_C^{-1}(a_k), k = 0, 1, 2, \ldots$. The points are pictured in Fig. 1.7.

One can see that $S_0 = [1 - \epsilon, 1], S_1 = [a_0, a_2]$, and $S_k = (a_{k-1}, a_{k+1}]$, if $k \geq 3$ is an odd positive integer, and $S_k = [a_{k+1}, a_{k-1})$ if $k \geq 2$ is an even positive integer. We have that $a_k \to v^*$ as $k \to \infty$. We shall call $S_k, k \geq 0$, the rate intervals.

From the preceding discussion it follows that there is no finite time in which *all points* of the unit square synchronize. The closer v is to the equilibrium v^*, the later is the moment of synchronization.

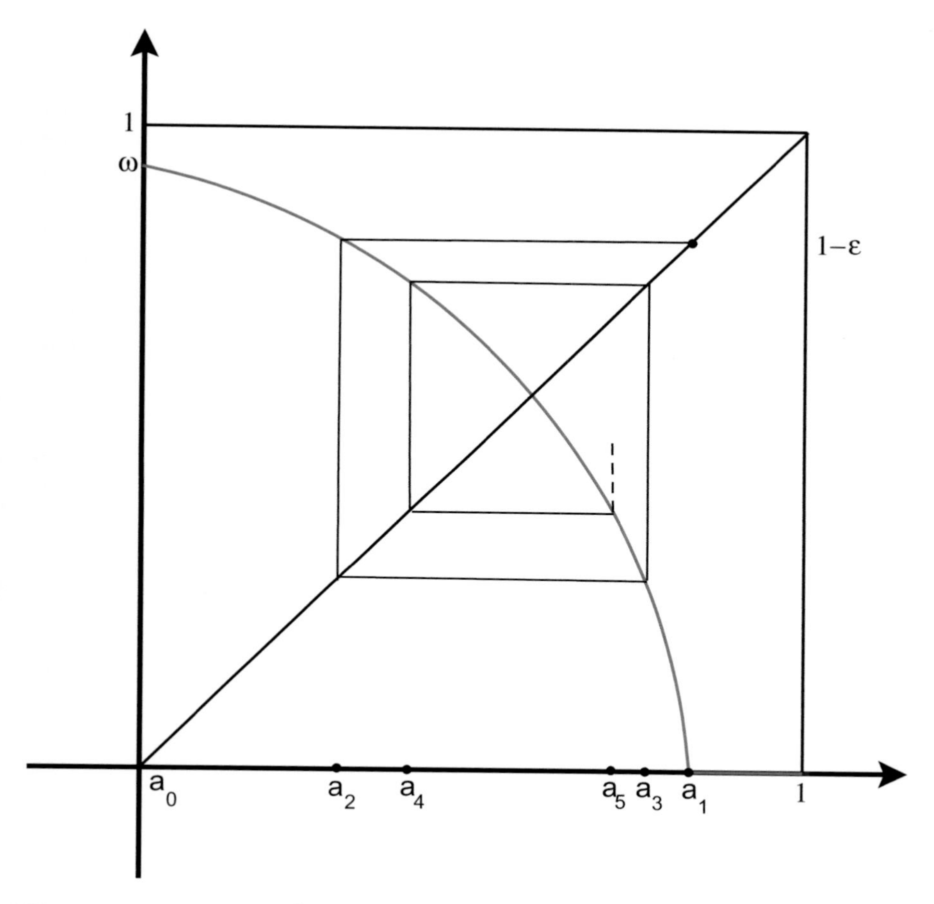

Fig. 1.7 The points $a_0 = L^{-1}(\eta) = 0, a_{k+1} = L^{-1}(a_k), k = 0, 1, 2, \ldots$

Denote by $T = \frac{1}{\gamma} \ln \frac{\kappa}{\kappa-1}$ the natural period of oscillators, that is, the period of motion without couplings, and denote by $\tilde{T}$ the time needed for the solution $u(t, 0, v^*)$ of the equation $u' = S - \gamma u$ to achieve threshold. Since both oscillators fire within an interval of length T and the distance between two firing moments of an oscillator is not less than $\tilde{T}$, the following assertion is valid.

Theorem 3. *Assume that* $t_1 \geq 0$ *is a firing moment,* $x_1(t_1) = 1, x_1(t_1+) = 0$. *If* $x_2(t_1) \in S_m$ *for some natural number* m*, then the couple* x_1, x_2 *of continuously coupled identical biological oscillators synchronizes within the time interval* $[t_0 + \frac{m}{2}\tilde{T}, t_0 + Tm]$.

1.3.2 Synchronization of an Ensemble of Identical Oscillators

Consider the integrate-and-fire model of continuously coupled identical biological oscillators $x_1, x_2, \ldots, x_n$. We intend to apply the map L_C defined in the last section to this model. Let us start with the synchronization of a pair of oscillators in the multioscillatory ensemble and prove that the synchrony occurs for this couple if the parameters are close to zero. Next, we prove the phenomenon for the whole model. Fix two of the oscillators, let's say x_l and x_r.

Lemma 2. *If $t_0 \geq 0$ is a firing moment, $x_l(t_0) = 1, x_l(t_0+) = 0$. If parameter η is sufficiently small, then the couple x_l, x_r synchronizes within the time interval $[t_0, t_0 + T]$ if $x_r(t_0) \notin [a_0, a_1)$ and within the time interval $[t_0 + \frac{m-1}{2}\tilde{T}, t_0 + (m+1)T]$ if $x_r(t_0) \in S_m, m \geq 1$.*

Proof. While the pair does not synchronize, there exists a sequence of firing moments, t_i, such that $0 \leq t_0 < t_1 < \ldots$, the oscillator x_l fires at t_i with even i, and x_r fires at t_i with odd i. For the sake of brevity, let $u_i = x_l(t_i), i = 2j+1, j \geq 0, u_i = x_r(t_i), i = 2j, j \geq 0$.

Let us fix an even i. There are $k, k \leq n-2$, distinct firing moments of the motion $x(t)$ on the interval (t_i, t_{i+1}). Denote by $t_i < \theta_1 < \theta_2 < \ldots < \theta_k < t_{i+1}$ the moments of firing, when at least one of the coordinates of $x(t)$ fires. We assume, without loss of generality, that the length of intervals $(t_i, \theta_1), (\theta_1, \theta_2), \ldots, (\theta_k, t_{i+1})$ is more than τ. Other cases can be considered similarly.

We have

$$
\begin{aligned}
x_r(t_i + \tau) &= x_r(t_i)\mathrm{e}^{-\gamma\tau} + (\kappa + \frac{\eta}{\gamma})(1 - \mathrm{e}^{-\gamma\tau}), \\
x_r(\theta_1) &= x_r(t_i + \tau)\mathrm{e}^{-\gamma(\theta_1 - t_i - \tau)} + \kappa(1 - \mathrm{e}^{-\gamma(\theta_1 - t_i - \tau)}), \\
x_r(\theta_1 + \tau) &= x_r(\theta_1)\mathrm{e}^{-\gamma\tau} + (\kappa + \frac{\eta}{\gamma})(1 - \mathrm{e}^{-\gamma\tau}), \\
x_r(\theta_2) &= x_r(\theta_1 + \tau)\mathrm{e}^{-\gamma(\theta_2 - \theta_1 - \tau)} + \kappa(1 - \mathrm{e}^{-\gamma(\theta_2 - \theta_1 - \tau)}), \\
&\ldots\ldots \\
x_r(\theta_j) &= x_r(\theta_{j-1} + \tau)\mathrm{e}^{-\gamma(\theta_j - \theta_{j-1} - \tau)} + \kappa(1 - \mathrm{e}^{-\gamma(\theta_j - \theta_{j-1} - \tau)}), \\
x_r(\theta_j + \tau) &= x_r(\theta_j)\mathrm{e}^{-\gamma\tau} + (\kappa + \frac{\eta}{\gamma})(1 - \mathrm{e}^{-\gamma\tau}), \\
&\ldots\ldots \\
x_r(t_{i+1}) &= x_r(\theta_k + \tau)\mathrm{e}^{-\gamma(t_{i+1} - \theta_k - \tau)} + \kappa(1 - \mathrm{e}^{-\gamma(t_{i+1} - \theta_k - \tau)}).
\end{aligned}
\tag{1.36}
$$

The moment t_{i+1} satisfies

$$1 - \epsilon \leq x_r(t_{i+1}) \leq 1. \tag{1.37}$$

We also have

$$
\begin{aligned}
x_l(\theta_1) &= \kappa(1 - \mathrm{e}^{-\gamma(\theta_1 - t_i)}),\\
x_l(\theta_1 + \tau) &= x_l(\theta_1)\mathrm{e}^{-\gamma\tau} + (\kappa + \frac{\eta}{\gamma})(1 - \mathrm{e}^{-\gamma\tau}),\\
&\cdots\cdots\\
x_l(\theta_k + \tau) &= x_l(\theta_k)\mathrm{e}^{-\gamma\tau} + (\kappa + \frac{\eta}{\gamma})(1 - \mathrm{e}^{-\gamma\tau}),\\
x_l(t_{i+1}) &= x_l(\theta_k)\mathrm{e}^{-\gamma(t_{i+1} - \theta_k - \tau)} + \kappa(1 - \mathrm{e}^{-\gamma(t_{i+1} - \theta_k - \tau)}).
\end{aligned}
\tag{1.38}
$$

The last three formulas determine the relation $u_{i+1} = K_i(u_i)$. A similar one can be found if i is odd. Evaluations in (1.36) and (1.38) bring us to the expressions

$$
\begin{aligned}
&x_r(t_{i+1}) = x_r(t_i)\mathrm{e}^{-\gamma(t_{i+1} - t_i)} + \kappa(1 - \mathrm{e}^{-\gamma(t_{i+1} - t_i)}) +\\
&\frac{\eta}{\gamma}(1 - \mathrm{e}^{-\gamma\tau})(\mathrm{e}^{-\gamma(t_{i+1} - t_i - \tau)} + \sum_{j=1}^{k} \mathrm{e}^{-\gamma(t_{i+1} - \theta_j - \tau)})
\end{aligned}
\tag{1.39}
$$

and

$$
\begin{aligned}
&x_l(t_{i+1}) = \kappa(1 - \mathrm{e}^{-\gamma(t_{i+1} - t_i)}) +\\
&\frac{\eta}{\gamma}\mathrm{e}^{-\gamma(t_{i+1} - \theta_k - \tau)}(1 - \mathrm{e}^{-\gamma\tau}) \sum_{j=1}^{k} \mathrm{e}^{-\gamma(\theta_k - \theta_j)}.
\end{aligned}
\tag{1.40}
$$

Recall the map L_C defined in the last section. We have

$$
\phi(t_i + \tau) = x_r(t_i)\mathrm{e}^{-\gamma\tau} + (\kappa + \frac{\eta}{\gamma})(1 - \mathrm{e}^{-\gamma\tau}),
$$

$$
\phi(\bar{t}_{i+1}) = \phi(t_i + \tau)\mathrm{e}^{-\gamma(\bar{t}_{i+1} - t_i - \tau)} + \kappa(1 - \mathrm{e}^{-\gamma(\bar{t}_{i+1} - t_i - \tau)}),
$$

or

$$
\begin{aligned}
&\phi(\bar{t}_{i+1}) = x_t(t_i)\mathrm{e}^{-\gamma(\bar{t}_{i+1} - t_i)} +\\
&\kappa(1 - \mathrm{e}^{-\gamma(\bar{t}_{i+1} - t_i)}) + \frac{\eta}{\gamma}\mathrm{e}^{-\gamma(\bar{t}_{i+1} - t_i - \tau)}(1 - \mathrm{e}^{-\gamma\tau}),
\end{aligned}
\tag{1.41}
$$

where $\bar{t}_{i+1}$ satisfies

$$
\phi(\bar{t}_{i+1}) = 1
\tag{1.42}
$$

and

$$\psi(\bar{t}_{i+1}) = \kappa(1 - \mathrm{e}^{-\gamma(\bar{t}_{i+1}-t_i)}), \tag{1.43}$$

to evaluate $L_C(u_i) = \psi(\bar{t}_{i+1})$.

We assume, without loss of generality, that $\bar{t}_{i+1} \leq t_{i+1}$. Then one can find that

$$x_r(\bar{t}_{i+1}) - \phi(\bar{t}_{i+1}) = x_r(\bar{t}_{i+1}) - 1 = \Phi(\eta, \gamma, \tau), \tag{1.44}$$

where

$$\Phi(\eta, \gamma, \tau) = \frac{\eta}{\gamma}(1 - \mathrm{e}^{-\gamma\tau}) \sum_{j=1}^{k} \mathrm{e}^{-\gamma(\bar{t}_{i+1}-\theta_j-\tau)},$$

and the last expression tends to zero as $\eta \to 0$. Next, by applying (1.37) and (1.44), we have

$$t_{i+1} - \bar{t}_{i+1} \leq \frac{|\Phi(\eta, \gamma, \tau)|}{S + \eta - \gamma}.$$

Now, consider

$$K_i(u_i) - L_C(u_i, \epsilon) = x_l(t_{i+1}) - \psi(\bar{t}_{i+1}) = \frac{\eta}{\gamma}\mathrm{e}^{-\gamma(t_{i+1}-\theta_k-\tau)}(1 - \mathrm{e}^{-\gamma\tau}) \sum_{j=1}^{k} \mathrm{e}^{-\gamma(\theta_k-\theta_j)} +$$

$$\kappa(\mathrm{e}^{-\gamma(\bar{t}_{i+1}-t_i)} - \mathrm{e}^{-\gamma(t_{i+1}-t_i)})$$

to see that $K_i(u_i) - L_C(u_i, \epsilon)$ can be made arbitrarily small if η is sufficiently small. This convergence is uniform with respect to u_0. We can also vary the number of points θ_i between 0 and $n-1$, as well as the distance between them. The convergence is indifferent with respect to these variations. Remember that the exciting strengths are not additive. Consider now the sequence $L^i_C(u_0, \epsilon)$. We have $1-\epsilon \leq L^m_C(u_0, \epsilon) \leq 1$. Now, since L_C is a continuous function, we can recurrently discuss inequalities

$$|u_i - L^i_C(u_0, \epsilon)| \leq |K_{i-1}(u_{i-1}) - L^i_C(u_{i-1}, \epsilon)| +$$

$$|L^i_C(u_{i-1}, \epsilon) - L_C(L^{i-1}_C(u_0, \epsilon))|, i = 1, 2, \ldots,$$

to conclude that either $1 - \epsilon \leq u_m \leq 1$ or $1 - \epsilon \leq u_{m+1} \leq 1$ if the parameters are sufficiently small. Both of these inequalities confirm synchronization.

In Fig. 1.8 one can see the sequence of maps K_i, and the synchronizing sequence u_i is constructed. In the figure we show not only u_i, but also the graphs of functions $w = K_i(u), u_{i+1} = K_i(u_i)$, in the neighborhood of u_i, to give a better geometrical visualization of the convergence.

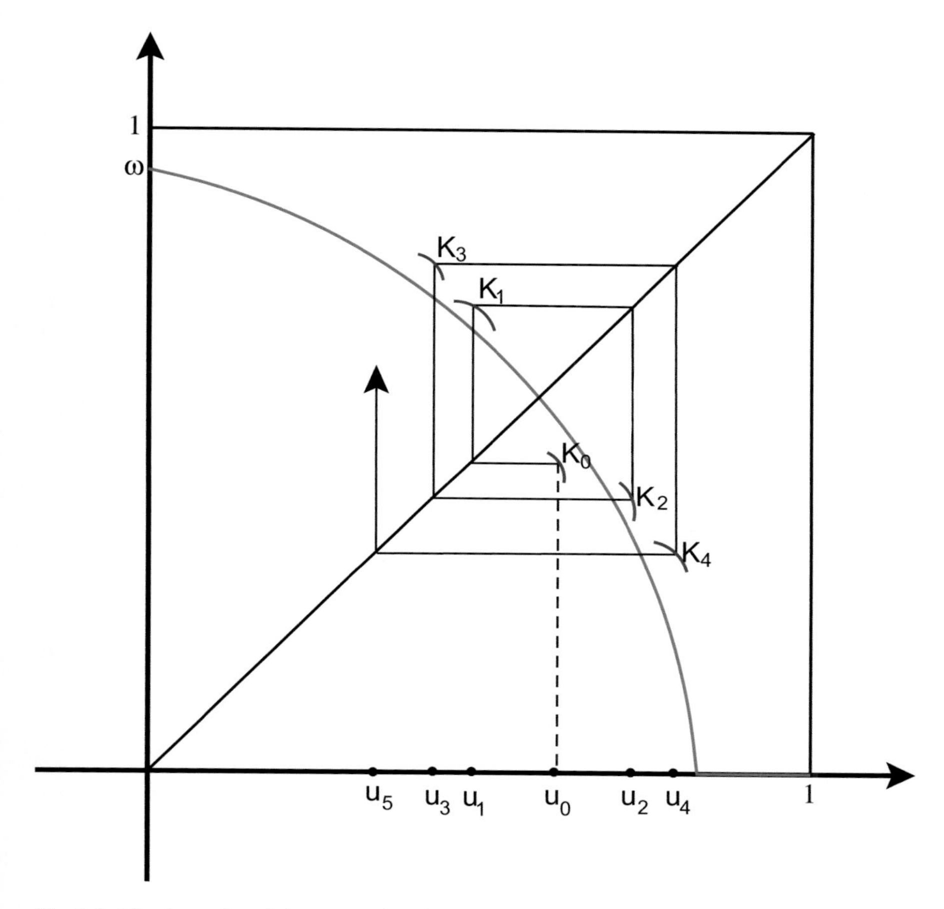

Fig. 1.8 The dynamics of the couple (x_l, x_r)

Since each of the iterations of the map L_C happens within an interval of length not more than T, and the distance between two firing moments of an oscillator is not smaller than $\tilde{T}$, we obtain that the couple x_l, x_r is synchronized no earlier than $t = t_0 + \frac{m-1}{2}\tilde{T}$, and no later than $t = t_0 + (m+1)T$.

Thus, one can conclude that if a couple of oscillators is synchronized at some moment of time, then since the oscillators are identical, the couple persists in firing in unison.

The lemma is proved.

Let us apply the last lemma to the entire ensemble.

Theorem 4. *Let $t_0 \geq 0$ be a firing moment such that $x_j(t_0) = 1, x_j(t_0+) = 0$. If parameter η is sufficiently small, then the motion $x(t)$ of the system synchronizes within the time interval $[t_0, t_0 + T]$, if $x_i(t_0) \notin [a_0, a_1), i \neq j$, and within the time* *interval* $[t_0 + \frac{\max_{i \neq j} k_i - 1}{2}\tilde{T}, t_0 + (\max_{i \neq j} k_i + 1)T]$ *if there exist* $x_s(t_0) \in [a_0, a_1)$ *for some* $s \neq j$ *and* $x_i(t_0) \in S_{k_i}, i \neq j$.

Proof. Apply the last lemma to each pair $(x_j, x_i), i \neq j$ to obtain that it synchronizes within the time interval. The theorem is proved.

On the basis of the last proof and the analysis of formulas (1.41)–(1.43) with (1.39), (1.37), and (1.40), one can conclude that the following assertion, which can be useful in applications and theory, is valid.

Theorem 5. *Assume that* $t_0 \geq 0$ *is a firing moment,* $x_j(t_0) = 1, x_j(t_0+) = 0$. *The motion* $x(t)$ *of the integrate-and-fire model of identical continuously coupled biological oscillators synchronizes within the time interval* $[t_0, t_0 + T]$ *if* $x_i(t_0+) \notin [a_0, a_1), i \neq j$, *and within the time interval* $[t_0 + \frac{\max_{i \neq j} k_i - 1}{2}\tilde{T}, t_0 + (\max_{i \neq j} k_i + 1)T]$ *if there exist* $x_s(t_0) \in [a_0, a_1)$ *for some* $s \neq j$ *and* $x_i(t_0) \in S_{k_i}, i \neq j$, *and if the delay* τ *is sufficiently small.*

1.3.3 Nonidentical Oscillators

Let us describe a more general system of oscillators such that the synchronization is still true.

Consider a system of n nonidentical oscillators $x_i, i = 1, 2, \ldots, n$, whose values are in $[0, 1 + \xi_i]$. We assume that the following conditions are valid:

*B*1). If several oscillators $x_{i_m}, m = 1, 2, \ldots, k$, fire at a moment $t = s$, such that $x_{i_m}(s) = 1 + \xi_{i_m}$ and $x_{i_m}(s+) = 0$, then all other oscillators $x_{i_p}, p = k + 1, k + 1, \ldots, n$, exhibit the following behavior near the moment of firing:

- If $x_{i_p}(s) + \epsilon + \epsilon_{i_p} \geq 1 + \xi_{i_p}$, then $x_{i_p}(s+) = 0$.
- Otherwise,

$$x'_{i_p} = (S + s_{i_p} + \eta + \sum_{m=1}^{k} s_{i_p i_m}(t - \theta_{i_p i_m}(t))) - (\gamma + \gamma_{i_p})x_{i_p}, \qquad (1.45)$$

for all $t \in [s, s + \tau + \tau_{i_p}]$ that belong to the same continuity interval of x_{i_p} as s. Functions s_{ij} are piecewise continuous and $\theta_{i_p i_m}(t) > 0$ are bounded delays. There exist positive constants η_{ij} such that $|s_{ij}(t)| < \eta_{ij}$ for all i, j.

If x_j fires at a moment $t = s$, we say the interval $[s, s + \tau]$ is an e^j-interval. An oscillator x_i *is excited* at a moment t if the moment belongs to an e^j-interval with $j \neq i$, or $x_i(t) = 1 + \xi_i$.

*B*2). When the ith oscillator is not excited,

$$x'_i = (S + s_i) - (\gamma + \gamma_i)x_i. \qquad (1.46)$$

In $(B1)$ and $(B2)$ constants $S, \gamma, \epsilon, \eta$ are the same as in $(A1) - (A4)$, parameters $s_i, \gamma_i, \xi_i, \eta_{ij}, \tau_{i,j}, i, j = 1, 2, \ldots, n$, are fixed real numbers. Additionally, we require that

$B3)$. $\tau + \tau_{i_p} > 0, \eta - \sum_{s=1}^{k} \eta_{i_p i_s} > 0$, for all possible k, i_p and i_s.

We shall call the system of n oscillators with conditions $(B1) - (B3)$, $(A3)$ *the integrate-and-fire model of continuously coupled nonidentical biological oscillators.*

Theorem 6. *Let $t_0 \geq 0$ be a firing moment such that $x_j(t_0) = 1, x_j(t_0+) = 0$. If parameters $s_i, \gamma_i, \zeta_i, \eta_{ij}, \tau_{ij}$, and η are sufficiently small, then the motion $x(t)$ of the integrate-and-fire model of continuously coupled nonidentical biological oscillators synchronizes within the time interval $[t_0, t_0 + T]$ if $x_i(t_0) \notin [a_0, a_1), i \neq j$, and within the time interval $[t_0 + \frac{\max_{i \neq j} k_i - 1}{2}\tilde{T}, t_0 + (\max_{i \neq j} k_i + 1)T]$ if there exist $x_s(t_0) \in [a_0, a_1)$ for some $s \neq j$ and $x_i(t_0) \in S_{k_i}, i \neq j$.*

We decided to omit the proof of the last theorem since it is very similar to that of Theorem 4 with slight changes caused by newly introduced parameters. Still, one point in the proof deserves special attention. If two oscillators x_l and x_r are nonidentical and fire simultaneously at a moment $t = \theta$, how will they retain the state of firing in unison despite being different? To find the required conditions, let us denote by $\tau, \tau > \theta$ a moment when one of them, let's say x_l, fires. We have that $x_l(\theta+) = x_r(\theta+) = 0$. This time it is not necessary to have $x_l(t) = x_r(t), \theta \leq t \leq \tau$. It is clear that to satisfy $x_l(\tau+) = x_r(\tau+) = 0$, we need $x_r(\tau) + \epsilon + \epsilon_r \geq 1 + \xi_r$. By applying formulas similar to (1.36) and (1.37), this time with $t_i = \theta, t_{i+1} = \tau, x_r(\theta) = 0$, one can easily obtain that the inequality is correct if the parameters are sufficiently small. Thus, one can conclude that if a couple of oscillators is synchronized at some moment of time, then it will persistently fire in unison.

Remark 5. We do not impose any restriction on the delay functions $\theta_{i_p i_m}(t)$ in (1.45) except that they be bounded functions. Oscillators with a delayed excitatory interaction, without leakage, and their applications are discussed in [19]. Similar to the way it is done for pulse-coupled models in [3], all results can be extended to systems when the coupling is not all-to-all, and general types of thresholds and differential equations are considered. The parameter η is chosen as the main one to establish synchronization. It is obvious that the choice of the control can be varied, for example, by choosing τ, or both of them, instead.

1.3.4 Simulations

To illustrate the theory, consider a system of oscillators, $x_1, x_2, \ldots, x_{100}$, with random start values in $[0, 1]$. Choose, also randomly, numbers $\xi_i, \alpha_i, \beta_i, i = 1, 2, \ldots, 100$, from the interval $[0, 1]$. Assume that if $x_j(s) = 1 + 0.005\xi_j$ at some moment $t = s$, then the oscillator fires, $x_j(s+) = 0$, and other oscillators

$x_i, i \neq j$, change their behavior near the firing moment in the following way: if $x_i(s) + 0.03 \geq 1 + \xi_i$, then $x_i(s+) = 0$; otherwise,

$$x_i' = (13 + 0.01\alpha_i) - (2 + 0.01\beta_i)x_i, \tag{1.47}$$

for all $t \in [s, s + 0.01]$, until x_i fires.

If x_j fires at a moment $t = s$, then an oscillator x_i is excited at the moment t if either the moment belongs to the interval $[s, s + 0.01]$ with $j \neq i$ or $x_i(t) = 1 + 0.005\xi_i$.

When $x_i, i = 1, 2, \ldots, n$, is not excited, then

$$x_i' = (3 + 0.01\alpha_i) - (2 + 0.01\beta_i)x_i. \tag{1.48}$$

In Fig. 1.9 one can see the result of the simulation, where the upper left figure corresponds to the initial states, the upper right one shows the situations just before the 30th jump, the lower left one shows the conditions just before the 60th jump, and the final one depicts the state before the 90th jump of the system.

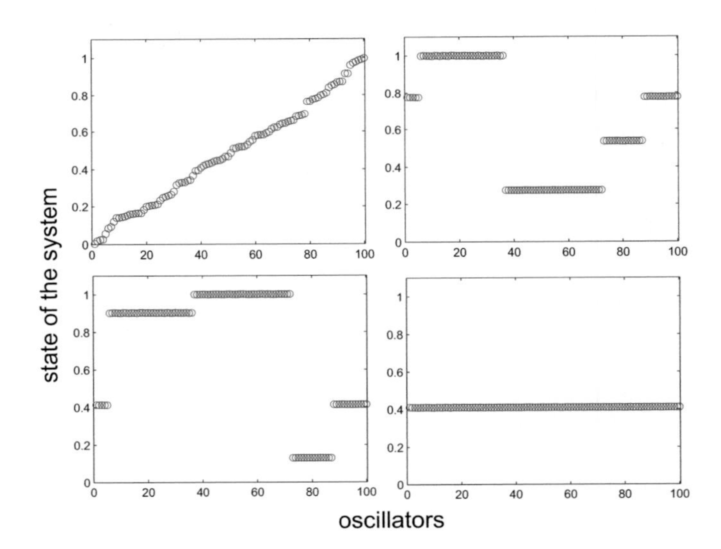

Fig. 1.9 The figure in the *upper left corner* depicts the initial positions, the one in the *upper right corner* depicts the situation just before the 30th jump, the one in the *lower left corner* depicts just before the 60th jump, and the final figure shows the positions before the 90th jump of the system. The flat sections of the graph are groups of synchronized oscillators

1.4 Integrate-and-Fire Oscillators with Delayed Couplings

1.4.1 A Couple of Identical Oscillators

Let us start with the analysis of two identical oscillators which satisfy, if they do not fire, the following differential equations:

$$x_i' = S - \gamma x_i, \tag{1.49}$$

where $0 \leq x_i \leq 1, i = 1,2$. It is assumed that S, γ are positive numbers and $\kappa = \frac{S}{\gamma} > 1$. In (1.49) each $x_i, i = 1,2$, is a voltage-like state variable, S is an external stimulus, and γ is the leakage coefficient.

When $x_j(t) = 1$, then the oscillator fires, $x_j(t+) = 0$. The firing changes the value of another oscillator, x_i, such that

$$x_i(t+) = 0 \tag{1.50}$$

if $x_i(t) \geq 1 - \epsilon$; otherwise,

$$x_i(t+\tau+) = \begin{cases} x_i(t+\tau) + \epsilon, \text{ if } x_i(t+\tau) < 1-\epsilon, \\ 0, \text{ otherwise.} \end{cases} \tag{1.51}$$

Thus, (1.51) implies that $t + \tau$ is a firing moment for x_i if the jump makes the value of the oscillator not smaller than 1.

In paper [50], the following coupling mechanism was introduced. If oscillator x_j fires at the moment t, then the firing changes the value of the other oscillator, x_i, such that

$$x_i(t+) = \begin{cases} x_i(t) + \epsilon, \text{ if } x_i(t) < 1-\epsilon, \\ 0, \text{ otherwise.} \end{cases} \tag{1.52}$$

That is, no delay was assumed for the pulse coupling.

In what follows, assume that

$$\frac{\kappa - 1}{\kappa - 1 + \epsilon} < \mathrm{e}^{-\gamma\tau}. \tag{1.53}$$

We have

$$x_i(s) = x_i(t)\mathrm{e}^{-\gamma(s-t)} + \int_t^s \mathrm{e}^{-\gamma(s-u)} S du$$

near t, where t is assumed again to be the firing moment for x_j, and

$$x_i(s) \leq (1-\epsilon)\mathrm{e}^{-\gamma\tau} + \kappa(1 - \mathrm{e}^{-\gamma\tau}).$$

Equation (1.53) implies that if $x_i(t) < 1 - \epsilon$, then $x_i(s) < 1$ for all $s \in [t, t+\tau]$. In other words, the oscillator x_i does not achieve the threshold within interval $[t, t+\tau]$ if the distance of $x_i(t)$ to threshold is more than ϵ. This is important for the construction of the prototype map and makes sense of condition (1.51).

One must emphasize that couplings of units are not only delayed in our model. By (1.50), oscillators interact instantaneously if they are near the threshold. This assumption is natural as firing provokes another oscillator, which, being close to the threshold, "is ready" to react instantaneously. Otherwise, the interaction is delayed.

Next, we shall construct the prototype map. Fix a moment $t = \zeta$ when x_1 fires, and suppose that oscillators are not synchronized. In interval $[\zeta, \zeta+\tau]$, the oscillator x_2 moves according to the law $x_2(t) = x_2(\zeta)e^{-\gamma(t-\zeta)} + \int_\zeta^t e^{-\gamma(t-u)}Sdu$, and

$$x_2(\zeta+\tau) = [x_2(\zeta) - \kappa]e^{-\gamma\tau} + \kappa. \tag{1.54}$$

Let us first handle the problem in the case that $x_2(\zeta+\tau) + \epsilon < 1$. One can verify that this is true if $x_2(\zeta) < \bar{v}$, where $\bar{v} = e^{\gamma\tau}(1 - \epsilon - \kappa) + \kappa$. It is important that $\bar{v} < 1 - \epsilon_1$, where $\epsilon_1 = \epsilon e^{\gamma\tau}$. Take $\tau > 0$ so small such that the inequality

$$e^{-\gamma\tau} > \epsilon \tag{1.55}$$

holds. Equation (1.55) implies that $\epsilon_1 < 1$. If we denote by $t = \eta$ the firing moment of x_2, then one can reveal that

$$x_2(\eta) = [x_2(\zeta+\tau) + \epsilon]e^{-\gamma(\eta-\zeta-\tau)} + \kappa[1 - e^{-\gamma(\eta-\zeta-\tau)}].$$

The equation $x_2(\eta) = 1$ implies the following:

$$e^{-\gamma(\eta-\zeta)} = \frac{1-\kappa}{x_2(\zeta) - \kappa + \epsilon_1}. \tag{1.56}$$

It follows from $x_1(\eta) = \kappa[1 - e^{-\gamma(\eta-\zeta)}]$ that

$$x_1(\eta) = \kappa\frac{1 - (x_2(\zeta) + \epsilon_1)}{\kappa - (x_2(\zeta) + \epsilon_1)}. \tag{1.57}$$

Let us introduce the following map:

$$L_D(v, \epsilon) = \kappa\frac{1 - (v + \epsilon_1)}{\kappa - (v + \epsilon_1)}, \text{ for } 0 < v < \bar{v}, \tag{1.58}$$

such that $x_1(\eta) = L_D(x_2(\zeta)), \epsilon)$.

Next, let us consider the case that $1 \leq x_2(\zeta+\tau) + \epsilon$. By (1.51) we have that $\eta = \zeta + \tau$, and $x_1(\eta) = \kappa[1 - e^{-\gamma\tau}]$. Set $\tilde{v} = \kappa[1 - e^{-\gamma\tau}]$, and introduce

$$L_D(v, \epsilon) = \tilde{v}, \quad \text{for } \bar{v} \leq v < 1 - \epsilon. \tag{1.59}$$

In what follows we assume that

$$e^{\gamma\tau} < \sqrt{\frac{\kappa}{\kappa - 1 + \epsilon_1}}. \tag{1.60}$$

Now, we will define an extension of L_D on $[0, 1]$ in the following way. Let

$$\omega = \kappa \frac{1 - \epsilon_1}{\kappa - \epsilon_1}. \tag{1.61}$$

One can see that $1 - \epsilon < \omega < 1$, provided that

$$e^{\gamma\tau} < \frac{\kappa}{\kappa - 1 + \epsilon_1}. \tag{1.62}$$

In the sequel, we assume that the number ϵ is sufficiently small such that (1.53) implies (1.62). We set $L_D(0, \epsilon) = \omega$, and define $L_D(v, \epsilon) = 0$ if $1 - \epsilon \leq v \leq 1$.

The derivatives of the map in $(0, \bar{v})$ satisfy

$$L'_D(v, \epsilon) = \kappa \frac{1 - \kappa}{(\kappa - (v + \epsilon_1))^2} < 0, \tag{1.63}$$

and

$$L''_D(v, \epsilon) = 2\kappa \frac{1 - \kappa}{(\kappa - (v + \epsilon))^3} < 0. \tag{1.64}$$

It is possible to verify that there is a fixed point of the map,

$$v^* = (\kappa - \frac{\epsilon_1}{2}) - \sqrt{\kappa^2 - \kappa + \frac{\epsilon_1^2}{4}}, \tag{1.65}$$

and that

$$L'_D(v^*, \epsilon) < -1. \tag{1.66}$$

In other words, the fixed point v^* is a repeller. The inequality $v^* < \bar{v}$ holds if condition (1.60) is valid; consequently, all our previous evaluations are justified.

Suppose, additionally, that

$$\kappa(1 - e^{-\gamma\tau}) < \frac{\epsilon\kappa}{\kappa - 1 + \epsilon} - \epsilon e^{\gamma\tau}. \tag{1.67}$$

Denote by $v = a_2$ the solution of the equation $L_D(v, \epsilon) = 1 - \epsilon$. We find that $a_2 = \frac{\epsilon\kappa}{\kappa - 1 + \epsilon} - \epsilon e^{\gamma\tau}$. From the last inequality, we have $\tilde{v} < a_2$.

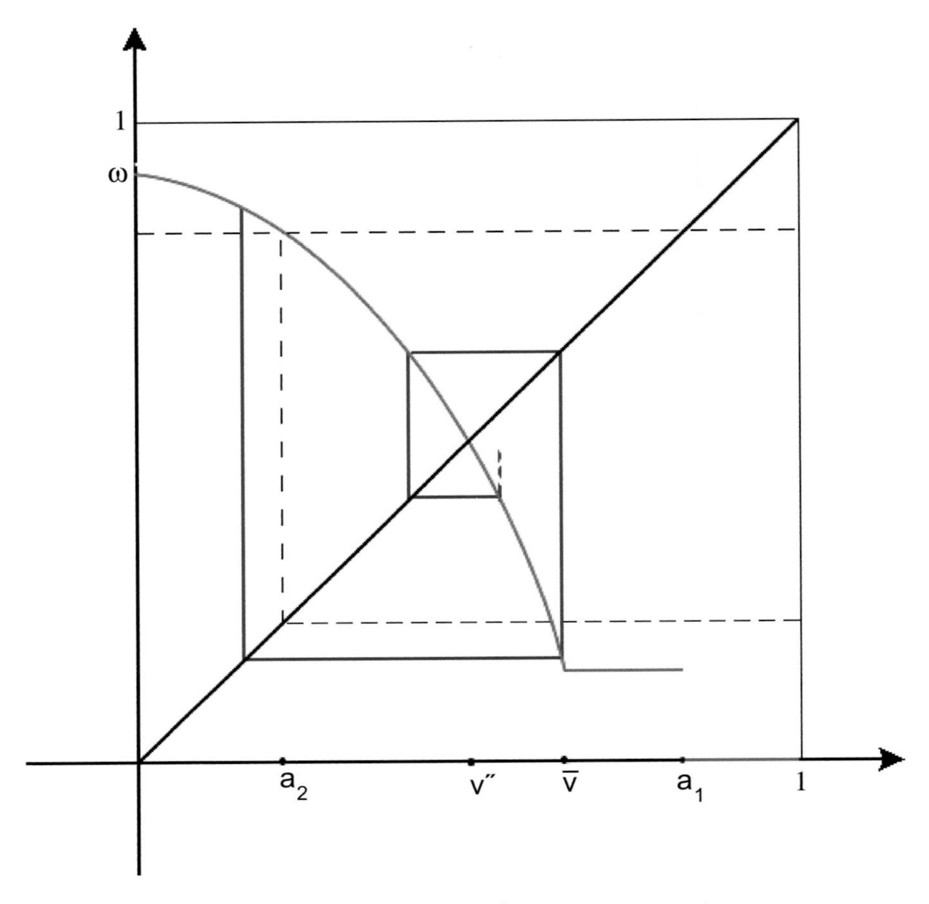

Fig. 1.10 The graph of map L_D in *red*, fixed point v^*, and stabilized trajectory are seen

We call τ the *small delay* since (1.53), (1.55), (1.60), (1.62), and (1.67) are assumed to be true. The graph of L_D (in red) under the above-mentioned conditions is illustrated in Fig. 1.10. One can see that L_D is a piecewise map. This is a curious fact, since in our previous papers for nondelayed pulse couplings or continuous couplings, the prototype map was continuous. Obviously, the discontinuity of the map gives more possibilities for various dynamical collective effects of integrate-and-fire oscillators.

To emphasize a significance of this map for the present analysis, let us see how iterations of it can help to observe the synchronization. Fix $t_0 \geq 0$, a firing moment, such that $x_1(t_0) = 1$ and $x_1(t_0+) = 0$. When the couple x_1 and x_2 are not in synchrony, there exists a sequence of moments $t_0 < t_1 < \ldots$ such that x_1 fires at t_i with even i and x_2 with odd indices. Denote $u_i = x_1(t_i)$ if i is odd, and $u_i = x_2(t_i)$ if i is even. One can easily see that $u_{i+1} = L_D(u_i, \epsilon), i \geq 0$. The pair synchronizes if and only if there exists $j \geq 1$ such that $x_1(t) \neq x_2(t)$ if $t \leq t_j$, and $x_1(t) = x_2(t)$ for $t > t_j$. In particular, both oscillators have to fire at t_j. In other words, the inequalities $1 - \epsilon \leq u_{j-1} < 1$ are valid. In particular, we have that $L_D(0) = \omega$ satisfies this condition. At the same time, if $1 - \epsilon_1 \leq u_{j-3} \leq 1$, then $u_{j-2} = 0 = L_D(u_{j-3})$ and $1 - \epsilon < u_{j-1} = \omega < 1$ again. That is, we have found that if there exists an integer $k \geq 0$ such that $1 - \epsilon \leq L_D^k(v) \leq 1$, then the motion $(x_1(t), x_2(t))$ with $x_1(t_0+) = v, x_2(t_0+) = 0$ synchronizes at the kth firing moment. Conversely, if a motion $(x_1(t), x_2(t))$ synchronizes, then one can find a firing moment, t_0, such that $x_1(t_0+) = 0, x_2(t_0+) = v, v \in [0, 1]$, and a number k with the property that $1 - \epsilon \leq L_D^k(v) \leq 1$.

Thus, the last discussion confirms that the analysis of synchronization is fully consistent with the dynamics of the introduced map $L_D(v, \epsilon)$ on $[0, 1]$, and the map L_D can be applied as the main instrument of the chapter. That is why we use this function as a prototype map in our investigations.

Now, with the help of the properties of the map L_D, and by analyzing self-compositions of the map, one can easily attain that for all $k \geq 0$ the functions L_D^k have only one fixed point, v^*, and $|[L_D^k(v^*, \epsilon)]'| > 1$. We skip the discussion as it is respectively simple and takes significant space. Since all the maps L_D^k have one and the same fixed point, v^*, there is not a k-periodic motion, $k > 1$, of the map. Consequently, for an arbitrary point $v \neq v^*$ one has a stabilized trajectory as presented in Fig. 1.10. The couple synchronizes when $L_D^k(v, \epsilon) \geq 1 - \epsilon$.

Next, we investigate the rate of synchronization. Set $a_0 = 0, a_1 = 1 - \epsilon$, and $a_{k+1} = L_D^{-1}(a_k), k = 2, 3, \ldots$ (see Fig. 1.11).

Denote by S_k the subset of the interval $[0, 1]$, consisting of the points v which are synchronized after exactly k iterations of the map L_D. It is easy to verify that $S_0 = [a_1, 1], S_1 = [a_0, a_2]$, and $S_k = (a_{k-1}, a_{k+1}]$ if $k \geq 3$ is an odd positive integer, and $S_k = [a_{k+1}, a_{k-1})$ if $k \geq 2$ is an even positive integer. One can observe that $a_k \to v^*$ as $k \to \infty$. We shall call $S_k, k \geq 0$, the rate intervals.

From the preceding discussion it follows that no finite time is available such that *all* points of the unit square synchronize at that moment. The closer v is to the equilibrium v^*, the later is the moment of synchronization.

Set $T = \frac{1}{\gamma} \ln \frac{\kappa}{\kappa - 1}$ and denote by $\tilde{T}$ the time needed for the solution $u(t, 0, v^*)$ of the equation $u' = S - \gamma u$, to achieve the threshold. Since all oscillators fire within an interval of length T and the distance between two firing moments of an oscillator is not less than $\tilde{T}$, we can conclude the validity of the following theorem.

Theorem 7. *Assume that conditions (1.53), (1.55), (1.60), (1.62), and (1.67) are valid. If $t_0 \geq 0$ is a firing moment, $x_1(t_0) = 1, x_1(t_0+) = 0$, and $x_2(t_0+) \in S_m$ for some natural number m, then the couple x_1, x_2 of continuously coupled identical biological oscillators synchronizes within the time interval* $[t_0 + \frac{m}{2}\tilde{T}, t_0 + Tm]$.

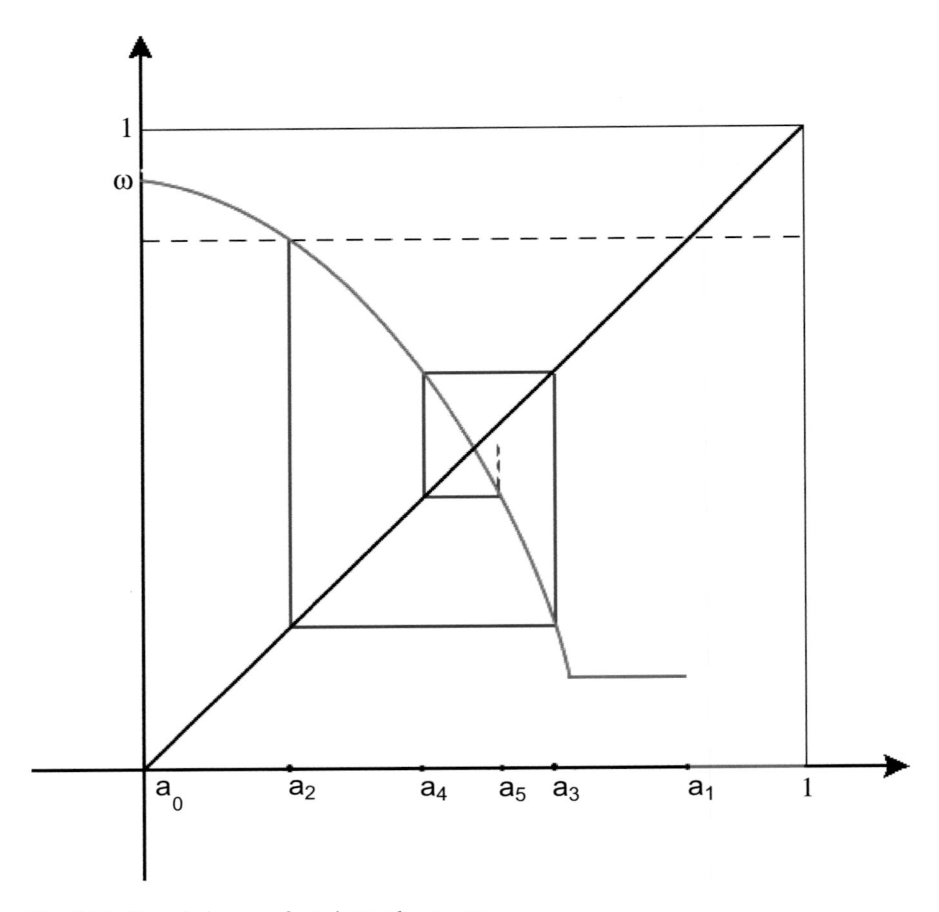

Fig. 1.11 Boundaries, a_i, of rate intervals are seen

1.4.2 Nonidentical Oscillators: The General Case

To make our investigation closer to real-world problems, one has to consider an ensemble of nonidentical oscillators. We will discuss the following system of equations:

$$x_i' = (S + \mu_i) - (\gamma + \zeta_i)x_i, \tag{1.68}$$

where $0 \le x_i \le 1 + \xi_i, i = 1, 2, \ldots, n$. The constants S and γ are the same as in the last section such that $\kappa = \frac{S}{\gamma} > 1$. Moreover, the constants μ_i and ζ_i are sufficiently small, satisfying $\kappa_i = \frac{S+\mu_i}{\gamma+\zeta_i} > 1$. When $x_j(t) = 1 + \xi_j$, the oscillator fires and $x_j(t+) = 0$. The firing changes the values of other oscillators $x_i, i \ne j$, such that

$$x_i(t+) = 0, \text{ if } x_i(t) \ge 1 - \epsilon, \tag{1.69}$$

and

$$x_i(t+\tau+) = \begin{cases} x_i(t+\tau)+\epsilon, & \text{if } x_i(t+\tau) < 1-\epsilon, \\ 0, & \text{otherwise.} \end{cases} \tag{1.70}$$

In what follows, we call the real numbers $\epsilon, \mu_i, \zeta_i, \xi_i, \epsilon_i$, *parameters*, assuming the first one is positive. Moreover, the constants $\mu_i, \zeta_i, \xi_i, \epsilon_i$ will be called *parameters of perturbation*. To achieve the model of *identical oscillators*, assume that the parameters are all zero. In our case, an exhibitory model is under discussion; that is, $\epsilon + \epsilon_i > 0$ for all i. Coupling is all-to-all such that each firing elicits jumps in all nonfiring oscillators. If several oscillators fire simultaneously, then other oscillators react as if just one oscillator fired. In other words, any firing acts only as a signal which abruptly provokes a state change, the intensity of the signal is not important, and pulse strengths are not additive. Moreover, we have

$$x_i(s) = x_i(t)\mathrm{e}^{-(\gamma+\zeta_i)(s-t)} + \int_t^s \mathrm{e}^{-(\gamma+\zeta_i)(s-u)}(S+\mu_i)du,$$

near t.

Under the circumstances that condition (1.53) is valid and the constants μ_i and ζ_i are sufficiently small such that

$$\frac{\kappa_i - 1}{\kappa_i - 1 + \epsilon} < \mathrm{e}^{-(\gamma+\zeta_i)\tau}, \tag{1.71}$$

we have $x_i(s) < 1$ for all $s \in [t, t+\tau]$, provided that $x(t) < 1-\epsilon$.

We begin this section by analyzing a couple of oscillators of the ensemble of n oscillators, finding that the couple synchronizes if the parameters are close to zero. After this, we will prove synchronization of the ensemble.

Consider the model of n nonidentical oscillators given by relations (1.49) and (1.51). Fix two of them, say x_l and x_r.

Lemma 3. *Assume that the inequalities (1.53), (1.55), (1.60), (1.62), and (1.67) are valid and $t_0 \geq 0$ is a firing moment such that $x_l(t_0) = 1+\xi_i, x_l(t_0+) = 0$. If the parameters are sufficiently close to zero and the absolute values of the parameters of perturbation are sufficiently small with respect to ϵ, then the couple x_l, x_r synchronizes within the time interval $[t_0, t_0+T]$ if $x_r(t_0+) \notin [a_0, a_1)$ and within the time interval $[t_0 + \frac{m-1}{2}\tilde{T}, t_0 + (m+1)T]$ if $x_r(t_0+) \in S_m, m \geq 1$.*

Proof. If $1+\xi_r - \varepsilon - \varepsilon_r \leq x_r(t_0) \leq 1+\xi_r$, then two oscillators fire simultaneously, and we have only to prove the persistence of the synchrony, which will be discussed later. Thus, fix another oscillator $x_r(t)$ such that $0 \leq x_r(t_0) < 1+\xi_r - \varepsilon - \varepsilon_r$. If the couple is not synchronized, then there is a sequence $\{t_i\}$ of firing moments such that $0 \leq t_0 < t_2 < \ldots$, and the oscillator x_l fires at t_i, with i even, and x_r fires at t_i with odd i. For the sake of brevity let's note that $u_i = x_l(t_i), i = 2j+1$, and $u_i = x_r(t_i), i = 2j, j \geq 0$. In what follows we shall show how one can evaluate u_{i+1}

through $L(u_i)$. Consider the case where i is even. There are $k \leq n-2$ distinct firing moments of the motion $x(t)$ in the interval (t_i, t_{i+1}). Denote by $t_i < \theta_1 < \theta_2 < \ldots < \theta_k < t_{i+1}$, the moments of firing, when at least one of the coordinates of $x(t)$ fires. We have

$$
\begin{aligned}
& x_r(\theta_1+\tau) = (x_r(t_i+\tau)+\epsilon+\epsilon_r)\mathrm{e}^{-(\gamma+\zeta_r)(\theta_1+\tau-t_i)} + \\
& \kappa_r(1-\mathrm{e}^{-(\gamma+\zeta_r)(\theta_1+\tau-t_i)}), \\
& x_r(\theta_2+\tau) = (x_r(\theta_1+\tau)+\epsilon+\epsilon_r)\mathrm{e}^{-(\gamma+\zeta_r)(\theta_2-\theta_1)} + \\
& \kappa_r(1-\mathrm{e}^{-(\gamma+\zeta_r)(\theta_2-\theta_1)}), \\
& \ldots\ldots \\
& x_r(\theta_j+\tau) = (x_r(\theta_{j-1}+\tau)+\epsilon+\epsilon_r)\mathrm{e}^{-(\gamma+\zeta_r)(\theta_j-\theta_{j-1})} + \\
& \kappa_r(1-\mathrm{e}^{-(\gamma+\zeta_r)(\theta_j-\theta_{j-1})}), \\
& \ldots\ldots \\
& x_r(t_{i+1}) = (x_r(\theta_k+\tau)+\epsilon+\epsilon_r)\mathrm{e}^{-(\gamma+\zeta_r)(t_{i+1}-\theta_k-\tau)} + \\
& \kappa_r(1-\mathrm{e}^{-(\gamma+\zeta_r)(t_{i+1}-\theta_k-\tau)}).
\end{aligned} \tag{1.72}
$$

The moment t_{i+1} satisfies the following:

$$
1+\xi_r-\epsilon-\epsilon_r \leq x_r(t_{i+1}) \leq 1+\xi_r \tag{1.73}
$$

and continuously depends on the parameters and $x_r(t_i)$.

We also have

$$
\begin{aligned}
& x_l(\theta_1+\tau) = \kappa_l(1-\mathrm{e}^{-(\gamma+\zeta_l)(\theta_1+\tau-t_i)}), \\
& x_l(\theta_2+\tau) = (x_l(\theta_1+\tau)+\epsilon+\epsilon_l)\mathrm{e}^{-\gamma(\gamma+\zeta_l)(\theta_2-\theta_1)} + \\
& \kappa_l(1-\mathrm{e}^{-(\gamma+\zeta_l)(\theta_2-\theta_1)}), \\
& \ldots\ldots \\
& x_l(\theta_j+\tau) = (x_l(\theta_{j-1}+\tau)+\epsilon+\epsilon_l)\mathrm{e}^{-(\gamma+\zeta_l)(\theta_j-\theta_{j-1})} + \\
& \kappa_l(1-\mathrm{e}^{-(\gamma+\zeta_l)(\theta_j-\theta_{j-1})}), \\
& \ldots\ldots \\
& x_l(t_{i+1}) = (x_l(\theta_k+\tau)+\epsilon+\epsilon_l)\mathrm{e}^{-(\gamma+\zeta_l)(t_{i+1}-\theta_k-\tau)} + \\
& \kappa_l(1-\mathrm{e}^{-(\gamma+\zeta_l)(t_{i+1}-\theta_k-\tau)}).
\end{aligned} \tag{1.74}
$$

The last two formulas describe the dependence of u_{i+1} on u_i. One can easily find a similar relation for the case when i is odd.

Set $\delta_i(\mu_i, \zeta_i) = \kappa_i - \kappa$. It is clear that $\delta_i(0,0) = 0$. By means of (1.72) and (1.74), it is possible to achieve

$$x_r(t_{i+1}) = (x_r(t_i + \tau) + \epsilon)\mathrm{e}^{-\gamma(t_{i+1}-t_i)}\mathrm{e}^{-\zeta_r(t_{i+1}-t_i)} +$$
$$\kappa(1 - \mathrm{e}^{-(\gamma+\zeta_r)(t_{i+1}-t_i)}) + \epsilon_r\mathrm{e}^{-\gamma(t_{i+1}-t_i)}\mathrm{e}^{-\zeta_r(t_{i+1}-t_i)} +$$
$$(\epsilon + \epsilon_r)\sum_{j=1}^{k}\mathrm{e}^{-(\gamma+\zeta_r)(t_{i+1}-\theta_j-\tau)} + \tag{1.75}$$
$$\delta_r(1 - \mathrm{e}^{-(\gamma+\zeta_r)(t_{i+1}-t_i)}), \tag{1.76}$$

and

$$x_l(t_{i+1}) = (\kappa + \delta_l)(1 - \mathrm{e}^{-(\gamma+\zeta_l)(t_{i+1}-t_i)}) +$$
$$(\epsilon + \epsilon_l)\sum_{j=1}^{k}\mathrm{e}^{-(\gamma+\zeta_l)(t_{i+1}-\theta_j-\tau)}. \tag{1.77}$$

Now, recall the map L_D defined in the last section. One can find out that

$$\phi(\bar{t}_{i+1}) = (x_r(t_i + \tau) + \epsilon)\mathrm{e}^{-\gamma(\bar{t}_{i+1}-t_i-\tau)} +$$
$$\kappa(1 - \mathrm{e}^{-\gamma(\bar{t}_{i+1}-t_i)}), \tag{1.78}$$

where $\bar{t}_{i+1}$ satisfies

$$\phi(\bar{t}_{i+1}) = 1 \tag{1.79}$$

and

$$\psi(\bar{t}_{i+1}) = \kappa(1 - \mathrm{e}^{-\gamma(\bar{t}_{i+1}-t_i)}). \tag{1.80}$$

With the help of the definition of L_D, we obtain $L_D(u_i) = \psi(\bar{t}_{i+1})$.
Without loss of generality, assume that $\bar{t}_{i+1} \leq t_{i+1}$. In this case, one has

$$\phi(\bar{t}_{i+1}) - x_r(\bar{t}_{i+1}) = 1 - x_r(\bar{t}_{i+1}) =$$
$$\Phi_1(\epsilon, \epsilon_r, \zeta_r, \delta_r, \tau), \tag{1.81}$$

where

$$\Phi_1(\epsilon, \epsilon_r, \zeta_r, \delta_r, \tau) =$$
$$\kappa(1 - \mathrm{e}^{-\gamma(\bar{t}_{i+1}-t_i)})(\mathrm{e}^{-\zeta_r(\bar{t}_{i+1}-t_i)} - 1) -$$
$$(x_r(t_i + \tau) + \epsilon)\mathrm{e}^{-\gamma(\bar{t}_{i+1}-t_i)}(\mathrm{e}^{-\zeta_r(\bar{t}_{i+1}-t_i)} - 1) -$$
$$\epsilon_r\mathrm{e}^{-\gamma(\bar{t}_{i+1}-t_i)}\mathrm{e}^{-\zeta_r(\bar{t}_{i+1}-t_i)} - (\epsilon + \epsilon_r)\sum_{j=1}^{k}\mathrm{e}^{-(\gamma+\zeta_r)(\bar{t}_{i+1}-\theta_j-\tau)} -$$
$$\delta_r(1 - \mathrm{e}^{-(\gamma+\zeta_r)(\bar{t}_{i+1}-t_i)}),$$

and the last expression tends to zero as all of its arguments tend to zero. Next, by utilizing (1.73) and (1.81) we achieve $t_{i+1} - \bar{t}_{i+1} \leq \Phi_2(\epsilon, \epsilon_r, \zeta_r, \delta_r)$, where

$$\Phi_2(\epsilon, \epsilon_r, \zeta_r, \delta_r, \tau) \equiv \frac{|\xi_r| + \epsilon + |\epsilon_r| + \Phi_1(\epsilon, \epsilon_r, \zeta_r, \delta_r, \tau)}{S - |\mu_r| - \gamma - |\zeta_r|}.$$

Now, from the last equation, (1.77), and (1.80), one can see that

$$|L_D(u_i) - K_i(u_i)| = |x_l(t_{i+1}) - \psi(\bar{t}_{i+1})| \leq |x_l(t_{i+1}) - x_l(\bar{t}_{i+1})| +$$

$$|x_l(\bar{t}_{i+1}) - \psi(\bar{t}_{i+1})| \leq \Phi_2(S + |\mu_l| + \gamma + |\zeta_l|) + \Phi_1.$$

That is, the difference $L_D(u_i, \epsilon) - u_{i+1}$ can be made arbitrarily small if the parameters are sufficiently close to zero. Moreover, we should assume smallness of absolute values of the parameters of perturbation with respect to ϵ, to satisfy (1.73). This convergence is uniform with respect to u_0. We can also vary the number of points θ_i and their location in the intervals (t_j, t_{j+1}) between 0 and $n-1$. The convergence is indifferent with respect to these variations, too.

Consider $L_D^i(u_0, \epsilon)$. It is true that $L_D^m(u_0, \epsilon) \in [1-\epsilon, 1]$. Assume, without loss of generality, that m is an even number. Since L_D is a continuous function, we can find recurrently, by applying the following sequence of inequalities, $|u_i - L_D^i(u_0, \epsilon)| \leq |u_i - L_D(u_{i-1}, \epsilon)| + |L_D(u_{i-1}, \epsilon) - L_D(L_D^{i-1}(u_0, \epsilon))|, i = 1, 2, \ldots,$ and that either $1 + \xi_r - \epsilon - \epsilon_r < u_m < 1 + \xi_r$ or $1 + \xi_l - \epsilon - \epsilon_l < u_{m+1} < 1 + \xi_l$ if the parameters are sufficiently small. The notation implies that each of the last two inequalities brings the couple to synchronization. Similarly, one can discuss relations connected to inequality (1.67).

Since each of the iterations of L_D is done within an interval whose length is not more than T, we now find that the couple x_l, x_r is synchronized not later than $t = t_0 + (m+1)T$.

We have found that oscillators x_l and x_r fire in unison at some moment $t = \theta$. Next, we show that they will save the state, being different. To find conditions for this, let us denote by $\tau > \theta$ the next moment the couple fires. For instance, let x_r fire at this moment. Thus, we have $x_l(\theta+) = x_r(\theta+) = 0$. Then $x_l(t) = x_r(t), \theta \leq t \leq \tau$. It is clear that to satisfy $x_l(\tau+) = x_r(\tau+) = 0$, we need $1 + \xi_r - \epsilon - \epsilon_r \leq x_l(\tau)$. By applying formula (1.73) again, this time with $t_i = \theta, t_{i+1} = \tau$, one can easily obtain that the inequality is correct if the parameters are close to zero and absolute values of the parameters of perturbation are small with respect to ϵ. Thus, one can conclude that if a couple of oscillators is synchronized at some moment of time, then it continues to fire in unison forever. The lemma is proved.

Let us extend the result of the last lemma for the whole ensemble.

Theorem 8. *Assume that conditions (1.53), (1.55), (1.60), (1.62), and (1.67) are valid, and* $t_0 \geq 0$ *is a firing moment such that* $x_j(t_0) = 1 + \xi_j, x_j(t_0+) = 0$. *If the parameters are sufficiently close to zero, and absolute values of parameters of the perturbation are sufficiently small with respect to* ϵ, *then the motion* $x(t)$ *of the system synchronizes within the time interval* $[t_0, t_0 + T]$, *if* $x_i(t_0+) \notin [a_0, a_1), i \neq j$, *and within the time interval* $[t_0 + \frac{\max_{i\neq j} k_i - 1}{2}\tilde{T}, t_0 + (\max_{i\neq j} k_i + 1)T]$ *if there exist* $x_s(t_0+) \in [a_0, a_1)$, *for some* $s \neq j$ *and* $x_i(t_0+) \in S_{k_i}, i \neq j$.

Proof. Consider the collection of couples $(x_i, x_j), i \neq j$. Each of these pairs synchronizes by the last lemma within the interval $[t_0 + \frac{\max_{i\neq j} k_i - 1}{2}\tilde{T}, t_0 + (\max_{i\neq j} k_i + 1)T]$. The theorem is proved.

Let us introduce a more general system of oscillators such that Theorem 8 is still true.

Consider a system of n oscillators such that if the ith oscillator does not fire or jump up, it satisfies the ith equation of system (1.49). If several oscillators $x_{i_s}, s = 1, 2, \ldots, k$, fire such that $x_{i_s}(t) = 1 + \phi(t, x(t), x(t - \tau_{i_s}))$, where $|\phi(t, x(t), x(t - \tau_i))| < \xi_i, i = 1, 2, \ldots, n$, and $x_{i_s}(t+) = 0$, then all other oscillators $x_{i_p}, p = k + 1, k + 1, \ldots, n$, change their coordinates by the law

$$x_i(t+) = 0, \text{ if } x_i(t) \geq 1 - \epsilon, \tag{1.82}$$

and if $x_i(t) < 1 - \epsilon$, then

$$x_i(t + \tau+) = x_i(t + \tau) + \epsilon + \sum_{s=1}^{k} \epsilon_{i_p i_s}. \tag{1.83}$$

One can easily see that the last theorem is correct for the model that was just described if $\epsilon + \sum_{s=1}^{k} \epsilon_{i_p i_s} > 0$, for all possible k, i_p and i_s, and we assume that ϵ_{ij} are also parameters of perturbation. Moreover, one can easily see that initial functions for threshold conditions can be chosen arbitrarily with values in the domain of the system.

Remark 6. Our preliminary analysis shows that the dynamics in a neighborhood of v^* can be very complex. We do not exclude that a chaos appearance can be observed, and trajectories may belong to a fractal, if parameters are not small. It does not contradict the zero Lebesgue measure of unsynchronized points. The analysis of nonidentical oscillators with not small parameters may be of significant interest to explore arrhythmias, earthquakes, chaotic flashing of fireflies, and so on.

Remark 7. The time of synchronization for a given initial point does not increase if the number of oscillators increases (but the parameters needed to be closer to zero). This property can possibly be accepted as a small-world phenomenon.

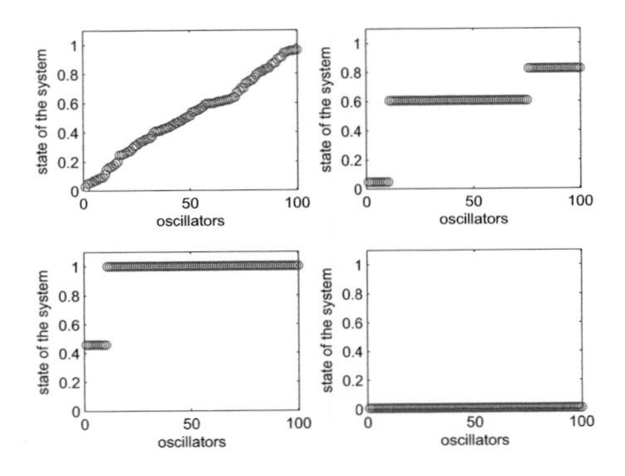

Fig. 1.12 The state of the model before the 1st, the 183rd, the 366th, and the 549th jumps. The flat fragments of the graph are groups of oscillators firing in unison

1.4.3 The Simulation Result

To demonstrate our main result numerically, let's consider a model of 100 oscillators, whose initial values are randomly uniformly distributed in [0, 1]. Their differential equations are of the form

$$x_i' = (4.1 + 0.01 * sort(rand(1,n)) - (3.2 + 0.01 * sort(rand(1,n))x_i,$$

and thresholds

$$1 + 0.005 * sort(rand(1,n)), i = 1, 2, \ldots, 100,$$

where deviations of coefficients of the threshold are also uniformly random in [0, 1]. We place the result of a simulation with $\epsilon = 0.06$ and $\tau = 0.002$ in Fig. 1.12, where the state of the system is shown at the initial moment, before the 183rd jump, before the 366th jump, and before the 549th jump. That is, it is obvious that eventually all oscillators fire in unison.

We verified that all conditions (1.53), (1.55), (1.60), (1.62), and (1.67) are valid.

1.5 Conclusion

There are two main approaches to analyzing the conjectures, which were first applied in [45] and [50]. The phase description method has dominated the field in the last decade, has been utilized in deterministic and indeterministic analyses, and has also been used in addressing various real-world problems. There is a rich collection of results on synchronization, obtained through experiments and simulations. The results of this chapter can give a theoretical background for them and form the basis for new ones. They can be applied, by using the theory of maps and their perturbations, not only to the problems of synchronization, but also to periodic, almost periodic motions, and complex behavior of biological models. New small-world phenomena can be discovered.

Peskin's [50] two famous conjectures were developed for further applications. One important additional question is whether continuous or piecewise continuous couplings synchronize the model. This chapter contains sufficient conditions to answer that question in the affirmative. The investigation is based on a specially constructed map. One can remark that the systems investigated in this chapter are, in fact, cooperative discontinuous systems [26–31] with monotone dynamics [54]. Consequently, by applying the methods of dynamical systems with discontinuities at variable moments [2], one can obtain more results concerning biological processes in the future.

The cardiac pacemaker model of identical and nonidentical oscillators with delayed pulse couplings is considered in the chapter. We apply the method developed in [3–5], which is based on a specially defined map. Sufficient conditions are found such that the involvement of delay in Peskin's model does not change the synchronization result for identical and nonidentical oscillators [5, 45, 50]. What we have done extends to biological processes since delay is often present therein, and if one proves that a phenomenon preserves even with delays, it makes us more confident that the model is adequate for the reality. Moreover, the method of treating models with delay can be useful for neural networks and earthquake fault [16, 19, 25, 28, 47] analysis. All the proved assertions are true with $\tau = 0$. Indeed, it is easy to see that conditions (1.53), (1.55), (1.60), (1.62), and (1.67) are valid with $\tau = 0$. Thus, the synchronization results for Peskin's model in [3, 5] are confirmed one more time. In our next papers, we plan to analyze models with delays that are not small. There are several interesting problems which can further develop results from this chapter, including the following: Suppose that condition (1.67) is violated; that is, $\tilde{v} > a_2$. Consider two identical oscillators. The corresponding graph of the map looks like that in Fig. 1.13.

One can see from the figure that the couple synchronizes after not more than three iterations if $v \notin [b, a_1]$. Otherwise, the pair moves periodically with period 2 ultimately. Considering this simple case of two identical oscillators, one can predict that for an ensemble of oscillators (identical or not quite identical) there should be two or more clusters of synchronized oscillators, and the clusters may move periodically if $\tilde{v}$ is near a_2. In our simulations, we observe clustering as well as periodicity in the motion of the clusters. Since the number of clusters changes with the variation of the parameters, one can investigate bifurcation of periodic solutions as well as the number of clusters.

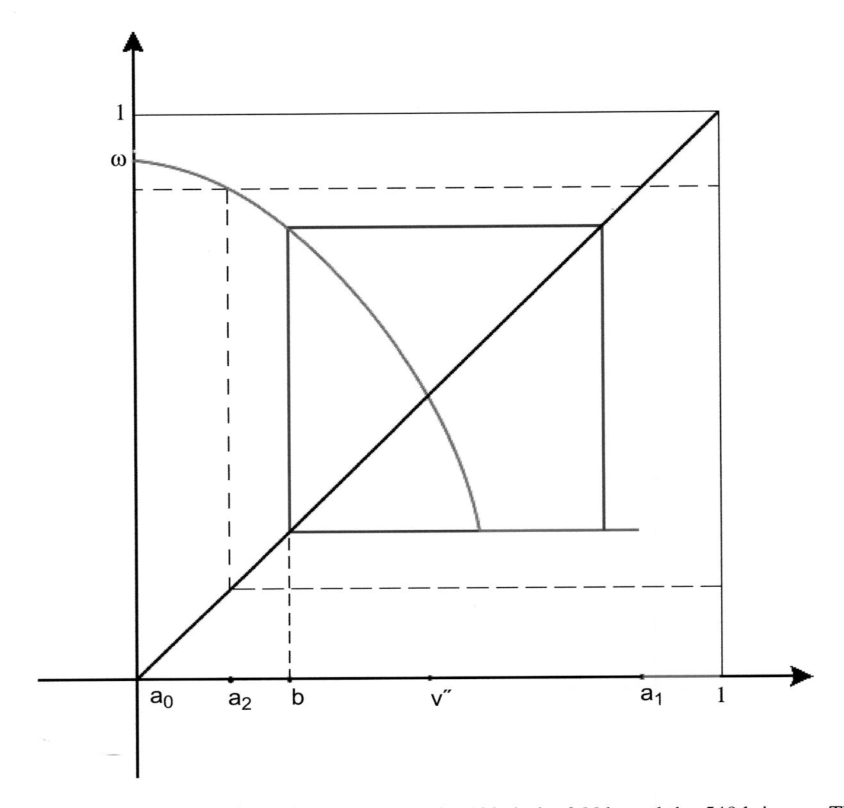

Fig. 1.13 The state of the model before the 1st, the 183rd, the 366th, and the 549th jumps. The flat fragments of the graph are groups of oscillators firing in unison

References

1. Akhmet, M.U.: Perturbations and Hopf bifurcation of the planar discontinuous dynamical system. Nonlinear Anal. Theory Methods Appl. **60**, 163–178 (2005)
2. Akhmet, M.U.: Principles of Discontinuous Dynamical Systems. Springer, New York (2010)
3. Akhmet, M.U.: Analysis of biological integrate-and-fire oscillators. Nonlinear Stud. **18**, 313–327 (2011)
4. Akhmet, M.U.: Nonlinear Hybrid Continuous/Discrete-Time Models. Atlantis Press, Amsterdam (2011)
5. Akhmet, M.U.: Self-synchronization of the integrate-and-fire pacemaker model with continuous couplings. Nonlinear Anal. Hybrid Syst. **6**, 730–740 (2012)
6. Akhmet, M.U.: Synchronization of the cardiac pacemaker model with delayed pulse-coupling. Discontinuity Nonlinearity Complex. **3**, 17–31 (2014)
7. Bottani, S.: Pulse-coupled relaxation oscillators: from biological synchronization to self-organized criticality. Phys. Rev. Lett. **74**, 4189–4192 (1995)
8. Bressloff, P.C.: Mean-field theory of globally coupled integrate-and-fire neural oscillators with dynamic synapses. Phys. Rev. E **60**, 2160–2170 (1999)
9. Brooks, C.M., Lu, H.H.: The Sinoatrial Pacemaker of the Heart. Thomas, Springfield, IL (1972)

10. Buck, J.: Synchronous rhythmic flashing of fireflies. II. Q. Rev. Biol. **63**(3), 265–290 (1988)
11. Buck, J., Buck, E.: Mechanism of rhythmic synchronous flashing of fireflies: fireflies of Southeast Asia may use anticipatory time-measuring in synchronizing their flashing. Science **159**, 1319–1327 (1968)
12. Devi Vasundara, J., Vatsala, A.S.: Generalized quasilinearization for an impulsive differential equation with variable moments of impulse. Dyn. Syst. Appl. **12**, 369–382 (2003)
13. Ermentrout, G.B.: The behavior of rings of coupled oscillators. J. Math. Biol. **23**, 55–74 (1985)
14. Ermentrout, G.B., Koppel, N.: Parabolic bursting in an excitable system coupled with a slow oscillators. SIAM J. Math. Anal. **15**, 233–253 (1986)
15. Ermentrout, G.B., Koppel, N.: Oscillator death in systems of coupled neural oscillators. SIAM J. Appl. Math. **50**(1), 125–146 (1990)
16. Ernst, U., Pawelzik, K., Geisel, T.: Delay-induced multi-stable synchronization of biological oscillators. Phys. Rev. E **57**, 2150–2162 (1998)
17. Feckan, M.: Bifurcation of periodic and chaotic solutions in discontinuous systems. Arch. Math. (Brno) **34**, 73–82 (1998)
18. Frigon, M., O'Regan, D.: Impulsive differential equations with variable times. Nonlinear Anal. Theory Methods Appl. **26**, 1913–1922 (1996)
19. Gerstner, W.: Rapid phase locking in systems of pulse-coupled oscillators with delays. Phys. Rev. Lett. **76**, 1755–1758 (1996)
20. Gerstner, W., Kistler, W.M.: Spiking Neuron Models: Single Neurons, Populations, Plasticity. Cambridge University Press, Cambridge (2002)
21. Glass, L., Mackey, M.C.: A simple model for phase locking of biological oscillators. J. Math. Biol. **7**, 339–367 (1979)
22. Glass, L., Mackey, M.C.: From Clocks to Chaos: The Rhythms of Life. Princeton University Press, Princeton (1988)
23. Goel, P., Ermentrout, B.: Synchrony, stability, and firing patterns in pulse-coupled oscillators. Phys. D **163**, 191–216 (2002)
24. Hanson, F.E., Case, J.F., Buck, E., Buck, J.: Synchrony and flash entrainment in a New Guinea firefly. Science **174**, 161–164 (1971)
25. Herz, A.V.M., Hopfield, J.J.: Earthquake cycles and neural perturbations: collective oscillations in systems with pulse-coupled thresholds elements. Phys. Rev. Lett. **75**, 1222–1225 (1995)
26. Hirsh, M.W.: Systems of differential equations which are competitive or cooperative 1: limit sets. SIAM J. Appl. Math. **13**, 167–179 (1982)
27. Hirsh, M.W.: Systems of differential equations that are competitive or cooperative II: convergence almost everywhere. SIAM J. Appl. Math. **16**, 423–439 (1985)
28. Hopfield, J.J.: Neurons, dynamics and computation. Phys. Today **47**, 40–46 (1994)
29. Hopfield, J.J., Herz, A.: Rapid local synchronization of action potentials: toward computation with coupled integrate-and-fire neurons. Proc. Natl. Acad. Sci. **92**, 6655–6662 (1995)
30. Hoppensteadt, F.C., Izhikevich, E.M.: Weakly Connected Neural Networks. Applied Mathematical Sciences, vol. 126. Springer, New York (1997)
31. Kamke, E.: Zur theorie der systeme gewohnlicher differentialgleichungen. II. Acta Math. **58**, 57–85 (1932) [German]
32. Knight, B.W.: Dynamics of encoding in a population of neurons. J. Gen. Physiol. **59**, 734–766 (1972)
33. Ko, T.W., Ermentrout, G.B.: Effects of axonal time delay on synchronization and wave formation in sparsely coupled neuronal oscillators. Phys. Rev. E. **76**, 1–8 (2007)
34. Koppel, N., Ermentrout, G.B., Williams, T.: On chains of oscillators forced at one end. SIAM J. Appl. Math. **51**(5), 1397–1417 (1991)
35. Kuramoto, Y.: Chemical Oscillators, Waves and Turbulence. Springer, Berlin (1984)
36. Kuramoto, Y.: Collective synchronization of pulse-coupled oscillators and excitable units. Phys. D **50**, 15–30 (1991)
37. Kuramoto, Y., Nishikawa, I.: Statistical macrodynamics of large dynamical systems. Case of a phase transition in oscillator communities. J. Stat. Phys.**49**, 569–605 (1987)

38. Lakshmikantham, V., Liu, X.: On quasistability for impulsive differential equations. Nonlinear Anal. Theory Methods Appl. **13**, 819–828 (1989)
39. Lakshmikantham, V., Bainov, D. D., Simeonov, P. S.: Theory of Impulsive Differential Equations. World Scientific, Singapore (1989)
40. Lakshmikantham, V., Leela, S., Kaul, S.: Comparison principle for impulsive differential equations with variable times and stability theory. Nonlinear Anal. Theory Methods Appl. **22**, 499–503 (1994)
41. Liu, X., Pirapakaran, R.: Global stability results for impulsive differential equations. Appl. Anal. **33**, 87–102 (1989)
42. Lloyd, J.E.: Fireflies of Melanesia: bioluminescence, mating behavior, and synchronous flashing (Coleoptera: Lampyridae). Environ. Entomol. **2**, 991–1008 (1973)
43. Luo, A.C.J.: Global Transversality, Resonance and Chaotic Dynamics. World Scientific, Hackensack, NJ (2008)
44. Mathar, R., Mattfeldt, J.: Pulse-coupled decentral synchronization. SIAM J. Appl. Math. **56**, 1094–1106 (1996)
45. Mirollo, R.E., Strogatz, S.H.: Synchronization of pulse-coupled biological oscillators. SIAM J. Appl. Math. **50**, 1645–1662 (1990)
46. Murray, J.D.: Mathematical Biology: I. An Introduction. Springer, New York (2002)
47. Olami, Z., Feder, H.J.S., Christensen, K.: Self-organized criticality in a continuous, nonconservative cellular automaton modeling earthquakes. Phys. Rev. Lett. **68**, 1244–1247 (1992)
48. Pavlidis, T.: A new model for simple neural nets and its application in the design of a neural oscillator. Bull. Math. Biophys. **27**, 215–229 (1965)
49. Pavlidis, T.: Biological Oscillators: Their Mathematical Analysis. Academic Press, New York (1973)
50. Peskin, C.S.: Mathematical Aspects of Heart Physiology. Courant Institute of Mathematical Sciences, pp. 268–278. New York University, New York (1975)
51. Pikovsky, A., Rosenblum, M., Kurths, J.: Synchronization: A Universal Concept in Nonlinear Sciences. Cambridge University Press, Cambridge (2001)
52. Senn, W., Urbanczik, R.: Similar non-leaky integrate-and-fire neurons with instantaneous couplings always synchronize. SIAM J. Appl. Math. **61**, 1143–1155 (2000)
53. Sherman, A., Rinzel, J., Keizer, J.: Emergence of organized bursting in clusters of pancreatic beta-cells by channel sharing. Biophys. J. **54**, 411–425 (1988)
54. Smith, H.L.: Monotone Dynamical Systems: An Introduction to the Theory of Competitive and Cooperative Systems. American Mathematical Society, Providence, RI, USA (1995)
55. Strogatz, S.: From Kuramoto to Crawford: exploring the onset of synchronization in populations of coupled oscillators. Phys. D **143**, 1–20 (2000)
56. Strogatz, S.: Nonlinear Dynamics and Chaos: With Applications to Physics, Biology, Chemistry and Engineering. Perseus Books Group, New York (2001)
57. Strogatz, S.: Exploring complex networks. Nature **410**, 268–276 (2001)
58. Strogatz, S.: Sync: The Emerging Science of Spontaneous Order. Hyperion, New York (2003)
59. Strogatz, S., Mirollo, R.: Collective synchronization in lattices of nonlinear oscillators with randomness. J. Phys. A **21**, L699–L705 (1988)
60. Timme, M., Wolf, F.: The simplest problem in the collective dynamics of neural networks: is synchrony stable? Nonlinearity **21**, 1579–1599 (2008)
61. Timme, M., Wolf, F., Geisel, T.: Prevalence of unstable attractors in networks of pulse-coupled oscillators. Phys. Rev. Lett. **89**, 154105 (2002)
62. Van Vreeswijk, C.: Partial synchronization in populations of pulse-coupled oscillators. Phys. Rev. E **54**, 5522–5537 (1996)
63. Winfree, A.T.: Biological rhythms and the behavior of populations of biological oscillators. J. Theor. Biol. **16**, 15–42 (1967)
64. Winfree, A.T.: The Geometry of Biological Time. Springer, New York (1980)

Chapter 2
On Periodic Motions in a Time-Delayed, Quadratic Nonlinear Oscillator with Excitation

Albert C.J. Luo and Hanxiang Jin

Abstract Analytical solutions of periodic motions in a time-delayed, quadratic nonlinear oscillator with periodic excitation are obtained through the finite Fourier series, and the stability and bifurcation analysis for periodic motions are discussed. The bifurcation trees of period-1 motion to chaos can be presented. Numerical illustration of periodic motion is given to verify the analytical solutions.

2.1 Introduction

The quadratic nonlinear oscillator is often used to describe boat motion under periodic ocean waves. To stabilize boat motions under waves, once the feedback is introduced, the boat motion equation will be a time-delayed dynamical system. In this chapter, the analytical solution of periodic motions in a time-delayed, quadratic nonlinear oscillator will be investigated for the stabilization of boat motion.

The study of periodic motions in dynamical systems dates back to the eighteenth century. In 1788, Lagrange [1] developed the standard Lagrange form to obtain the method of averaging and used this method for the periodic motions of three-body problems. In the nineteenth century, Poincaré [2] developed perturbation theory to determine the periodic motions of celestial bodies. In 1920, van der Pol [3] employed the method of averaging for the periodic solutions of oscillation systems in circuits. In 1928, Fatou [4] gave the first proof of the asymptotic validity of the method of averaging through the existing theorems of solutions of differential equations. In 1935, Krylov, Bogoliubov, and Mitropolsky [5] further developed the method of averaging and applied it to periodic motions in nonlinear oscillators. In 1961, Bogoliubov and Mitropolsky [6] summarized the asymptotic perturbation methods in nonlinear oscillations. In 1964, Hayashi [7] employed perturbation methods, the method of averaging, and the principle of harmonic balance for the approximate solutions of nonlinear oscillators, and the stability of approximate

---

A.C.J. Luo (✉) • H. Jin
Southern Illinois University Edwardsville, Edwardsville, IL 62026-1805, USA
e-mail: aluo@siue.edu; hjin@siue.edu

A.C.J. Luo, H. Merdan (eds.), *Mathematical Modeling and Applications in Nonlinear Dynamics*, Nonlinear Systems and Complexity 14,
DOI 10.1007/978-3-319-26630-5_2

periodic solutions in nonlinear oscillators was determined by the improved Mathieu equation. In 1973, Nayfeh [8] presented multiscale methods for approximate solutions of periodic motions in nonlinear structural dynamics (also see Nayfeh and Mook [9]). In 1990, Coppola and Rand [10] developed the method of averaging with elliptic functions for the approximate of limit cycle. In 2012, Luo [11] developed a methodology for analytical solutions of periodic motions in nonlinear dynamical systems. In 2012, Luo and Huang [12] applied such a generalized harmonic balance method to the Duffing oscillator for approximate solutions of periodic motions, and Luo and Huang [13] gave the analytical bifurcation trees of period-m motions to chaos in the Duffing oscillator. In 2013, Luo [14] systematically proposed a methodology for periodic motions in time-delayed, nonlinear dynamical systems. In 2014, Luo and Jin [15] used such a technique to investigate periodic motion in a quadratic nonlinear oscillator with time delay.

In this chapter, the analytical solutions of period-m motions for such a time-delayed, quadratic nonlinear oscillator will be presented and the stability and bifurcation of period-m motions in the time-delayed nonlinear oscillator will be discussed. From the bifurcation trees of period-1 motion to chaos, numerical simulations will be carried out for comparison of analytical and numerical solutions of periodic motions.

2.1.1 *Analytical Solutions*

As in Luo and Jin [15], consider a periodically forced, time-delayed, quadratic nonlinear oscillator as

$$\ddot{x} + \delta\dot{x} + \alpha_1 x - \alpha_2 x^{\tau} + \beta x^2 = Q_0 \cos\Omega t, \tag{2.1}$$

where $x^{\tau} = x(t - \tau)$ and $\dot{x}^{\tau} = \dot{x}(t - \tau)$. The coefficients in Eq. (2.1) are δ for linear damping, α_1 and α_2 for linear springs, β for quadratic nonlinearity, and Q_0 and Ω for excitation amplitude and frequency, respectively. The standard form of Eq. (2.1) is written as

$$\ddot{x} + f(x, \dot{x}, x^{\tau}, \dot{x}^{\tau}, t) = 0, \tag{2.2}$$

where

$$f(x, \dot{x}, x^{\tau}, \dot{x}^{\tau}, t) = \delta\dot{x} + \alpha_1 x - \alpha_2 x^{\tau} + \beta x^2 - Q_0 \cos\Omega t. \tag{2.3}$$

The analytical solution of period-m motion for the preceding equation is

$$\begin{aligned}
x^{(m)*} &= a_0^{(m)}(t) + \sum_{k=1}^{N} b_{k/m}(t)\cos\left(\frac{k}{m}\Omega t\right) + c_{k/m}(t)\sin\left(\frac{k}{m}\Omega t\right),\\
x^{\tau(m)*} &= a_0^{\tau(m)}(t) + \sum_{k=1}^{N}\left[b_{k/m}^{\tau}(t)\cos\left(\frac{k}{m}\Omega\tau\right) - c_{k/m}^{\tau}(t)\sin\left(\frac{k}{m}\Omega\tau\right)\right]\cos\left(\frac{k}{m}\Omega t\right)\\
&\quad + \left[b_{k/m}^{\tau}(t)\cos\left(\tfrac{k}{m}\Omega\tau\right) + c_{k/m}^{\tau}(t)\sin\left(\tfrac{k}{m}\Omega\tau\right)\right]\sin\left(\tfrac{k}{m}\Omega t\right).
\end{aligned}\tag{2.4}$$

where $a_0^{\tau(m)}(t) = a_0^{(m)}(t-\tau)$, $b_{k/m}^{\tau}(t) = b_{k/m}(t-\tau)$, $c_{k/m}^{\tau}(t) = c_{k/m}(t-\tau)$. The coefficients $a_0^{(m)}(t)$, $b_{k/m}(t)$, $c_{k/m}(t)$ vary with time. The first and second order of derivatives of $x^{(m)*}(t)$ and $x^{\tau(m)*}(t)$ are

$$\begin{aligned}
\dot{x}^{(m)*} &= \dot{a}_0^{(m)}(t) + \sum_{k=1}^{N}\left[\left(\dot{b}_{k/m}(t) + \frac{k}{m}\Omega c_{k/m}(t)\right)\cos\left(\frac{k}{m}\Omega t\right)\right.\\
&\quad \left. + \left(\dot{c}_{k/m}(t) - \tfrac{k}{m}\Omega b_{k/m}(t)\right)\sin\left(\tfrac{k}{m}\Omega t\right)\right],\\
\dot{x}^{\tau(m)*} &= \dot{a}_0^{\tau(m)}(t) + \sum_{k=1}^{N}\left\{\left[\left(\dot{b}_{k/m}^{\tau}(t) + \frac{k}{m}\Omega c_{k/m}^{\tau}(t)\right)\cos\left(\frac{k}{m}\Omega\tau\right)\right.\right.\\
&\quad \left. - \left(\dot{c}_{k/m}^{\tau}(t) - \tfrac{k}{m}\Omega b_{k/m}^{\tau}(t)\right)\sin\left(\tfrac{k}{m}\Omega\tau\right)\right]\cos\left(\tfrac{k}{m}\Omega t\right)\\
&\quad + \left[\left(\dot{b}_{k/m}^{\tau}(t) + \tfrac{k}{m}\Omega c_{k/m}^{\tau}(t)\right)\sin\left(\tfrac{k}{m}\Omega\tau\right)\right.\\
&\quad \left.\left. + \left(\dot{c}_{k/m}^{\tau}(t) - \tfrac{k}{m}\Omega b_{k/m}^{\tau}(t)\right)\cos\left(\tfrac{k}{m}\Omega\tau\right)\right]\sin\left(\tfrac{k}{m}\Omega t\right)\right\},
\end{aligned}\tag{2.5}$$

$$\begin{aligned}
\ddot{x}^{(m)}(t) &= \ddot{a}_0^{(m)}(t) + \sum_{k=1}^{N}\left[\ddot{b}_{k/m} + 2\left(\frac{k}{m}\Omega\right)\dot{c}_{k/m} - \left(\frac{k}{m}\Omega\right)^2 b_{k/m}\right]\cos\left(\tfrac{k}{m}\Omega t\right)\\
&\quad + \left[\ddot{c}_{k/m} - 2\left(\tfrac{k}{m}\Omega\right)\dot{b}_{k/m} - \left(\tfrac{k}{m}\Omega\right)^2 c_{k/m}\right]\sin\left(\tfrac{k}{m}\Omega t\right).
\end{aligned}\tag{2.6}$$

Substitution of Eqs. (2.4)–(2.6) into Eq. (2.1) and averaging for the harmonic terms of $\cos(k\Omega t/m)$ and $\sin(k\Omega t/m)$ $(k = 0, 1, 2, \ldots)$ gives

$$\begin{aligned}
&\ddot{a}_0^{(m)} + F_0^{(m)}\left(\mathbf{z}^{(m)}, \dot{\mathbf{z}}^{(m)}; \mathbf{z}^{\tau(m)}\dot{\mathbf{z}}^{\tau(m)}\right) = 0,\\
&\ddot{b}_{k/m} + 2\tfrac{k\Omega}{m}\dot{c}_{k/m} - \left(\tfrac{k\Omega}{m}\right)^2 b_{k/m} + F_{1k}^{(m)}\left(\mathbf{z}^{(m)}, \dot{\mathbf{z}}^{(m)}; \mathbf{z}^{\tau(m)}\dot{\mathbf{z}}^{\tau(m)}\right) = 0,\\
&\ddot{c}_{k/m} - 2\tfrac{k\Omega}{m}\dot{b}_{k/m} - \left(\tfrac{k\Omega}{m}\right)^2 c_{k/m} + F_{2k}^{(m)}\left(\mathbf{z}^{(m)}, \dot{\mathbf{z}}^{(m)}; \mathbf{z}^{\tau(m)}\dot{\mathbf{z}}^{\tau(m)}\right) = 0,\\
&k = 1, 2, \ldots, N,
\end{aligned}\tag{2.7}$$

where

$$\begin{aligned}
\mathbf{z}^{(m)} &= \left(a_0^{(m)}, \mathbf{b}^{(m)}, \mathbf{c}^{(m)}\right)^{\mathrm{T}} \text{ and } \dot{\mathbf{z}}^{(m)} = \left(\dot{a}_0^{(m)}, \dot{\mathbf{b}}^{(m)}, \dot{\mathbf{c}}^{(m)}\right)^{\mathrm{T}},\\
\mathbf{z}^{\tau(m)} &= \left(a_0^{\tau(m)}, \mathbf{b}^{\tau(m)}, \mathbf{c}^{\tau(m)}\right)^{\mathrm{T}} \text{ and } \dot{\mathbf{z}}^{\tau(m)} = \left(\dot{a}_0^{\tau(m)}, \dot{\mathbf{b}}^{\tau(m)}, \dot{\mathbf{c}}^{\tau(m)}\right)^{\mathrm{T}};\\
\mathbf{b}^{(m)} &= \left(b_1^{(m)}, b_2^{(m)}, \cdots, b_N^{(m)}\right)^{\mathrm{T}} \text{ and } \mathbf{b}^{\tau(m)} = \left(b_1^{\tau(m)}, b_2^{\tau(m)}, \cdots, b_N^{\tau(m)}\right)^{\mathrm{T}},\\
\mathbf{c}^{(m)} &= \left(c_1^{(m)}, c_2^{(m)}, \cdots, c_N^{(m)}\right)^{\mathrm{T}} \text{ and } \mathbf{c}^{\tau(m)} = \left(c_1^{\tau(m)}, c_2^{\tau(m)}, \cdots, c_N^{\tau(m)}\right)^{\mathrm{T}};
\end{aligned} \tag{2.8}$$

$$\begin{aligned}
F_0^{(m)}\left(\mathbf{z}^{(m)}, \dot{\mathbf{z}}^{(m)}; \mathbf{z}^{\tau(m)}, \dot{\mathbf{z}}^{\tau(m)}\right) &= \delta \dot{a}_0^{(m)} + \alpha_1 a_0^{(m)} - \alpha_2 a_0^{\tau(m)} + \beta f_0^{(m)},\\
F_{1k}^{(m)}\left(\mathbf{z}^{(m)}, \dot{\mathbf{z}}^{(m)}; \mathbf{z}^{\tau(m)}, \dot{\mathbf{z}}^{\tau(m)}\right) &= \delta\left(\dot{b}_{k/m} + \tfrac{k\Omega}{m} c_{k/m}\right) + \alpha_1 b_{k/m} - \alpha_2 \Big[b_{k/m}^{\tau} \cos\left(\tfrac{k}{m}\Omega\tau\right)\\
&\quad - c_{k/m}^{\tau} \sin\left(\tfrac{k}{m}\Omega\tau\right)\Big] + \beta f_{k/m}^{(c)} + Q_0 \delta_k^m\\
F_{2k}^{(m)}\left(\mathbf{z}^{(m)}, \dot{\mathbf{z}}^{(m)}; \mathbf{z}^{\tau(m)}, \dot{\mathbf{z}}^{\tau(m)}\right) &= \delta\left(\dot{c}_{k/m} - \tfrac{k\Omega}{m} b_{k/m}\right) + \alpha_1 c_{k/m} - \alpha_2 \Big[c_{k/m}^{\tau} \cos\left(\tfrac{k}{m}\Omega\tau\right)\\
&\quad + b_{k/m}^{\tau} \sin\left(\tfrac{k}{m}\Omega\tau\right)\Big] + \beta f_{k/m}^{(s)},
\end{aligned} \tag{2.9}$$

and

$$\begin{aligned}
f_0^{(m)} &= \left(a_0^{(m)}\right)^2 + \tfrac{1}{2}\sum_{i=1}^{N}\left(b_{i/m}^2 + c_{i/m}^2\right),\\
f_{k/m}^{(c)} &= 2a_0^{(m)} b_{k/m} + \tfrac{1}{2}\sum_{i=1}^{N}\sum_{j=1}^{N} b_{i/m} b_{j/m}\left(\delta_{i+j}^k + \delta_{j-i}^k + \delta_{i-j}^k\right)\\
&\quad + c_{i/m} c_{j/m}\left(\delta_{j-i}^k - \delta_{i+j}^k + \delta_{i-j}^k\right),\\
f_{k/m}^{(s)} &= 2a_0^{(m)} c_{k/m} + \sum_{i=1}^{N}\sum_{j=1}^{N} b_{i/m} c_{j/m}\left(\delta_{i+j}^k + \delta_{j-i}^k - \delta_{i-j}^k\right).
\end{aligned} \tag{2.10}$$

Equation (2.7) can be expressed in the form of a vector field as

$$\dot{\mathbf{z}}^{(m)} = \mathbf{z}_1^{(m)} \text{ and } \dot{\mathbf{z}}_1^{(m)} = \mathbf{g}^{(m)}\left(\mathbf{z}^{(m)}, \mathbf{z}_1^{(m)}; \mathbf{z}^{\tau(m)}, \mathbf{z}_1^{\tau(m)}\right), \tag{2.11}$$

where

$$\begin{aligned}
&\mathbf{g}^{(m)}\left(\mathbf{z}^{(m)}, \mathbf{z}_1^{(m)}; \mathbf{z}^{\tau(m)}, \mathbf{z}_1^{\tau(m)}\right)\\
&= \begin{pmatrix}
-F_0^{(m)}\left(\mathbf{z}^{(m)}, \mathbf{z}_1^{(m)}; \mathbf{z}^{\tau(m)}, \mathbf{z}_1^{\tau(m)}\right)\\
-\mathbf{F}_1^{(m)}\left(\mathbf{z}^{(m)}, \mathbf{z}_1^{(m)}; \mathbf{z}^{\tau(m)}, \mathbf{z}_1^{\tau(m)}\right) - 2\mathbf{k}_1 \tfrac{\Omega}{m} \dot{\mathbf{c}}^{(m)} + \mathbf{k}_2\left(\tfrac{\Omega}{m}\right)^2 \mathbf{b}^{(m)}\\
-\mathbf{F}_2^{(m)}\left(\mathbf{z}^{(m)}, \mathbf{z}_1^{(m)}; \mathbf{z}^{\tau(m)}, \mathbf{z}_1^{\tau(m)}\right) + 2\mathbf{k}_1 \tfrac{\Omega}{m} \dot{\mathbf{b}}^{(m)} + \mathbf{k}_2\left(\tfrac{\Omega}{m}\right)^2 \mathbf{c}^{(m)}
\end{pmatrix}
\end{aligned} \tag{2.12}$$

and

$$\begin{array}{l}\mathbf{k}_1 = \operatorname{diag}(1, 2, \ldots, N) \text{ and } \mathbf{k}_2 = \operatorname{diag}\left(1, 2^2, \ldots, N^2\right)\\ \mathbf{F}_1^{(m)} = \left(F_{11}^{(m)}, F_{12}^{(m)}, \ldots, F_{1N}^{(m)}\right)^{\mathrm{T}} \text{ and } \mathbf{F}_2^{(m)} = \left(F_{21}^{(m)}, F_{22}^{(m)}, \ldots, F_{2N}^{(m)}\right)^{\mathrm{T}}\\ \text{for } N = 1, 2, \ldots, \infty.\end{array} \tag{2.13}$$

Setting

$$\mathbf{y}^{(m)} \equiv \left(\mathbf{z}^{(m)}, \mathbf{z}_1^{(m)}\right),\ \mathbf{y}^{\tau(m)} \equiv \left(\mathbf{z}^{\pi(m)}, \mathbf{z}_1^{\pi(m)}\right), \text{ and } \mathbf{f}^{(m)} = \left(\mathbf{z}_1^{(m)}, \mathbf{g}^{(m)}\right)^{\mathrm{T}}, \tag{2.14}$$

equation (2.11) becomes

$$\dot{\mathbf{y}}^{(m)} = \mathbf{f}^{(m)}\left(\mathbf{y}^{(m)}, \mathbf{y}^{\tau(m)}\right). \tag{2.15}$$

The steady-state solutions for periodic motion in Eq. (2.1) can be obtained by setting

$$\begin{array}{l}F_0^{(m)}\left(\mathbf{z}^{(m)}, \mathbf{0}; \mathbf{z}^{\tau(m)}, \mathbf{0}\right) = 0,\\ \mathbf{F}_1^{(m)}\left(\mathbf{z}^{(m)}, \mathbf{0}; \mathbf{z}^{\tau(m)}, \mathbf{0}\right) - \mathbf{k}_2\left(\frac{\Omega}{m}\right)^2\mathbf{b}^{(m)} = \mathbf{0},\\ \mathbf{F}_2^{(m)}\left(\mathbf{z}^{(m)}, \mathbf{0}; \mathbf{z}^{\tau(m)}, \mathbf{0}\right) - \mathbf{k}_2\left(\frac{\Omega}{m}\right)^2\mathbf{c}^{(m)} = \mathbf{0}.\end{array} \tag{2.16}$$

The $(2N + 1)$ nonlinear equations in Eq. (2.16) are solved by the Newton–Raphson method. In Luo [11, 14], the linearized equation at the equilibrium point is given by

$$\Delta\dot{\mathbf{y}}^{(m)} = D\mathbf{f}\left(\mathbf{y}^{(m)*}, \mathbf{y}^{\tau(m)*}\right)\Delta\mathbf{y}^{(m)} + D^{\tau}\mathbf{f}\left(\mathbf{y}^{(m)*}, \mathbf{y}^{\tau(m)*}\right)\Delta\mathbf{y}^{\tau(m)}. \tag{2.17}$$

The corresponding eigenvalues are determined by

$$\left|\mathbf{A} + \mathbf{B}e^{-\lambda\tau} - \lambda\mathbf{I}_{2(2N+1)\times 2(2N+1)}\right| = 0, \tag{2.18}$$

where

$$\begin{array}{l}\mathbf{A} = D\mathbf{f}\left(\mathbf{y}^{(m)*}, \mathbf{y}^{\tau(m)*}\right) = \partial\mathbf{f}\left(\mathbf{y}^{(m)}, \mathbf{y}^{\tau(m)}\right)/\partial\mathbf{y}^{(m)}\big|_{\left(\mathbf{y}^{(m)*}, \mathbf{y}^{\tau(m)*}\right)},\\ \mathbf{B} = D^{\tau}\mathbf{f}\left(\mathbf{y}^{(m)*}, \mathbf{y}^{\tau(m)*}\right) = \partial\mathbf{f}\left(\mathbf{y}^{(m)}, \mathbf{y}^{\tau(m)}\right)/\partial\mathbf{y}^{\tau(m)}\big|_{\left(\mathbf{y}^{(m)*}, \mathbf{y}^{\tau(m)*}\right)}.\end{array} \tag{2.19}$$

The corresponding submatrices are

$$\begin{array}{l}\mathbf{A} = \begin{bmatrix}\mathbf{0}_{(2N+1)\times(2N+1)} & \mathbf{I}_{(2N+1)\times(2N+1)}\\ \mathbf{G} & \mathbf{H}\end{bmatrix},\\ \mathbf{B} = \begin{bmatrix}\mathbf{0}_{(2N+1)\times(2N+1)} & \mathbf{I}_{(2N+1)\times(2N+1)}\\ \mathbf{G}^{\tau} & \mathbf{H}^{\tau}\end{bmatrix},\end{array} \tag{2.20}$$

where

$$\mathbf{G} = \frac{\partial \mathbf{g}^{(m)}}{\partial \mathbf{z}^{(m)}} = \left(\mathbf{G}^{(0)}, \mathbf{G}^{(c)}, \mathbf{G}^{(s)}\right)^{\mathrm{T}}, \mathbf{G}^{\tau} = \frac{\partial \mathbf{g}}{\partial \mathbf{z}^{\tau}} = \left(\mathbf{G}^{\tau(0)}, \mathbf{G}^{\tau(c)}, \mathbf{G}^{\tau(s)}\right)^{\mathrm{T}}, \quad (2.21)$$

$$\begin{aligned} \mathbf{G}^{(0)} &= \left(G_0^{(0)}, G_1^{(0)}, \cdots, G_{2N}^{(0)}\right), \\ \mathbf{G}^{(c)} &= \left(\mathbf{G}_1^{(c)}, \mathbf{G}_2^{(c)}, \cdots, \mathbf{G}_N^{(c)}\right)^{\mathrm{T}}, \\ \mathbf{G}^{(s)} &= \left(\mathbf{G}_1^{(s)}, \mathbf{G}_2^{(s)}, \cdots, \mathbf{G}_N^{(s)}\right)^{\mathrm{T}}; \\ \mathbf{G}^{\tau(0)} &= \left(G_0^{\tau(0)}, G_1^{\tau(0)}, \cdots, G_{2N}^{\tau(0)}\right), \\ \mathbf{G}^{\tau(c)} &= \left(\mathbf{G}_1^{\tau(c)}, \mathbf{G}_2^{\tau(c)}, \cdots, \mathbf{G}_N^{\tau(c)}\right)^{\mathrm{T}}, \\ \mathbf{G}^{\tau(s)} &= \left(\mathbf{G}_1^{\tau(s)}, \mathbf{G}_2^{\tau(s)}, \cdots, \mathbf{G}_N^{\tau(s)}\right)^{\mathrm{T}} \end{aligned} \quad (2.22)$$

for $N = 1, 2, \ldots, \infty$ with

$$\begin{aligned} \mathbf{G}_k^{(c)} &= \left(G_{k0}^{(c)}, G_{k1}^{(c)}, \cdots, G_{k(2N)}^{(c)}\right), \ \mathbf{G}_k^{(s)} = \left(G_{k0}^{(s)}, G_{k1}^{(s)}, \cdots, G_{k(2N)}^{(s)}\right), \\ \mathbf{G}_k^{\tau(c)} &= \left(G_{k0}^{\tau(c)}, G_{k1}^{\tau(c)}, \cdots, G_{k(2N)}^{\tau(c)}\right), \ \mathbf{G}_k^{\tau(s)} = \left(G_{k0}^{\tau(s)}, G_{k1}^{\tau(s)}, \cdots, G_{k(2N)}^{\tau(s)}\right) \end{aligned} \quad (2.23)$$

for $k = 1, 2, \ldots, N$. The corresponding components are

$$\begin{aligned} G_r^{(0)} &= -\alpha_1 \delta_r^0 - \beta g_r^{(0)}, \\ G_{kr}^{(c)} &= \left(\tfrac{k\Omega}{m}\right)^2 \delta_k^r - \delta \tfrac{k\Omega}{m} \delta_{k+N}^r - \alpha_1 \delta_k^r - \beta g_{kr}^{(c)}, \\ G_{kr}^{(s)} &= \left(\tfrac{k\Omega}{m}\right)^2 \delta_{k+N}^r + \delta \tfrac{k\Omega}{m} \delta_k^r - \alpha_1 \delta_{k+N}^r - \beta g_{kr}^{(s)}, \end{aligned} \quad (2.24)$$

where

$$\begin{aligned} g_r^{(0)} &= 2a_0^{(m)} \delta_r^0 + \sum_{i=1}^{N} \left(b_{i/m} \delta_i^r + c_{i/m} \delta_{i+N}^r\right), \\ g_{kr}^{(c)} &= 2\left(b_{k/m} \delta_r^0 + a_0^{(m)} \delta_k^r\right) + \sum_{i=1}^{N} \sum_{j=1}^{N} b_{j/m} \delta_i^r \left(\delta_{i+j}^k + \delta_{j-i}^k + \delta_{i-j}^k\right) \\ &\quad + c_{i/m} \delta_{i+N}^r \left(\delta_{j-i}^k - \delta_{i+j}^k + \delta_{i-j}^k\right), \\ g_{kr}^{(s)} &= 2\left(c_{k/m} \delta_r^0 + a_0^{(m)} \delta_{k+N}^r\right) + \sum_{i=1}^{N} \sum_{j=1}^{N} \left(c_{j/m} \delta_i^r + b_{i/m} \delta_{j+N}^r\right) \left(\delta_{i+j}^k + \delta_{j-i}^k - \delta_{i-j}^k\right), \end{aligned} \quad (2.25)$$

for $r = 0, 1, \ldots, 2N$. The components relative to the time delay for $r = 0, 1, \ldots, 2N$ are

$$
\begin{aligned}
G_{r}^{\tau(0)} &= \alpha_2 \delta_r^0, \\
G_{kr}^{\tau(c)} &= \alpha_2 \left[\delta_k^r \cos\left(\tfrac{k}{m}\Omega\tau\right) - \delta_{k+N}^r \sin\left(\tfrac{k}{m}\Omega\tau\right)\right], \\
G_{kr}^{\tau(s)} &= \alpha_2 \left[\delta_{k+N}^r \cos\left(\tfrac{k}{m}\Omega\tau\right) + \delta_k^r \sin\left(\tfrac{k}{m}\Omega\tau\right)\right].
\end{aligned}
\tag{2.26}
$$

The matrices relative to the velocity are

$$
\begin{aligned}
\mathbf{H} &= \frac{\partial \mathbf{g}^{(m)}}{\partial \mathbf{z}_1{}^{(m)}} = \left(\mathbf{H}^{(0)}, \mathbf{H}^{(c)}, \mathbf{H}^{(s)}\right)^{\mathrm{T}}, \\
\mathbf{H}^{\tau} &= \frac{\partial \mathbf{g}^{(m)}}{\partial \mathbf{z}_1^{\tau(m)}} = \left(\mathbf{H}^{\tau(0)}, \mathbf{H}^{\tau(c)}, \mathbf{H}^{\tau(s)}\right)^{\mathrm{T}},
\end{aligned}
\tag{2.27}
$$

where

$$
\begin{aligned}
\mathbf{H}^{(0)} &= \left(H_0^{(0)}, H_1^{(0)}, \cdots, H_{2N}^{(0)}\right), \\
\mathbf{H}^{(c)} &= \left(\mathbf{H}_1^{(c)}, \mathbf{H}_2^{(c)}, \cdots, \mathbf{H}_N^{(c)}\right)^{\mathrm{T}}, \\
\mathbf{H}^{(s)} &= \left(\mathbf{H}_1^{(s)}, \mathbf{H}_2^{(s)}, \cdots, \mathbf{H}_N^{(s)}\right)^{\mathrm{T}}; \\
\mathbf{H}^{\tau(0)} &= \left(H_0^{\tau(0)}, H_1^{\tau(0)}, \cdots, H_{2N}^{\tau(0)}\right), \\
\mathbf{H}^{\tau(c)} &= \left(\mathbf{H}_1^{\tau(c)}, \mathbf{H}_2^{\tau(c)}, \cdots, \mathbf{H}_N^{\tau(c)}\right)^{\mathrm{T}}, \\
\mathbf{H}^{\tau(s)} &= \left(\mathbf{H}_1^{\tau(s)}, \mathbf{H}_2^{\tau(s)}, \cdots, \mathbf{H}_N^{\tau(s)}\right)^{\mathrm{T}}
\end{aligned}
\tag{2.28}
$$

for $N = 1, 2, \ldots, \infty$, with

$$
\begin{aligned}
\mathbf{H}_k^{(c)} &= \left(H_{k0}^{(c)}, H_{k1}^{(c)}, \cdots, H_{k(2N)}^{(c)}\right), \\
\mathbf{H}_k^{(s)} &= \left(H_{k0}^{(s)}, H_{k1}^{(s)}, \cdots, H_{k(2N)}^{(s)}\right); \\
\mathbf{H}_k^{\tau(c)} &= \left(H_{k0}^{\tau(c)}, H_{k1}^{\tau(c)}, \cdots, H_{k(2N)}^{\tau(c)}\right), \\
\mathbf{H}_k^{\tau(s)} &= \left(H_{k0}^{\tau(s)}, H_{k1}^{\tau(s)}, \cdots, H_{k(2N)}^{\tau(s)}\right).
\end{aligned}
\tag{2.29}
$$

for $k = 1, 2, \ldots, N$. The corresponding components are

$$
\begin{aligned}
&H_r^{(0)} = -\delta\delta_0^r, H_{kr}^{(c)} = -2k\Omega\delta_{k+N}^r - \delta\delta_k^r,\ H_{kr}^{(s)} = 2k\Omega\delta_k^r - \delta\delta_{k+N}^r; \\
&H_r^{\tau(0)} = 0,\ H_{kr}^{\tau(c)} = 0,\ H_{kr}^{\tau(s)} = 0
\end{aligned}
\tag{2.30}
$$

for $r = 0, 1, \ldots, 2N$.

From Luo [11, 14], the eigenvalues of Eq. (2.17) are classified as

$$
\left(n_1, n_2, n_3 \middle| n_4, n_5, n_6\right), \tag{2.31}
$$

where n_1 is the total number of negative real eigenvalues, n_2 is the total number of positive real eigenvalues, n_3 is the total number of negative zero eigenvalues;

n_4 is the total pair number of complex eigenvalues with negative real parts, n_5 is the total pair number of complex eigenvalues with positive real parts, n_6 is the total pair number of complex eigenvalues with zero real parts. If $\mathrm{Re}(\lambda_k) < 0$ $(k = 1, 2, \cdots, 2(2N+1))$, the approximate steady-state solution $\mathbf{y}^*$ with truncation of $\cos(N\Omega t)$ and $\sin(N\Omega t)$ is stable. If $\mathrm{Re}(\lambda_k) > 0$ $(k \in \{1, 2, \cdots, 2(2N+1)\})$, the truncated approximate steady-state solution is unstable. The corresponding boundary between the stable and unstable solutions is given by the saddle-node bifurcation and Hopf bifurcation.

The harmonic amplitude and phase are defined by

$$A_{k/m} \equiv \sqrt{b_{k/m}^2 + c_{k/m}^2} \text{ and } \varphi_{k/m} = \arctan\frac{c_{k/m}}{b_{k/m}}. \tag{2.32}$$

The corresponding solution in Eq. (2.4) becomes

$$\begin{aligned} x^*(t) &= a_0^{(m)} + \sum_{k=1}^{N} A_{k/m}\cos\left(\frac{k}{m}\Omega t - \varphi_{k/m}\right), \\ x^{\tau *}(t) &= a_0^{(m)} + \sum_{k=1}^{N} A_{k/m}\cos\left[\frac{k}{m}\Omega(t-\tau) - \varphi_{k/m}\right]. \end{aligned} \tag{2.33}$$

Consider system parameters as

$$\delta = 0.05, \alpha_1 = 15.0, \alpha_2 = 5.0, \beta = 5.0, Q_0 = 4.5, \tau = T/4. \tag{2.34}$$

2.2 Numerical Illustrations

To verify the approximate analytical solutions of periodic motion in the time-delayed, quadratic nonlinear oscillator, numerical simulations will be completed through the midpoint discrete scheme. The initial conditions and the initial time-delay values in the range of $t \in (-\tau, 0)$ for numerical simulation are computed from the approximate analytical solutions. The numerical results are depicted by solid curves, but the analytical solutions are given by red circular symbols. The big filled circular symbols are initial conditions and initial time-delay response values. The initial starting and final points of the time delay are represented by the acronyms D.I.S. and D.I.F., respectively.

The displacement, velocity, trajectory, and amplitude spectrum of stable period-1 motion for the time-delayed, quadratic nonlinear oscillator are presented in Fig. 2.1 for $\Omega = 7.767$ with initial condition ($x_0 \approx -0.100171$, $\dot{x}_0 \approx 0.089894$) with initial time-delayed responses. This analytical solution is based on 20 harmonic terms (HB20) in the Fourier series solution of period-1 motion. In Fig. 2.1a, b, for over 100 periods, the analytical and numerical solutions of the period-1 motion in the time-delayed, quadratic nonlinear oscillator match very well. The

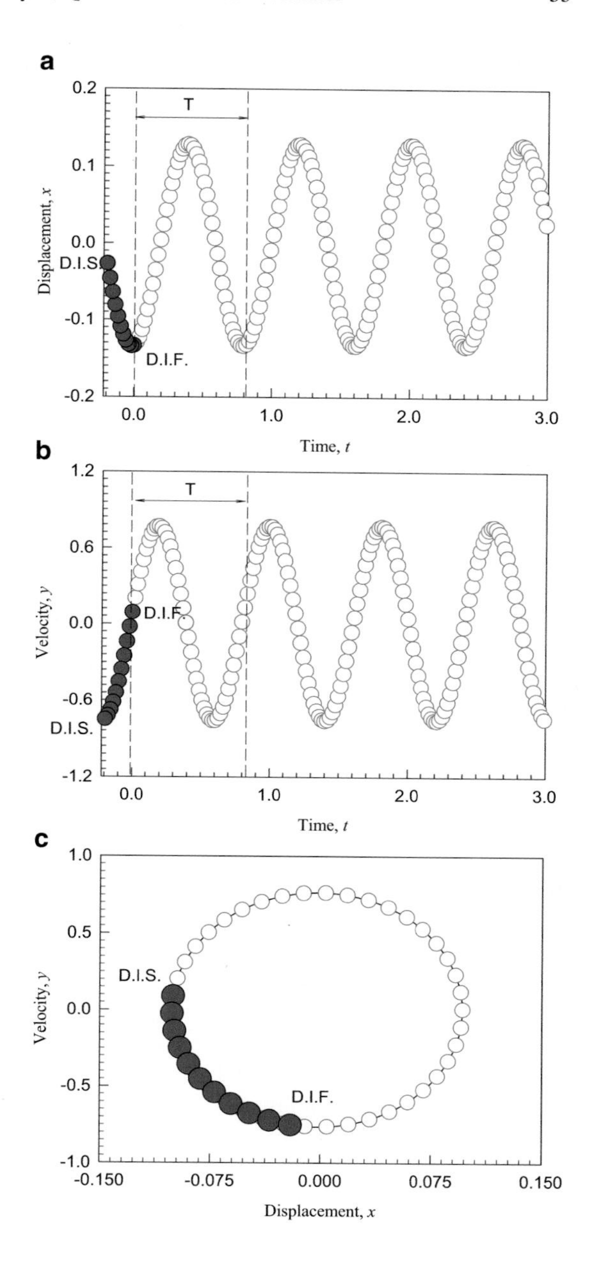

Fig. 2.1 Analytical and numerical solutions of stable period-1 motion based on 20 harmonic terms (HB20) ($\Omega = 7.767$): (**a**) displacement, (**b**) velocity, (**c**) phase plane, and (**d**) amplitude spectrum. Initial condition ($x_0 \approx -0.100171, \dot{x}_0 \approx 0.089894$). Parameters: ($\delta = 0.05, \alpha_1 = 15.0, \alpha_2 = 5.0, \beta = 5.0, Q_0 = 4.5, \tau = T/4$)

Fig. 2.1 (continued)

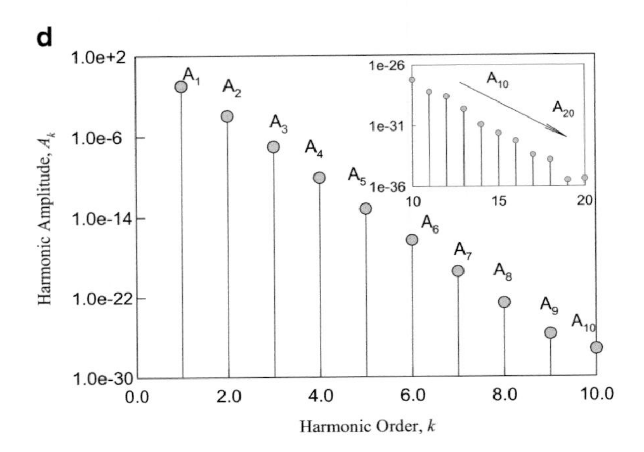

initial time-delayed displacement and velocity are represented by the large circular symbols for the initial delay period of $t \in (-\tau, 0)$. In Fig. 2.1c, analytical and numerical trajectories match very well, and the initial time-delay response in the phase plane is clearly depicted. In Fig. 2.1d, the amplitude spectrum versus the harmonic order is presented. The corresponding quantity levels of the harmonic amplitudes are given as follows: $a_0 \approx -2.4302\text{e-}3$, $A_1 \approx 0.0985$, and $A_k \in (10^{-36}, 10^{-4})$ $(k = 2, 3, \ldots, 20)$. For the distribution of harmonic amplitudes, the harmonic amplitudes decrease with harmonic order nonuniformly. The main contribution for this periodic motion is from the primary harmonics. The truncated harmonic amplitude is $A_{20} \sim 10^{-36}$. For this periodic motion, one can use a harmonic term to get an accurate enough analytical solution.

From the bifurcation tree of period-1 motion to chaos in Luo and Jin [15], the stable period-1, period-2, period-4, and period-8 motions are presented in Fig. 2.2 at $\Omega = 1.897, 1.8965, 1.8920, 1.88906$ for illustrations of the complexity of periodic motions. The initial conditions for such stable periodic motions are listed in Table 2.1.

In Fig. 2.2a, the analytical and numerical trajectories of period-1 motion are presented. Such period-1 motion possesses two cycles and the initial time-delay conditions are presented. The harmonic amplitude distribution is presented in Fig. 2.2b. The main amplitudes of the period-1 motion in such a time-delayed, nonlinear system are $a_0 \approx -0.618722$, $A_1 \approx 0.309591$, $A_2 \approx 1.264949$, $A_3 \approx 0.086255$, $A_4 \approx 0.076064$, and $A_k \in (10^{-14}, 10^{-2})$ for $k = 5, 6, \ldots, 20$. The second harmonic amplitude plays an important role in the period-1 motion.

In Fig. 2.2c, the analytical and numerical trajectories of period-1 motion are presented. Such period-1 motion possesses two cycles and the initial time-delay conditions are presented. The harmonic amplitude distribution is presented in Fig. 2.2d. The main amplitudes of the period-2 motion in such a time-delayed,

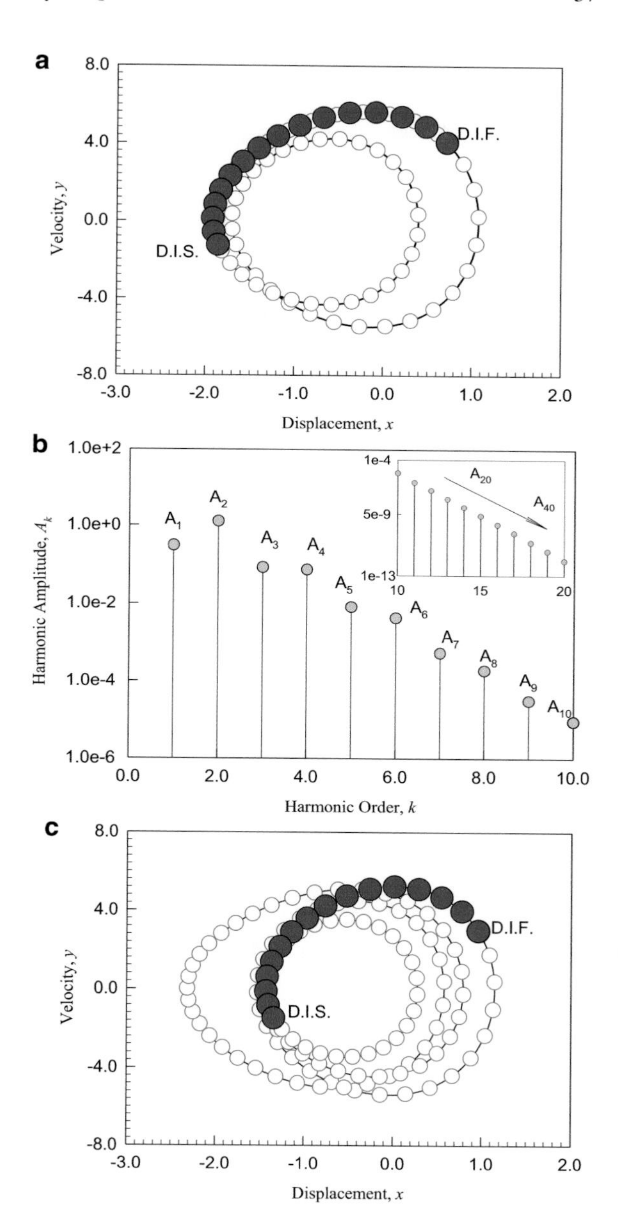

Fig. 2.2 Phase plane and amplitude spectrum: (**a**) and (**b**) period-1 motion ($\Omega = 1.8970$, HB20); (**c**) and (**d**) period-2 motion ($\Omega = 1.8965$, HB40); (**e**) and (**f**) period-4 motion ($\Omega = 1.8920$, HB80); (**g**) and (**h**) period-4 motion ($\Omega = 1.88906$, HB80). Parameters: ($\delta = 0.05, \alpha_1 = 15.0, \alpha_2 = 5.0, \beta = 5.0, Q_0 = 4.5$, $\tau = T/4$)

Fig. 2.2 (continued)

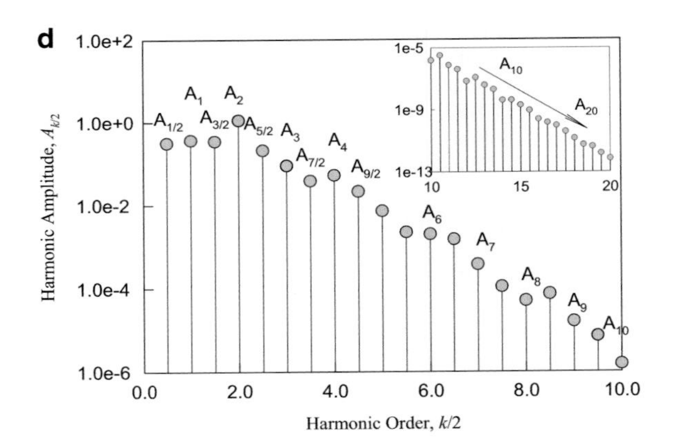

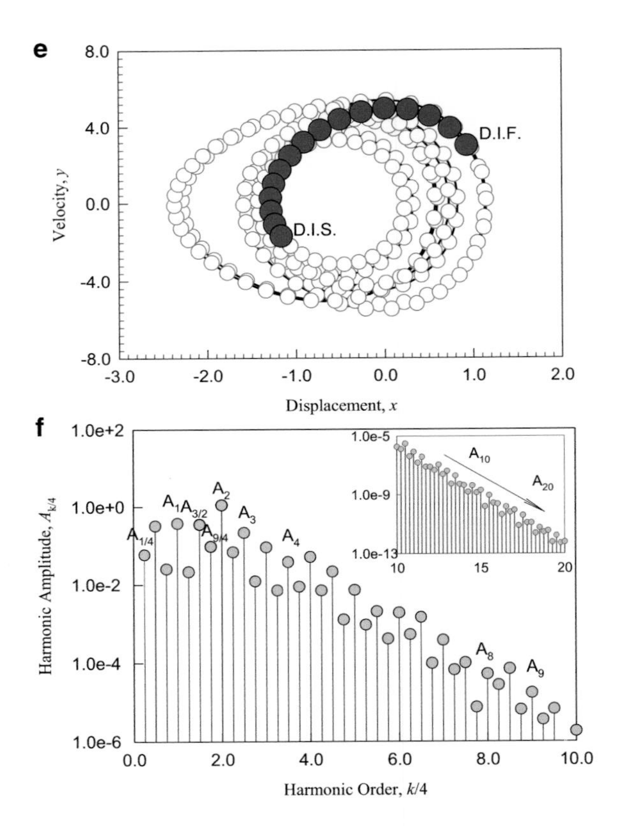

Fig. 2.2 (continued)

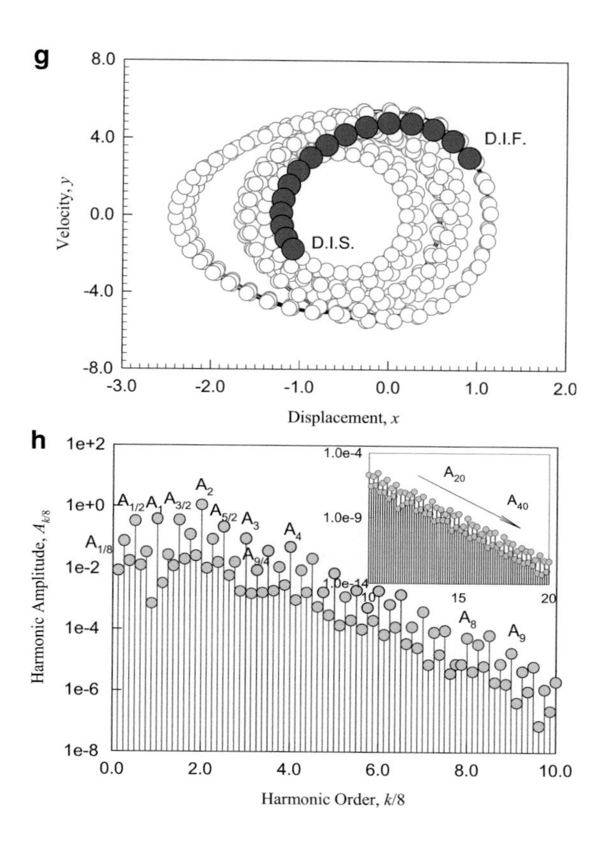

nonlinear system are $a_0^{(2)} \approx -0.589080$, $A_{1/2} \approx 0.312662$, $A_1 \approx 0.366173$, $A_{3/2} \approx 0.345472$, $A_2 \approx 1.120050$, $A_{5/2} \approx 0.209455$, $A_3 \approx 0.089404$, $A_{7/2} \approx 0.038283$, $A_4 \approx 0.052349$, $A_{9/2} \approx 0.021267$, and $A_{k/2} \in (10^{-14}, 10^{-2})$ for $k = 10, 11, \ldots, 40$. The biggest contribution is from the harmonic term of $A_2 \approx 1.120050$.

In Fig. 2.2e, the analytical and numerical trajectories of period-4 motion are presented. Such period-4 motion possesses eight cycles and the initial time-delay conditions are presented. The harmonic amplitude distribution is presented in Fig. 2.2f. The main amplitudes of the period-4 motion are $a_0^{(4)} \approx -0.591813$, $A_{1/4} \approx 0.058286$, $A_{1/2} \approx 0.322076$, $A_{3/4} \approx 0.025289$, $A_1 \approx 0.373248$, $A_{5/4} \approx 0.021254$, $A_{3/2} \approx 0.351173$, $A_{7/4} \approx 0.094394$, $A_2 \approx 1.106125$, $A_{9/4} \approx 0.067732$, $A_{5/2} \approx 0.214359$, $A_{11/4} \approx 0.012157$, $A_3 \approx 0.090130$, $A_{13/4} \approx 7.042438\text{E-}3$, $A_{7/2} \approx 0.037581$, $A_{15/4} \approx 8.784526\text{E-}3$, $A_4 \approx 0.050681$, $A_{17/4} \approx 7.035358\text{E-}3$, $A_{9/2} \approx 0.021354$, $A_{19/4} \approx 1.263319\text{E-}3$, and $A_{k/4} \in (10^{-14}, 10^{-2})$ for $k = 20, 21, \ldots, 80$.

The analytical and numerical trajectories of period-8 motion are presented in Fig. 2.2g. Such period-8 motion possesses 16 cycles and the initial time-

Table 2.1 Input data for numerical illustrations ($\delta = 0.05, \alpha_1 = 15.0, \alpha_2 = 5.0, \beta = 5.0, Q_0 = 4.5, \tau = T/4$)

Figure no.	Ω	Initial condition $(x_0, \dot{x}_0)$	Types	Harmonics terms
Figure 2.2a, b	1.8970	(0.713984, 4.045130)	P-1	HB20 (stable)
Figure 2.2c, d	1.8965	(0.959465, 2.965047)	P-2	HB40 (stable)
Figure 2.2e, f	1.8920	(0.926914, 3.026495)	P-4	HB80 (stable)
Figure 2.2g, h	1.88876	(0.904503, 3.045649)	P-8	HB160 (stable)

delay conditions are presented clearly. As presented before, the harmonic amplitude spectrum is presented in Fig. 2.2h. The main amplitudes of the period-8 motion are $a_0^{(8)} \approx -0.594919$, $A_{1/8} \approx 8.668953\text{e-}3$, $A_{1/4} \approx 0.075480$, $A_{3/8} \approx 0.017434$, $A_{1/2} \approx 0.324209$, $A_{5/8} \approx 0.012676$, $A_{3/4} \approx 0.033521$, $A_{7/8} \approx 6.822809\text{e-}4$, $A_1 \approx 0.376686$, $A_{9/8} \approx 3.278184\text{e-}3$, $A_{5/4} \approx 0.027110$, $A_{11/8} \approx 0.012213$, $A_{3/2} \approx 0.351086$, $A_{13/8} \approx 0.019842$, $A_{7/4} \approx 0.122173$, $A_{15/8} \approx 0.025622$, $A_2 \approx 1.099997$, $A_{17/8} \approx 0.010327$, $A_{9/4} \approx 0.087137$, $A_{19/8} \approx 0.015794$, $A_{5/2} \approx 0.214882$, $A_{21/8} \approx 5.998294\text{e-}3$, $A_{11/4} \approx 0.016157$, $A_{23/8} \approx 1.775930\text{e-}3$, $A_3 \approx 0.090622$, $A_{25/8} \approx 1.485620\text{e-}3$, $A_{13/4} \approx 8.904199\text{e-}3$, $A_{27/8} \approx 1.592552\text{e-}3$, $A_{7/2} \approx 0.036887$, $A_{29/8} \approx 1.829681\text{e-}3$, $A_{15/4} \approx 0.011286$, $A_{31/8} \approx 2.891636\text{e-}3$, $A_4 \approx 0.050091$, $A_{33/8} \approx 9.021719\text{e-}4$, $A_{17/4} \approx 8.953262\text{e-}3$, $A_{35/8} \approx 1.640158\text{e-}3$, $A_{9/2} \approx 0.021173$, and $A_{k/8} \in (10^{-14}, 10^{-2})$ for $k = 37, 38, \ldots, 160$. The biggest contribution of the period-8 motion is still from the harmonic amplitude of $A_2 \approx 1.099997$.

2.3 Conclusion

In this chapter, the analytical solutions of period-m motions in the time-delayed, quadratic nonlinear oscillator were obtained from the finite Fourier series expression. Based on such analytical solutions, the stability and bifurcation of period-m motions of the time-delayed nonlinear oscillator were discussed. From the bifurcation trees of period-1 motion to chaos, numerical simulations were carried out to compare analytical and numerical solutions of periodic motions. The numerical and analytical solutions of periodic motions are well matched in such a time-delayed, quadratic nonlinear oscillator once enough harmonic terms are included in the finite Fourier series expression.

References

1. Lagrange, J.L.: Mécanique Analytique, 2 vol. (édition Albert Blanchard Paris, 1965) (1788) [French]
2. Poincaré, H.: Méthodes Nouvelles de la Mécanique Celeste, vol. 3. Gauthier-Villars, Paris (1899) [French]
3. van der Pol, B.: A theory of the amplitude of free and forced triode vibrations. Radio Rev. **1**, 701–710 (1920). 754–762
4. Fatou, P.: Sur le mouvement d'un système soumis à des forces à courte periode. Bull. Soc. Math. **56**, 98–139 (1928) [French]
5. Krylov, N.M., Bogolyubov, N.N.: Methodes approchées de la mécanique non-linéaire dans leurs application à l'éetude de la perturbation des mouvements périodiques de divers phénomènes de résonance s'y rapportant. Academie des Sciences d'Ukraine, Kiev (1935) [French]
6. Bogoliubov, N.N., Mitropolsky, Y.A.: Asymptotic Methods in the Theory of Nonlinear Oscillations. Gordon and Breach, New York (1961)
7. Hayashi, C.: Nonlinear Oscillations in Physical Systems. McGraw-Hill, New York (1964)
8. Nayfeh, A.H.: Perturbation Methods. Wiley, New York (1973)
9. Nayfeh, A.H., Mook, D.T.: Nonlinear Oscillation. Wiley, New York (1979)
10. Coppola, V.T., Rand, R.H.: Averaging using elliptic functions: approximation of limit cycle. Acta Mech. **81**, 125–142 (1990)
11. Luo, A.C.J.: Continuous Dynamical Systems. Higher Education Press/L&H Scientific, Beijing/Glen Carbon (2012)
12. Luo, A.C.J., Huang, J.Z.: Approximate solutions of periodic motions in nonlinear systems via a generalized harmonic balance. J. Vib. Control. **18**, 1661–1674 (2012)
13. Luo, A.C.J., Huang, J.Z.: Analytical dynamics of period-*m* flows and chaos in nonlinear systems. Int. J. Bifurcation Chaos **22**, 29 p (2012). Article No. 1250093
14. Luo, A.C.J.: Analytical solutions of periodic motions in dynamical systems with/without time-delay. Int. J. Dyn. Control **1**, 330–359 (2013)
15. Luo, A.C.J., Jin, H.X.: Bifurcation trees of period-*m* motions in a time-delayed, quadratic nonlinear oscillator under a periodic excitation. Discontin. Nonlinearity Complex. **3**, 87–107 (2014)

Chapter 3
Mathematical Analysis of a Delayed Hematopoietic Stem Cell Model with Wazewska–Lasota Functional Production Type

Radouane Yafia, M.A. Aziz Alaoui, Abdessamad Tridane, and Ali Moussaoui

Abstract In this chapter, we consider a more general model describing the dynamics of a hematopoietic stem cell (HSC) model with Wazewska–Lasota functional production type describing the cycle of proliferating and quiescent phases. The model is governed by a system of two ordinary differential equations with discrete delay. Its dynamics are studied in terms of local stability and Hopf bifurcation. We prove the existence of the possible steady state and their stability with respect to the time delay and the apoptosis rate of proliferating cells. We show that a sequence of Hopf bifurcations occurs at the positive steady state as the delay crosses some critical values. We illustrate our results with some numerical simulations.

R. Yafia (✉)
Polydisciplinary Faculty of Ouarzazate, Ibn Zohr University, B.P. 638, Ouarzazate, Morocco
e-mail: yafia1@yahoo.fr

M.A.A. Alaoui
Normandie University, Le Havre, France

ULH, LMAH, 76600 Le Havre, France

FR CNRS 3335, 25 rue Philippe Lebon, 76600 Le Havre, France
e-mail: aziz.alaoui@univ-lehavre.fr

A. Tridane
Department of Mathematical Sciences, United Arab Emirates University, P.O. Box 15551, Al Ain, United Arab Emirates
e-mail: a-tridane@uaeu.ac.ae

A. Moussaoui
Department of Mathematics, University of Tlemcen, 13000 Tlemcen, Algeria
e-mail: moussaouidz@yahoo.fr

A.C.J. Luo, H. Merdan (eds.), *Mathematical Modeling and Applications in Nonlinear Dynamics*, Nonlinear Systems and Complexity 14,
DOI 10.1007/978-3-319-26630-5_3

3.1 Biological Background

Hematopoietic stem cells (HSCs) are found in adult bone marrow, which is found in femurs, hips, ribs, sternum, and other bones. HSCs are precursor cells which give rise to all types of both the myeloid and lymphoid lineages of blood cells. HSCs have the ability to form multiple cell types (multipotency) and an ability to self-renew.

Multipotency: Individual HSCs can give rise to all of the end-stage blood cell types.

During differentiation, daughter cells derived from HSCs undertake a series of commitment decisions, retaining differentiation potential for some lineages while losing others. Intermediate cells become progressively more restrictive in their lineage potential until eventually, at the end stage, the cells are lineage-committed.

Self-Renewal: Some kinds of stem cells are thought to undertake asymmetric cell division to generate one daughter cell that remains a stem cell and one daughter cell that is differentiated. However, it is not known with certainty whether or not asymmetric cell division occurs during self-renewal. An alternative possibility is that hematopoiesis occurs via symmetric divisions that sometimes give rise to two HSC daughter cells, and sometimes to two daughter cells that are committed to differentiate. The balance between self-renewal and differentiation would then be determined by the control of these two distinct kinds of symmetric cell divisions (see Fig. 3.1).

HSCs are either proliferating or nonproliferating (quiescent or resting) cells. The majority of HSCs are actually in a quiescent stage [14].

Quiescent HSCs represent a pool of stem cells that are used to produce new blood cells.

Proliferating HSCs are actively involved in cell division (growth, DNA synthesis, etc.).

After entering the proliferating phase, a cell is committed to undergo cell division at a fixed time τ later. The generation time τ is assumed to consist of four phases: G_1, the presynthesis phase; S, the DNA synthesis phase; G_2, the postsynthesis phase; and M, the mitotic phase.

Just after the division, both daughter cells go into the resting (quiescent) phase called the G_0-phase. Once in this phase, they can either return to the proliferating phase and complete the cycle or die before ending the cycle (see Fig. 3.3).

The first mathematical model was introduced by Mackey [19] and Burns and Tannock [8]. Mackey's model is governed by a system of delay differential equations taking into account the proliferating and quiescent phases and the necessary time delay of cell division. It was also proposed to describe some periodic hematological diseases, such as periodic autoimmune hemolytic anemia [6, 22], cyclical thrombocytopenia [26, 28], cyclical neutropenia [17, 18], and periodic chronic myelogenous leukemia [14]. Periodic hematological disorders are classic examples of dynamic diseases. Because of their dynamic properties, they offer an almost unique opportunity to understand the nature of the regulatory processes involved in hematopoiesis. Periodic hematological disorders are characterized by

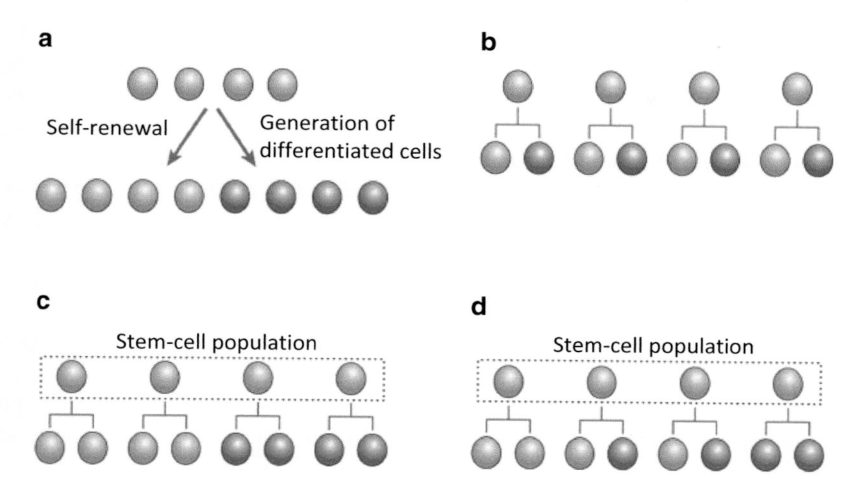

Fig. 3.1 (**a**) Stem cells (*orange*) must accomplish the dual task of self-renewal and generation of differentiated cells (*green*). (**b**)–(**d**) Possible stem cell strategies that maintain a balance of stem cells and differentiated progeny. (**b**) Asymmetric cell division: Each stem cell generates one daughter stem cell and one daughter destined to differentiate. (**c**), (**d**) Population strategies. A population strategy provides dynamic control over the balance between stem cells and differentiated cells—a capacity that is necessary for repair after injury or disease. In this scheme, stem cells are defined by their "potential" to generate both stem cells and differentiated daughters, rather than their actual production of a stem cell and a differentiated cell at each division. (**c**) Symmetric cell division: Each stem cell can divide symmetrically to generate either two daughter stem cells or two differentiated cells. (**d**) Combination of cell divisions: Each stem cell can divide either symmetrically or asymmetrically (courtesy of www.nature.com)

oscillations in the number of one or more of the circulating blood cells with periods on the order of days to months (see the figures in [17] for examples of experimental data for four hematological diseases. AIHA: Reticulocyte numbers ($\times 10^4$cells/μL) in an AIHA subject. Adapted from Orr et al. [23]. CT: Cyclical fluctuations in platelet counts ($\times 10^3$cells/μL). From Yanabu et al. [30]. CN: Circulating neutrophils ($\times 10^3$cells/μL), platelets ($\times 10^5$cells/μL), and reticulocytes ($\times 10^4$cells/μL) in a cyclical neutropenic patient. From Guerry et al. [15]. PCML: White blood cell (top) ($\times 10^4$cells/μL), platelet (middle) ($\times 10^5$cells/μL), and reticulocyte (bottom) ($\times 10^4$cells/μL) counts in a PCML patient. From Chikkappa et al. [9]. AIHA: Autoimmune hemolytic anemia. CT: cyclical thrombocytopenia. CN: cyclical neutropenia. PCML: periodic chronic myelogenous leukemia).

Recently, many authors have tried to reintroduce Mackey's model in the unstructured and structured versions. In the unstructured version with discrete and distributed time delays, the model was intensively studied by Adimy et al. [2]. They studied the dynamics of the model with respect to the time delay and occurrence and direction of Hopf bifurcation. It was also studied by Alaoui and Yafia [4] and Alaoui et al. [5] in terms of local stability, occurrence, and direction of Hopf bifurcation by proposing an approachable model. In recent years, Adimy et al. [1, 2] proposed the structured model of HSC dynamics in which the cell cycle duration depends on the cell maturity by reducing the model to a system of delay differential equations by the characteristic method. This is a way of indicating that cell cycles can be shortened for some types of cells, or in particular situations such as diseases or anemia.

In 2010, Adimy et al. [3] proposed the same Mackey model with a system of differential equations with state-dependent delay; they proved the global stability and the Hopf bifurcation occurrence.

Such stem cells are released by the marrow to help with the regeneration of damaged bone and tissue. "Techniques already exist to increase the numbers of blood cell producing stem cells from the bone marrow, but the study focuses on two other types-endothelial, which produce the cells which make up our blood vessels, and mesenchymal, which can become bone or cartilage cells." The scientists hope that the increased production rate could be used to greatly speed tissue repair and to allow recovery from wounds that would otherwise be too severe. "There are also hopes that the technique could help damp down autoimmune diseases such as rheumatoid arthritis, where the body's immune system attacks its own tissues. Mesenchymal stem cells are known to have the ability to damp down the immune system (see Pitchford et al. [25]).

It is generally agreed that the production rate is a decreasing function over a wide range of cells levels. Indeed, we would expect the production rate to increase when the number of cells decreases. There are many functions that fit this description, for example the Hill function type $\beta(x) = \beta_0 \frac{\theta^n}{\theta^n + x^n}$ (see Mackey [19]) and the Lasota function type $l(x) = e^{-\gamma x}$ [29].

In this work, we focus on the influence of the necessary time delay (duration) of division and the apoptosis rate of the proliferating cells and the production rate of HSCs.

3.2 Description of Hematopoietic Stem Cells

The classic model of HSCs is as follows (see [8, 21, 27]):

$$\begin{cases} \frac{dN}{dt} = -\delta N - \beta(N)N + 2e^{-\gamma\tau}\beta(N_\tau)N_\tau \\ \\ \frac{dP}{dt} = -\gamma P + \beta(N)N - e^{-\gamma\tau}\beta(N_\tau)N_\tau, \end{cases} \tag{3.1}$$

where β is a monotone decreasing function of N which has the explicit form of a Hill function (see [7, 13, 19, 24]):

$$\beta(N) = \beta_0 \frac{\theta^n}{\theta^n + N^n}. \tag{3.2}$$

The symbols in Eq. (3.1) have the following interpretation. N is the number of cells in the nonproliferating phase, $N_\tau = N(t-\tau)$, P the number of cycling proliferating cells, γ the rate of cell loss from the proliferating phase (apoptosis rate), δ the rate of cell loss from the nonproliferating phase, τ the time spent in the proliferating phase, β the feedback function, the rate of recruitment from nonproliferating phase, $\beta_0 > 0$ the maximal rate of reentry in the proliferating phase, and $\theta \geq 0$ the number of resting cells at which β has its maximum rate of change with respect to the resting phase population; $n > 0$ describes the sensitivity of the reintroduction rate with changes in the population, and $e^{-\gamma\tau}$ accounts for the attenuation due to apoptosis (programmed cell death) at rate γ (or the survival function).

Low cell counts lead to quick reactions of the organism, in order to produce enough cells to return to a normal state, and this can then induce shorter cell cycles and a small rate of apoptosis (this is observed for red cells, where, following an anemia, immature cells enter the bloodstream and replace mature cells very quickly) [11]. To control this low cell count and increase the speed of production of HSCs, we replace the quantity $e^{-\gamma\tau}$ by the Wazewska–Lasota function $e^{-\gamma N_\tau}$ (Fig. 3.2). Let's denote the change in the levels of quiescent cells between $t-\tau$ and $t-\tau+\Delta t$ as

$$\Delta N(t-\tau) = N(t+\Delta t-\tau) - N(t-\tau).$$

The production stimulated of level of proliferating cells between $t-\tau$ and $t-\tau+\Delta t$ is given by

$$\Delta P(t) = P(t+\Delta t) - P(t).$$

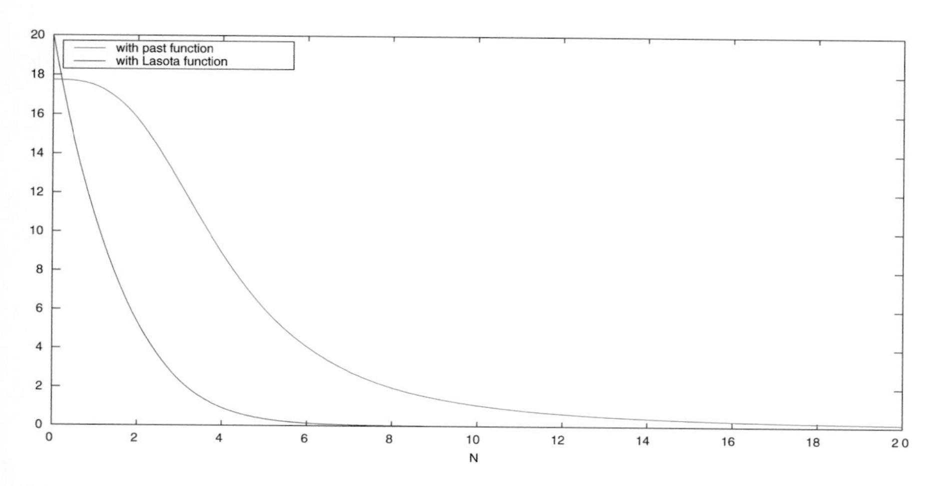

Fig. 3.2 From this figure, we observe that the new function $e^{-\gamma N}\beta(N)$ is much more decreasing than the old function $e^{-\gamma\tau}\beta(N)$

The number of quiescent cells is decreasing, and the production increases after the time delay τ. Therefore we look for a nonnegative function $l(t, \tau)$ such that

$$\Delta P(t) = -l(t, \tau)\Delta N(t - \tau).$$

We suppose there exists some kind of per capita increase. Therefore we choose simply $l(t, \tau) = \xi P(t)$:

$$P(t + \Delta t) - P(t) = -\xi P(t)N(t + \Delta t - \tau) - N(t - \tau),$$

where ξ characterizes the excitability of the HSCs. After dividing by Δt and choosing $\Delta t \longrightarrow 0^+$, we have

$$\frac{d}{dt}P(t) = -\xi P(t)\frac{d}{dt}N(t - \tau).$$

The solution of this equation with some constant υ is

$$P(t) = \upsilon e^{-\xi N(t-\tau)}.$$

υ is a medical constant.

We consider the case when $\upsilon = \gamma$; without loss of generality, we suppose that the survival function of the active cells takes the form $e^{-\gamma N(t-\tau)}$ instead of $e^{-\gamma\tau}$.

The model that is under consideration is governed by the following schematic representation (see Fig. 3.3):

The mathematical model is as follows:

$$\begin{cases} \frac{dN}{dt} = -\delta N - \beta(N)N + 2e^{-\gamma N_\tau}\beta(N_\tau)N_\tau \\ \\ \frac{dP}{dt} = -\gamma P + \beta(N)N - e^{-\gamma N_\tau}\beta(N_\tau)N_\tau. \end{cases} \quad (3.3)$$

Parameter estimation and their references appear in the following table.

Parameters	Value used	Unit	Sources
β_0	3–3.5,	day^{-1}	Mackey et al. [20], Colijn et al. [10]
θ	1.38×10^8–0.5×10^6	cells kg^{-1}	Mackey et al. [20], Colijn et al. [10]
n	3–4		Mackey et al. [20], Colijn et al. [10]
δ	0.16	day^{-1}	Mackey et al. [20]
γ	0.1–0.36	day^{-1}	Mackey et al. [20], Colijn et al. [10]
τ	0.83–0.88	day^{-1}	Mackey et al. [20], Colijn et al. [10]

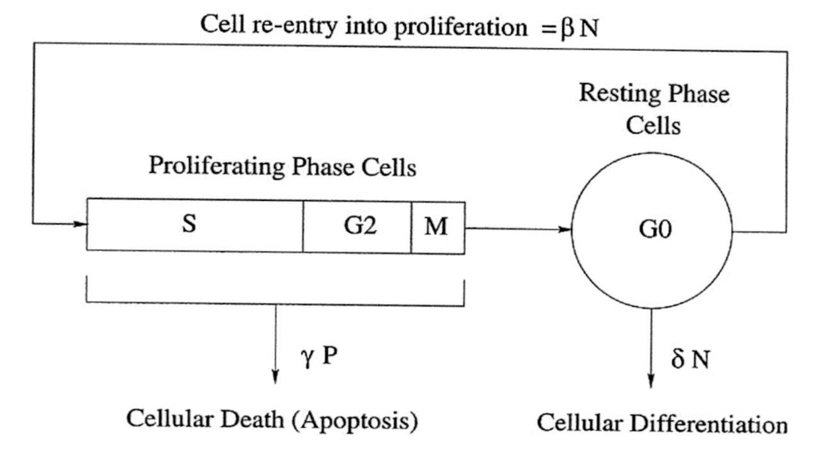

Fig. 3.3 A schematic representation of the $G0$ stem cell model. Proliferating phase cells P include those cells in S (DNA synthesis), $G2$, and M (mitosis), while the resting phase N cells are in the $G0$ phase. δ is the rate of differentiation into all of the committed stem cell populations, while γ represents a loss of proliferating phase cells due to apoptosis. $G(\gamma, N)$ is the rate of cell reentry from $G0$ into the proliferating phase, and τ is the duration of the proliferating phase. See Mackey [19] for further details

Remark by M. C. Mackey This term just tries to capture the fact that the production of erythrocytes is a decreasing function of the number of erythrocytes in the circulation. The delay τ takes into account the fact that it requires a number of days τ between the time the signal to produce erythrocyte precursors is felt in the bone marrow and when mature blood cells are ready for circulation.

This work is organized as follows. In Sect. 3.3, we prove the existence and stability of the possible steady states both with and without delay. Section 3.4 is devoted to the occurrence of Hopf bifurcation by considering the delay as a parameter bifurcation; we prove the occurrence of a sequence of Hopf bifurcation. In Sect. 3.5, we give an algorithm determining the stability and instability of periodic solutions bifurcating from the nontrivial steady state and the direction of bifurcation. At the end we illustrate our result with numerical simulations.

3.3 Steady States and Stability

In this section, we establish the conditions of the existence of the possible steady states. We prove their stability for the model without and with delay and show the influence of the delay and the rate of the apoptosis of the proliferating cells on the stability of the positive steady state.

3.3.1 Existence of Possible Steady States

Consider the following system:

$$\begin{cases} \frac{dN}{dt} = -\delta N - \beta(N)N + 2e^{-\gamma N_\tau}\beta(N_\tau)N_\tau \\ \\ \frac{dP}{dt} = -\gamma P + \beta(N)N - e^{-\gamma N_\tau}\beta(N_\tau)N_\tau. \end{cases} \tag{3.4}$$

The equilibrium points are given by resolving the equations

$$\begin{cases} \frac{dN}{dt} = 0 \\ \\ \frac{dP}{dt} = 0. \end{cases} \tag{3.5}$$

Let $d = \frac{\ln(2)}{\gamma}$ and define the function $F(N) = \beta(N)(2e^{-\gamma N} - 1)$.

As $F(0) = \beta_0$ and $F(d) = 0$, we have that F is a positive decreasing function on $]0, d[$ (Fig. 3.4).

From Eq. $(3.5)_1$, there exists $N^* \in]0, d[$ such that $F(N^*) = \delta$ if and only if (iff) $\delta \in]0, \beta_0[$, where $N^* = F^{-1}(\delta)$ (see, Fig. 3.4), and from Eq. $(3.5)_2$ we obtain

$$P^* = \frac{1}{\gamma}(1 - e^{-\gamma N^*})\beta(N^*)N^*.$$

Let

(H_1): $\delta > \beta_0$,
(H_2): $\delta \in]0, \beta_0[$,

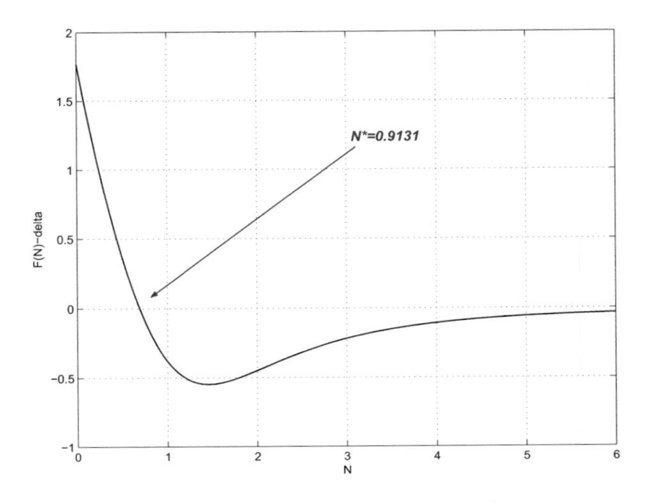

Fig. 3.4 The curve of the functional F showing the existence of N^*

(H_3): $\beta_0 > 0$,

(H_4): $N^* < \inf\left(d = \frac{\ln(2)}{\gamma}, (\frac{\gamma}{2})^{\frac{1}{n-1}}\right)$.

Proposition 1. *(1) If (H_1) is satisfied, system (3.4) has a unique trivial equilibrium point $E_0 = (0, 0)$.*

(2) If (H_2) is satisfied, system (3.4) has two equilibrium points: The first is trivial, $E_0 = (0, 0)$, and the second is nontrivial (positive), given by $E^* = (N^*, P^*)$, where $N^* = F^{-1}(\delta)$ and $P^* = \frac{1}{\gamma}(1 - e^{-\gamma N^*})\beta(N^*)N^*$.

The previous proposition gives a condition of the existence of two different equilibria. In fact, by definition δ is the differentiation rate of cells and β_0 is the maximal proliferation rate of reentry into the proliferating phase. Therefore, if the proliferation rate is small, then in addition to the trivial equilibrium we get another nontrivial equilibrium. The normal step is to investigate the condition of stability of each equilibrium; for this we will first use the case without delay $\tau = 0$; second, we will study the effect of increasing the delay $\tau > 0$ on the stability of our model.

3.3.2 Stability of Steady States for $\tau = 0$

For $\tau = 0$, system (3.4) becomes a system of ordinary differential equations (ODEs) given by the following system:

$$\begin{cases} \frac{dN}{dt} = -\delta N - \beta(N)N + 2e^{-\gamma N}\beta(N)N \\ \frac{dP}{dt} = -\gamma P + \beta(N)N - e^{-\gamma N}\beta(N)N. \end{cases} \tag{3.6}$$

Proposition 2. *(1) If (H_1) is satisfied, the trivial equilibrium point $E_0 = (0, 0)$ is asymptotically stable.*

(2) If (H_2) is satisfied, the equilibrium point $E_0 = (0, 0)$ is unstable and the nontrivial (positive) $E^* = (N^*, P^*)$ is asymptotically stable.

Proof. (1) The steady states are the same given in Proposition 1. To study the stability of $E_0 = (0, 0)$, we linearize system (3.6) around the concerned steady state E_0.

The linearized equation is given as follows:

$$\begin{cases} \frac{dN}{dt} = -\delta N + \beta(0)N \\ \frac{dP}{dt} = -\gamma P, \end{cases} \tag{3.7}$$

and the characteristic equation associated to E_0 is

$$(\lambda + \delta - \beta(0))(\lambda + \gamma) = 0. \tag{3.8}$$

Then the characteristic roots are as follows: $\lambda_1 = -\gamma$ and $\lambda_2 = -\delta + \beta_0$.

(2) Suppose now that $0 < \delta < \beta_0$ and let $N = x+N^*$ and $P = y+P^*$. We linearize system (3.6) around the equilibrium point E^* and the linearized system is given as follows:

$$\begin{cases} \frac{dx}{dt} = -\delta x + \left\{-\beta'(N^*)N^* - 2\gamma e^{-\gamma N^*}\beta(N^*)N^* + 2e^{-\gamma N^*}\beta'(N^*)N^*\right\} x \\ \frac{dy}{dt} = -\gamma y + \left\{\gamma \frac{P^*}{N^*} + \gamma \frac{P^*\beta'(N^*)}{\beta(N^*)} + \gamma e^{-\gamma N^*}\beta(N^*)N^*\right\} x. \end{cases} \tag{3.9}$$

The characteristic equation is given by

$$(\lambda+\gamma)(\lambda+\beta'(N^*)N^* + 2\gamma e^{-\gamma N^*}\beta(N^*)N^* - 2e^{-\gamma N^*}\beta'(N^*)N^*) = 0 \tag{3.10}$$

and the associated characteristic roots are $\lambda_1 = (2e^{-\gamma N^*} - 1)\beta'(N^*)N^* - 2\gamma e^{-\gamma N^*}\beta(N^*)N^*$ and $\lambda_2 = -\gamma$. As β is a decreasing positive function and $2e^{-\gamma N^*} - 1 > 0$, we have $\lambda_i < 0$, $i = 1, 2$.

Then the steady states E^* are asymptotically stable.

It is clear from the previous results for a nondelay model that when the trivial equilibrium exists and is unique, then it is asymptotically stable; otherwise, the nontrivial equilibrium exists and is asymptotically stable. Next, we will study the stability of our delay model and the effect of the delay on the stability of these equilibria.

3.3.3 Stability of Steady States for $\tau > 0$

Proposition 3. *(1) If (H_1) is satisfied, the trivial equilibrium point $E_0 = (0, 0)$ is asymptotically stable for all $\tau > 0$.*

(2) If (H_2)–(H_4) are satisfied, there exists $\tau_0 > 0$ such that the nontrivial (positive) steady state $E^* = (N^*, P^*)$ is asymptotically stable for $\tau < \tau_0$ and unstable for $\tau > \tau_0$ and the equilibrium point $E_0 = (0, 0)$ is unstable for all $\tau > 0$.

Proof. (1) By linearizing system (3.3) around the steady state E_0, we obtain the following linearized equation:

$$\begin{cases} \frac{dN}{dt} = -\delta N - \beta_0 N + 2\beta_0 N_\tau \\ \frac{dP}{dt} = -\gamma P + \beta_0 N - \beta_0 N_\tau. \end{cases} \tag{3.11}$$

The characteristic equation is

$$(\lambda+\gamma)(\lambda+\delta+\beta_0 - 2\beta_0 e^{-\lambda\tau}) = 0. \tag{3.12}$$

For the stability of E_0, one needs to study the position of the characteristic roots of the following equation:

$$(\lambda + \delta + \beta_0 - 2\beta_0 e^{-\lambda\tau}) = 0. \tag{3.13}$$

From Proposition 3, E_0 is asymptotically stable. For a change of stability, replacing $\lambda = i\omega$ in (3.13) and separating the real and imaginary parts gives us

$$\begin{cases} \delta + \beta_0 - 2\beta_0 \cos(\omega\tau) = 0 \\ \omega + 2\beta_0 \sin(\omega\tau) = 0. \end{cases} \tag{3.14}$$

From (3.14), we have $\omega^2 = (\beta_0 - \delta)(3\beta_0 + \delta)$. As $\beta_0 < \delta$, there exists any value of τ in which E_0 changes the stability. Then we conclude that E_0 is asymptotically stable for all $\tau > 0$.

(2) Suppose now that $\tau > 0$ and $\delta < \beta_0$, and by linearizing system (3.3) around the nontrivial steady state we have the following linearized system:

$$\begin{cases} \frac{dx(t)}{dt} = -\delta x(t) - h(N^*)x(t) + 2g(N^*)x(t-\tau) \\ \frac{dy(t)}{dt} = -\gamma y(t) + h(N^*)x(t) - g(N^*)x(t-\tau), \end{cases} \tag{3.15}$$

where

$$h(N^*) = \beta(N^*) + \beta'(N^*)N^* = (\beta(N)N)'_{N=N*} = H'(N)_{/N=N*},$$

$$\begin{aligned} g(N^*) &= e^{-\gamma N^*}\beta(N^*) - \gamma e^{-\gamma N^*}\beta(N^*)N^* + e^{-\gamma N^*}\beta'(N^*)N^* \\ &= (e^{-\gamma N}\beta(N)N)'_{N=N*} == G'(N)_{/N=N*}, \end{aligned}$$

and

$$x = N - N^* \qquad y = P - P^*.$$

The characteristic equation is

$$\Delta(\lambda, \tau) = (\lambda + \gamma)(\lambda + \delta + h(N^*) - g(N^*)e^{-\lambda\tau}) = 0. \tag{3.16}$$

To study the change of stability, replacing $\lambda = i\omega$ and separating the real and imaginary parts gives us $\delta + h(N^*) - g(N^*)\cos(\omega\tau) = 0$ and $\omega + g(N^*)\sin(\omega\tau) = 0$.

Then

$$\omega^2 = g(N^*)^2 - (\delta + h(N^*))^2 = (g(N^*) - \delta - h(N^*)).$$

From the expressions of h and g, we have

$$g(N^*) - \delta - h(N^*) = (2e^{-\gamma N^*} - 1)\beta'(N^*)N^* - 2\gamma e^{-\gamma N^*}\beta(N^*)N^* < 0.$$

By calculations, we obtain

$$g(N^*) + \delta + h(N^*) = 2e^{-\gamma N^*}\beta(N^*)(2 - \gamma N^*) + \beta'(N^*)N^* + 2e^{-\gamma N^*}\beta'(N^*)N^*.$$

From the expression of β, we have

$$\begin{aligned} &2e^{-\gamma N^*}\beta(N^*)(2 - \gamma N^*) + 2e^{-\gamma N^*}\beta'(N^*)N^* \\ &= 2e^{-\gamma N^*}\beta(N^*)(2 - \gamma N^* - \frac{\beta_0\theta^n}{\theta^n + N^{*n}}) \\ &= 2e^{-\gamma N^*}\beta(N^*)(2N^{*n} - \gamma N^* + (2 - \beta_0)\theta^n). \end{aligned}$$

As $\beta_0 < 2$ and $N^* < \inf\left(\frac{\ln(2)}{2}, (\frac{\gamma}{2})^{\frac{1}{n-1}}\right)$, and from the expression of the function β, we have

$$g(N^*) + \delta + h(N^*) < 0$$

and the quantity of ω^2 is positive.

As

$$\mid \frac{\delta + h(N^*)}{g(N^*)} \mid < 1,$$

let

$$\tau_k = \frac{1}{\omega_0}\left\{\arccos\left(\frac{\delta + h(N^*)}{g(N^*)}\right) + 2k\pi\right\}, k = 0, 1, 2, 3, \ldots, \tag{3.17}$$

and

$$\omega_0 = \sqrt{g(N^*)^2 - (\delta + h(N^*))^2}. \tag{3.18}$$

Then Eq. (3.16) has a pair of purely imaginary roots $\pm i\omega_0$ at $\tau = \tau_k$, $k = 0, 1, 2, 3, \ldots$.

Let $\lambda(\tau) = \eta(\tau) + \omega(\tau)$ denote a root of (3.16) near $\tau = \tau_k$ such that $\eta(\tau_k) = 0$, $\omega(\tau_k) = \omega_0$.

Then we deduce the result.

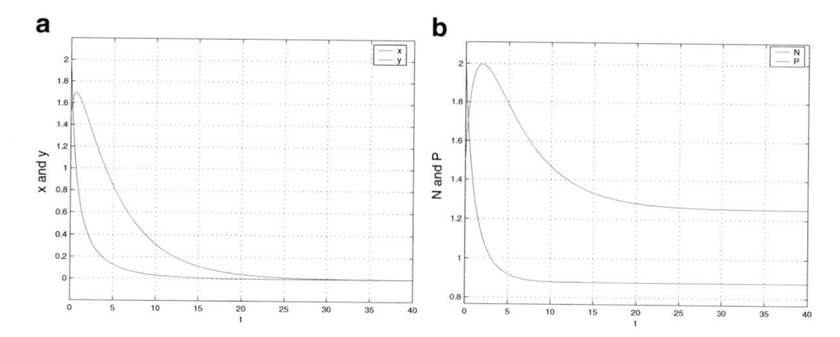

Fig. 3.5 (**a**) Stability of $E_0 = (0, 0)$ and the nonexistence of E^* for $\delta > \beta_0$. (**b**) Instability of $E_0 = (0, 0)$ and stability of E^* for $\tau = 0$ and $\delta < \beta_0$

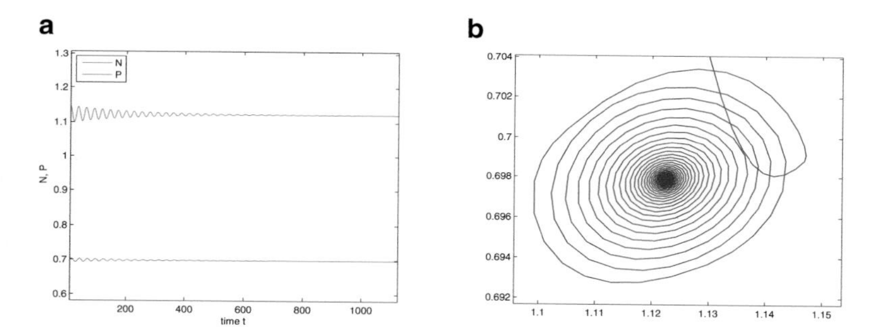

Fig. 3.6 Stability of E^* for $n = 3$, $\tau = 10$ in (t,P) and (t,N) planes (a) and in (P,N) plane

From this result, we showed that the condition of stability of the trivial solution is the same for the delay and nondelay models (see Fig. 3.5). On the other hand, we have additional conditions for the stability of the nontrivial solution (see, Figs. 3.6 and 3.8); there exists a threshold delay τ_0 under which the local asymptotic stability holds if $0 < \sup(\delta, 2) < \beta_0$ ($2 < \beta_0$ means that the maximal rate of proliferation is greater than the rate of division of one cell into two daughters) and $N^* < \inf\left(\frac{\ln(2)}{\gamma}, (\frac{\gamma}{2})^{\frac{1}{n-1}}\right)$ and beyond this threshold the system goes to Hopf bifurcation and becomes unstable (see, Figs. 3.7, 3.9, and 3.10).

It is worth mentioning that $\inf\left(\frac{\ln(2)}{\gamma}, (\frac{\gamma}{2})^{\frac{1}{n-1}}\right)$ is determined by the order of $\gamma^{\frac{n}{n-1}}$ and $2^{\frac{1}{n-1}}\ln(2)$. This can be determined by knowing the range of possible values of γ.

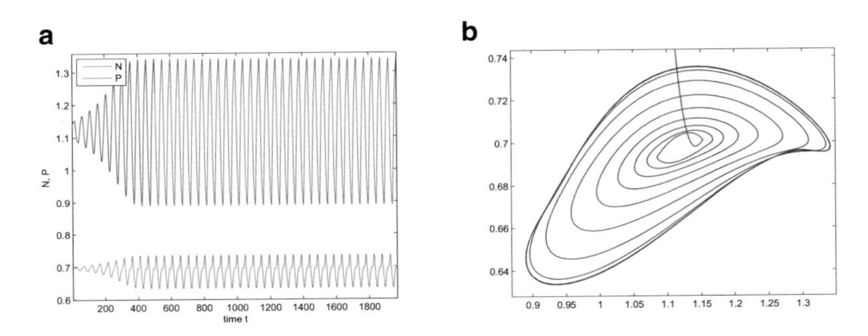

Fig. 3.7 Periodic solutions for $n = 3$, $\tau = 20$ in (t,P) and (t,N) planes (a) and in (P,N) plane

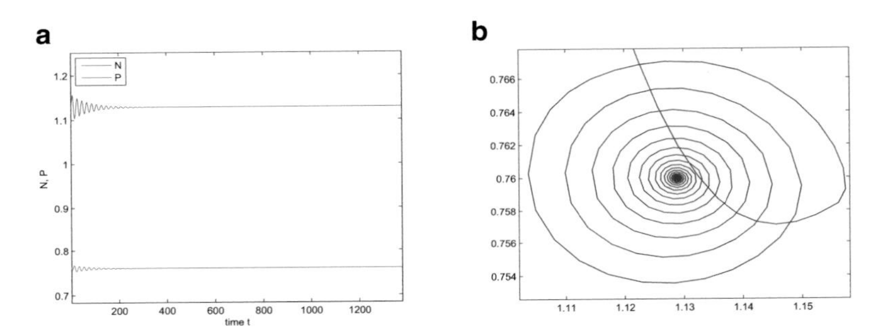

Fig. 3.8 Stability of E^* for $n = 4$, $\tau = 7$ in (t,P) and (t,N) planes (a) and in (P,N) plane

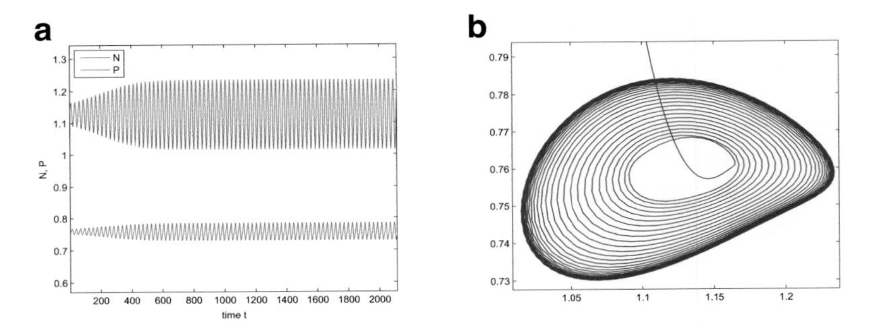

Fig. 3.9 Periodic solutions for $n = 4$, $\tau = 10$ in (t,P) and (t,N) planes (a) and in (P,N) plane

3.4 Branch of Bifurcating Periodic Solutions

We apply the Hopf bifurcation theorem to show the existence of a nontrivial periodic solution of system (3.4), for suitable values of parameter delay, used as a bifurcation parameter. Therefore, the periodicity is a result of changing the type of stability, from a stable stationary solution to a limit cycle.

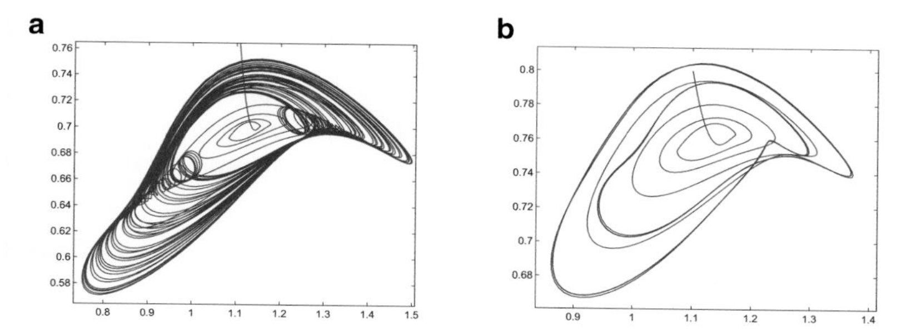

Fig. 3.10 Chaotic solutions for $n = 3$, $\tau = 30$ (**a**) and $n = 4$, $\tau = 16$ (**b**)

In what follows, we recall the formulation of the Hopf bifurcation theorem for delayed differential equations. Let

$$\frac{dx(t)}{dt} = F(\alpha, x_t), \tag{3.19}$$

with $F : \mathbb{R} \times C \longrightarrow \mathbb{R}^n$, F of class $\mathscr{C}^k$, $k \geq 2$, $F(\alpha, 0) = 0$ for all $\alpha \in \mathbb{R}$, and $C = C([-r, 0], \mathbb{R}^n)$ the space of continuous functions from $[-r, 0]$ into $\mathbb{R}^n$. As usual, x_t is the function defined from $[-r, 0]$ into $\mathbb{R}^n$ by $x_t(\theta) = x(t + \theta)$, $r \geq 0$, and $n \in \mathbb{N}^*$.

The following assumptions are stated:

(M_0) F of class $\mathscr{C}^k$, $k \geq 2$, $F(\alpha, 0) = 0$ for all $\alpha \in \mathbb{R}$, and the map $(\alpha, \varphi) \longrightarrow D^k_\varphi F(\alpha, \varphi)$ sends bounded sets into bounded sets.

(M_1) The characteristic equation

$$\Delta(\alpha, \lambda) = \det(\lambda Id - D_\varphi F(\alpha, 0) \exp(\lambda(.)Id)) \tag{3.20}$$

of the linearized equation of (3.19) around the equilibrium $v = 0$,

$$\frac{dv(t)}{dt} = D_\varphi F(\alpha, 0) v_t, \tag{3.21}$$

has in $\alpha = \alpha_0$ a simple imaginary root $\lambda_0 = \lambda(\alpha_0) = i$. All others roots λ satisfy $\lambda \neq m\lambda_0$ for $m \in \mathbb{Z}$.

As [(M_2)] $\lambda(\alpha)$ is the branch of roots passing through λ_0, we have

$$\frac{\partial}{\partial \alpha} Re\lambda(\alpha)_{/\alpha=\alpha_0} \neq 0. \tag{3.22}$$

Theorem 1 ([16]). *Under the assumptions* (M_0), (M_1), *and* (M_2), *there exist constants* $\varepsilon_0 > 0$ *and* δ_0 *and functions* $\alpha(\varepsilon)$, $T(\varepsilon)$, *and a* $T(\varepsilon)$*-periodic function* $x^*(\varepsilon)$ *such that*

(a) *All of these functions are of class* $\mathscr{C}^{k-1}$ *with respect to* ε, *for* $\varepsilon \in [0, \varepsilon_0[$, $\alpha(0) = \alpha_0$, $T(0) = 2\pi$, $x^*(0) = 0$.
(b) $x^*(\varepsilon)$ *is a* $T(\varepsilon)$*-periodic solution of* (3.19), *for the parameter values equal* $\alpha(\varepsilon)$.
(c) *For* $\mid \alpha - \alpha_0 \mid < \delta_0$ *and* $\mid T - 2\pi \mid < \delta_0$, *any* T*-periodic solution* p, *with* $\parallel p \parallel < \delta_0$, *of* (3.19) *for the parameter value* α, *there exists* $\varepsilon \in [0, \varepsilon_0[$ *such that* $\alpha = \alpha(\varepsilon)$, $T = T(\varepsilon)$, *and* p *is up to a phase shift equal to* $x^*(\varepsilon)$.

Normalizing the delay τ by the time scaling $t \to \frac{t}{\tau}$, effecting the change of variables $u(t) = N(t\tau)$ and $v(t) = P(t\tau)$, system (3.3) is transformed into

$$\begin{cases} \dot{u}\ (t) = \tau[-\delta u(t) - \alpha(u(t)) + 2e^{-\gamma u(t-1)}\alpha(u(t-1))] \\ \dot{v}\ (t) = \tau[-\gamma v(t) + \alpha(u(t)) - e^{-\gamma u(t-1)}\alpha(u(t-1))], \end{cases} \tag{3.23}$$

where $\alpha(x) = \beta(x)x$.

By the translation $z(t) = (u(t), v(t)) - (N^*, P^*)$, system (3.23) is written as a functional differential equation (FDE) in $C := C([-1, 0], \mathbb{R}^2)$:

$$\dot{z}\ (t) = L(\tau)z_t + f_0(z_t, \tau), \tag{3.24}$$

where $L(\tau) : C \longrightarrow \mathbb{R}^2$ is a linear operator and $f_0 : C \times \mathbb{R} \longrightarrow \mathbb{R}^2$ are respectively given by

$$L(\tau)\varphi = \tau \begin{pmatrix} -(\delta + h(N^*))\varphi_1(0) + 2g(N^*)\varphi_1(-1) \\ -\gamma\varphi_2(0) + h(N^*)\varphi_1(0) - g(N^*)\varphi_1(-1) \end{pmatrix}$$

$$f_0(\varphi, \tau) = \tau \begin{pmatrix} -H(\varphi_1(0)+N^*)+h(N^*)\varphi_1(0)+2G(\varphi_1(-1)+N^*)-\delta N^*-2g(N^*)\varphi_1(-1) \\ H(\varphi_1(0)+N^*)-h(N^*)\varphi_1(0)-G(\varphi_1(-1)+N^*)-\gamma P^*+g(N^*)\varphi_1(-1). \end{pmatrix}$$

for $\varphi = (\varphi_1, \varphi_2) \in C$.

The following theorem gives the existence of bifurcating periodic solutions.

Theorem 2. *Suppose* (H_2)*–*(H_4). *Then Eq.* (3.23) *has a family of periodic solutions* $p_l(\varepsilon)$ *with period* $T_l = T_l(\varepsilon)$ *for the parameter values* $\tau = \tau(\varepsilon)$ *such that* $p_l(0) = 0$ ($p_l(0) = (N^*, P^*)$ *for system* (3.3)), $T_l(0) = \frac{2\pi}{\omega_0}$, *and* $\tau(0) = \tau_k$, $k = 0, 1, 2, \ldots$. *In this case* τ_k, $k = 0, 1, 2, \ldots$, *and* ω_0 *are respectively given by Eqs.* (3.17) *and* (3.18).

Proof. We apply the Hopf bifurcation theorem. From the expression of f in (3.24), we have

$$f(0,\tau) = 0 \qquad \text{and} \qquad \frac{\partial f(0,\tau)}{\partial \varphi} = 0, \text{ for all } \tau > 0.$$

From (3.16), we have

$$\Delta(i\omega,\tau) = 0 \qquad \Leftrightarrow \qquad \begin{cases} \omega = \omega_0 \\ \text{and} \\ \tau = \tau_k, k = 0,1,2,\ldots. \end{cases}$$

Thus, characteristic equation (3.16) has a pair of simple imaginary roots $\lambda_0 = i\omega_0$ and $\overline{\lambda}_0 = -i\omega_0$ at $\tau = \tau_k$, $k = 0,1,2,\ldots$.

Lastly, we need to verify the transversality condition.

From (3.16), $\Delta(\lambda_0,\tau_k) = 0$ and $\frac{\partial}{\partial\lambda}\Delta(\lambda_0,\tau_k) = (\lambda_0+\gamma)(1-\tau_k g(N^*)e^{-\lambda\tau_k}) \neq 0$. According to the implicit function theorem, there exists a complex function $\lambda = \lambda(\tau)$ defined in a neighborhood of τ_k such that $\lambda(\tau_k) = \lambda_0$ and $\Delta(\lambda(\tau),\tau) = 0$ and

$$\lambda^{'}(\tau) \;=\; -\frac{\partial\Delta(\lambda,\tau)/\partial\tau}{\partial\Delta(\lambda,\tau)/\partial\lambda}, \text{ for } \tau \text{ in a neighborhood of } \tau_k,\ k = 0,1,2,\ldots. \tag{3.25}$$

Let $\lambda(\tau) = \eta(\tau) + \omega(\tau)$. From (3.25) we have

$$\eta(\tau)^{'}(\tau)_{/\tau=\tau_k} = -\frac{\omega_0^2}{\cos(\omega_0\tau_k) + \tau_k g(N^*))^2 + \sin^2(\omega_0\tau_k)} \text{ for } k = 0,1,2,\ldots.$$

By the continuity property, we conclude that $\eta^{'}(\tau)_{/\tau=\tau_k} < 0$, for $k = 0,1,2,\ldots$.

3.5 Direction of Hopf Bifurcation

In this section we follow methods presented in [12], where the direction and stability of the bifurcating branch are obtained by the Taylor expansion of the delay function τ that describes the parameter of bifurcation near the critical value τ_0 (see Theorem 2). Namely, this direction and stability are determined by the sign of the first nonzero term of Taylor expansion, that is,

$$\tau(\varepsilon) = \tau_0 + \tau_2\varepsilon^2 + o(\varepsilon^2), \tag{3.26}$$

and the sign of τ_2 determines that either the bifurcation is supercritical (if $\tau_2 > 0$) and periodic orbits exist for $\tau > \tau_0$, or it is subcritical (if $\tau_2 < 0$) and periodic orbits exist for $\tau < \tau_0$. The term τ_2 may be calculated (see [12]) using the formula

$$\tau_2 = \frac{Re(c)}{Re(qD_2M_0(i\zeta_0, \tau_0)p)}, \tag{3.27}$$

where M_0 is the characteristic matrix of (3.24) given by

$$M_0(\lambda, \tau) = \begin{pmatrix} \lambda - \tau a(\tau) - \tau b(\tau)e^{-\lambda} & 0 \\ -\tau h(N^*) + \tau g(N^*)e^{-\lambda}) & \lambda + \gamma\tau \end{pmatrix},$$

where $a = a(\tau) = -(\delta + h(N^*))$ and $b = b(\tau) = 2g(N^*)$.

$D_2M_0(i\zeta_0, \tau_0)$ denotes the derivative of M_0 with respect to τ at the critical point $(i\zeta_0, \tau_0)$, and the constant c is defined as follows:

$$c = \frac{1}{2}qD_1^3f_0(0, \tau_0)(P^2(\theta), \overline{P}(\theta))$$

$$+qD_1^2f_0(0, \tau_0)(e^{0\cdot}M_0^{-1}(0, \tau_0)D_1^2f_0(0, \tau_0)(P(\theta), \overline{P}(\theta)), P(\theta))$$

$$+\frac{1}{2}qD_1^2f_0(0, \tau_0)(e^{2i\zeta_0\cdot}M_0^{-1}(2i\zeta_0, \tau_0)D_1^2f_0(0, \tau_0)(P(\theta), P(\theta)), \overline{P}(\theta)),$$

where f_0 is the nonlinear part of (3.24), $D_1^if_0, i = 2, 3$, denotes the ith derivative of f_0 with respect to φ, $P(\theta)$ denotes the eigenvector of A, $\overline{P}(\theta)$ denotes the conjugate eigenvector, and p and q are defined later.

Now, we will describe all the preceding operators and vectors precisely. Let $L := L(\tau_0) : C([-1, 0], \mathbb{R}^2) \longrightarrow \mathbb{R}^2$ denote the linear part of (3.24). Using the Riesz representation theorem, one obtains (see [16])

$$L\varphi = \int_{-1}^{0} d\eta(\theta)\varphi(\theta), \tag{3.28}$$

where

$$d\eta(\theta) = \tau_0 \begin{pmatrix} -(\delta + h(N^*))\delta(\theta) + 2g(N^*)\delta(\theta + 1) & 0 \\ h(N^*)\delta(\theta) - g(N^*)\delta(\theta + 1) & -\gamma\delta(\theta); \end{pmatrix} \tag{3.29}$$

δ denotes the Dirac function and $u^* = u^*(\tau_0)$.

Let A denote the generator of a semigroup generated by the linear part of (3.24). Then

$$A\varphi(\theta) = \begin{cases} \frac{d\varphi}{d\theta}(\theta) \text{ for } \theta \in [-1, 0) \\ \\ L\varphi \text{ for } \theta = 0, \end{cases} \tag{3.30}$$

where $\varphi \in C([-1, 0], \mathbb{R}^2)$.

To study the direction of Hopf bifurcation, one needs to calculate the second and third derivatives of the nonlinear part of (3.24):

$$D_1^2 f_0(\varphi, \tau)\psi\chi = \tau \begin{pmatrix} -H''(N^* + \varphi_1(0))\psi_1(0)\chi_1(0) \\ +2G''(N^* + \varphi_1(-1))\psi_1(-1)\chi_1(-1) \\ \\ H''(N^* + \varphi_1(0))\psi_1(0)\chi_1(0) \\ -G''(N^* + \varphi_1(-1))\psi_1(-1)\chi_1(-1) \end{pmatrix} \tag{3.31}$$

and

$$D_1^3 f_0(\varphi, \tau)\psi\chi\upsilon = \tau \begin{pmatrix} -H'''(N^* + \varphi_1(0))\psi_1(0)\chi_1(0)\upsilon_1(0) \\ +2G'''(N^* + \varphi_1(-1))\psi_1(-1)\chi_1(-1)\upsilon_1(-1) \\ \\ H'''(N^* + \varphi_1(0))\psi_1(0)\chi_1(0)\upsilon_1(0) \\ -G'''(N^* + \varphi_1(-1))\psi_1(-1)\chi_1(-1)\upsilon_1(-1). \end{pmatrix} \tag{3.32}$$

Then

$$\begin{aligned} D_1^2 f_0(0, \tau_0)\psi\chi = {} & \tau_0 H''(N^*)\psi_1(0)\chi_1(0)\binom{-1}{1} \\ & +\tau_0 G''(N^*)\psi_1(-1)\chi_1(-1)\binom{2}{-1} \end{aligned} \tag{3.33}$$

and

$$\begin{aligned} D_1^3 f_0(0, \tau_0)\psi\chi\upsilon = {} & \tau_0 H'''(N^*)\psi_1(0)\chi_1(0)\upsilon_1(0)\binom{-1}{1} \\ & +\tau_0 G'''(N^*)\psi_1(-1)\chi_1(-1)\upsilon_1(-1)\binom{2}{-1}; \end{aligned} \tag{3.34}$$

$\psi = (\psi_1, \psi_2)$, $\chi = (\chi_1, \chi_2)$, $\upsilon = (\upsilon_1, \upsilon_2) \in C([-1, 0], \mathbb{R}^2)$.

As $(i\zeta_0, \tau_0)$ is a solution of (3.16), then $i\zeta_0$ is an eigenvalue of A and there is an eigenvector of the form $P(\theta) = pe^{i\zeta_0\theta}$ and $p_i, i = 1, 2$ are complex numbers which satisfy the following system of equations:

$$Mp = 0$$

with

$$M = M_0(i\zeta_0, \tau_0) = \begin{pmatrix} 0 & 0 \\ -\tau_0 h(N^*) + \tau_0 g(N^*)e^{-i\zeta_0} & i\zeta_0 + \gamma\tau_0. \end{pmatrix}. \tag{3.35}$$

Then one may assume

$$p_1 = 1$$

and calculate

$$p_2 = \tau_0 \frac{h(N^*) - g(N^*)e^{-i\zeta_0}}{i\zeta_0 + \gamma\tau_0}.$$

Now, consider A^*, namely, an operator conjugated to A, $A^* : C([0, 1], \mathbb{R}^2) \longrightarrow \mathbb{R}^2$, defined by

$$A^*\psi(s) = \begin{cases} -\frac{d\psi}{ds}(s) \text{ for } s \in (0, 1] \\ -\int_{-1}^{0} \psi(-s)d\eta(s) \text{ for } s = 0, \end{cases} \tag{3.36}$$

and $\psi = (\psi_1, \psi_2) \in C([0, 1], \mathbb{R}^2)$.

Let $Q(s) = qe^{i\zeta_0 s}$ be the eigenvector for A^* associated to eigenvalue $i\zeta_0$, $q = (q_1, q_2)^T$. One needs to choose q such that the inner product (see [16])

$$< Q, \overline{P} >= Q(0)\overline{P}(0) - \int_{-1}^{0} \int_{0}^{\theta} Q(\xi - \theta)d\eta(\theta)\overline{P}(\xi)d\xi$$

is equal to 1. Therefore

$$q_2 = 0$$

leads to

$$q_1 = \frac{1}{1 - 2\tau_0 a + i\zeta_0}.$$

From (3.33) and (3.34) we have

$$D_1^2 f_0(0,\tau_0)(P(\theta),\overline{P}(\theta)) = \tau_0\left[H''(N^*)\begin{pmatrix}-1\\1\end{pmatrix} + G''(N^*)\begin{pmatrix}2\\-1\end{pmatrix}\right] \tag{3.37}$$

$$D_1^2 f_0(0,\tau_0)(P(\theta),P(\theta)) = \tau_0\left[H''(N^*)\begin{pmatrix}-1\\1\end{pmatrix} + G''(N^*)e^{-2i\zeta_0}\begin{pmatrix}2\\-1\end{pmatrix}\right] \tag{3.38}$$

and

$$D_1^3 f_0(0,\tau_0)(P^2(\theta),\overline{P}(\theta)) = \tau_0\left[H'''(N^*)\begin{pmatrix}-1\\1\end{pmatrix} + e^{-i\zeta_0}G'''(N^*)\begin{pmatrix}2\\-1\end{pmatrix}\right]. \tag{3.39}$$

and

$$\frac{1}{2}qD_1^3 f_0(0,\tau_0)(P^2(\theta),\overline{P}(\theta)) = \frac{\tau_0 q_1}{2}\left(H'''(N^*) + 2e^{-i\zeta_0}G'''(N^*)\right). \tag{3.40}$$

From the expression of M_0, we have

$$M_0^{-1}(0,\tau_0) = -\frac{1}{\gamma\tau_0(a+b)}\begin{pmatrix}\gamma & 0\\ h(N^*)-g(N^*) & -(a+b)\end{pmatrix} \tag{3.41}$$

and

$$M_0^{-1}(2i\zeta_0,\tau_0) = \Delta^{-1}(2i\zeta_0,\tau_0)\begin{pmatrix}2i\zeta_0+\gamma\tau_0 & 0\\ \tau_0 h(N^*)+\tau_0 g(N^*)e^{-2i\zeta_0} & 2i\zeta_0-\tau_0 a-\tau_0 be^{-2i\zeta_0}\end{pmatrix}. \tag{3.42}$$

From (3.33), (3.37), (3.38), (3.41), and (3.42) we have

$$qD_1^2 f_0(0,\tau_0)(e^0 M_0^{-1}(0,\tau_0)D_1^2 f_0(0,\tau_0)(P(\theta),\overline{P}(\theta)),P(\theta)) = q_1\tau_0 M_1\left(-H''(N^*)+2G''(N^*)e^{-i\zeta_0}\right), \tag{3.43}$$

where

$$M_1 = -\frac{1}{\gamma\tau_0(a+b)}\gamma\tau_0(-H''(N^*)+2G''(N^*))$$

and

$$qD_1^2 f_0(0,\tau_0)(e^{2i\zeta_0}M_0^{-1}(2i\zeta_0,\tau_0)D_1^2 f_0(0,\tau_0)(P(\theta),P(\theta)),\overline{P}(\theta)) = q_1\tau_0 N_1\left(-H''(N^*)+2G''(N^*)e^{-i\zeta_0}\right), \tag{3.44}$$

where

$$N_1 = \tau_0 \Delta^{-1}(2i\zeta_0, \tau_0)(2i\zeta_0 + \gamma\tau_0)(-H^{''}(N^*) + 2G^{''}(N^*)).$$

Then

$$c = \frac{\tau_0 q_1}{2}\left(-H^{'''}(N^*) + 2e^{-i\zeta_0}G^{'''}(N^*)\right) + q_1\tau_0 M_1\left(-H^{''}(N^*) + 2G^{''}(N^*)e^{-i\zeta_0}\right)$$

$$\frac{q_1\tau_0 N_1}{2}\left(-H^{''}(N^*) + 2G^{''}(N^*)e^{-i\zeta_0}\right)$$

and

$$Re(c) = \frac{\tau_0}{2}\left(\frac{(1-\tau_0 a)}{(1-2\tau_0 a)^2 + \zeta_0^2}X + \frac{-\zeta_0}{(1-2\tau_0 a)^2 + \zeta_0^2}Y\right), \tag{3.45}$$

where

$$\begin{aligned}
X = & -H^{'''}(N^*) + 2\cos(\zeta_0)G^{'''}(N^*) - 2M_1H^{''}(N^*) + 4\cos(\zeta_0)M_1G^{''}(N^*) \\
& + \tau_0\frac{4\zeta_0^2 + \gamma^2\tau_0^2}{\|\Delta(2i\zeta_0, \tau_0)\|^2}\Big(-(\tau_0 a - \tau_0 b\cos(\zeta_0))(-H^{''}(N^*) \\
& + 2\cos(\zeta_0)M_1G^{''}(N^*)) - 2\sin(\zeta_0)G^{''}(N^*)(2\zeta_0 + \tau_0 b\sin(\zeta_0))\Big) \\
Y = & -2\sin(\zeta_0)G^{'''}(N^*) - 4\sin(\zeta_0)M_1G^{''}(N^*) \\
& + \tau_0\frac{4\zeta_0^2 + \gamma^2\tau_0^2}{\|\Delta(2i\zeta_0, \tau_0)\|^2}\Big((\tau_0 a - \tau_0 b\cos(\zeta_0))\sin(\zeta_0)G^{''}(N^*) \\
& - (2\zeta_0 + \tau_0 b\sin(\zeta_0))(-H^{''}(N^*) + 2\cos(\zeta_0)M_1G^{''}(N^*)).\Big)
\end{aligned}$$

Then we deduce the following result:

Theorem 3. *Let $Re(c)$ be given in (3.45) and γ sufficiently small. Then*

(a) The Hopf bifurcation occurs as τ crosses τ_0 to the right (supercritical Hopf bifurcation) if $Re(c) > 0$ and to the left (subcritical Hopf bifurcation) if $Re(c) < 0$.
(b) Also, the bifurcating periodic solutions are stable if $Re(c) > 0$ and unstable if $Re(c) < 0$.

Note that Theorem 3 provides an explicit algorithm for detecting the direction and stability of Hopf bifurcation (Figs. 3.7, 3.9, and 3.10).

Acknowledgements We are very grateful to the editors and to Professor M. C. Mackey for their valuable discussions.

References

1. Adimy, M., Crauste, F.: Existence, positivity and stability for a nonlinear model of cellular proliferation. Nonlinear Anal. Real World Appl. **6**, 337–366 (2005)
2. Adimy, M., Crauste, F., Pujo-Menjouet, L.: On the stability of a nonlinear maturity structured model of cellular proliferation. Discret. Cont. Dyn. Syst. **12**, 501–522 (2005)
3. Adimy, M., Crauste, F., Hbid, M.Y.L., Qesmi, R.: Stability and Hopf bifurcation for a cell population model with state-dependent delay. SIAM J. Appl. Math. **70**(5), 1611–1633 (2010)
4. Alaoui, H.T., Yafia, R.: Stability and Hopf bifurcation in approachable hematopoietic stem cells model. Math. Biosci. **206**, 176–184 (2007)
5. Alaoui, H.T., Yafia, R., Aziz Alaoui, M.A.: Dynamics and Hopf bifurcation analysis in a delayed haematopoietic stem cells model. Arab J. Math. Math. Sci. **1**(1), 35–49 (2007)
6. Bélair, J., Mahaffy, J.M., Mackey, M.C.: Age structured and two delay models for erythropoiesis. Math. Biosci. **128**, 317–346 (1995)
7. Bullough, W.S.: Mitotic control in adult mammalian tissues. Biol. Rev. **50**, 99–127 (1975)
8. Burns, F., Tannock, I.: On the existence of a G_0 phase in the cell cycle. Cell Tissue Kinet. **3**, 321–334 (1970)
9. Chikkappa, G., Burlington, H., Borner, G., Chanana, A.D., Cronkite, E.P., Ohl, S., Pavelec, M., Robertso, J.S.: Periodic oscillation of blood leukocytes, platelets, and reticulocytes in a patient with chronic myelocytic leukemia. Blood **47**, 1023–1030 (1976)
10. Colijn, C., Mackey, M.C.: A mathematical model of hematopoiesis: Periodic chronic myelogenous leukemia, part I, J. Theor. Biol. **237**, 117–132 (2005)
11. Crauste, F., Pujo-Menjouet, L., Genieys, S., Molina, C., Gandrillon, O.: Adding self-renewal in committed erythroid progenitors improves the biological relevance of a mathematical model of erythropoiesis. J. Theor. Biol. **250**, 322–338 (2008)
12. Diekmann, O., Van Giles, S., Verduyn Lunel, S., Walter, H.: Delay Equations. Springer, New York (1995)
13. Ferrell, J.J.: Tripping the switch fantastic: how protein kinase cascade convert graded into switch-like outputs. Trends Biochem. Sci. **21**, 460–466 (1996)
14. Fortin, P., Mackey, M.C.: Periodic chronic myelogenous leukemia: spectral analysis of blood cell counts and etiological implications. Br. J. Haematol. **104**, 336–345 (1999)
15. Guerry, D., Dale, D., Omine, D.C., Perry, S., Wol, S.M.: Periodic hematopoiesis in human cyclic neutropenia. J. Clin. Inves. **52**, 3220–3230 (1973)
16. Hale, J.K., Lunel, S.M.V.: Introduction to Functional Differential Equations. Springer, New York (1993)
17. Haurie, C., Dale, D.C., Mackey, M.C.: Occurrence of periodic oscillations in the differential blood counts of congenital, idiopathic and cyclical neutropenic patients before and during treatment with G-CSF. Exp. Hematol. **27**, 401–409 (1999)
18. Haurie, C., Dale, D.C., Rudnicki, R., Mackey, M.C.: Mathematical modeling of complex neutrophil dynamics in the grey collie. J. Theor. Biol. **204**, 505–519 (2000)
19. Mackey, M.C.: Unified hypothesis for the origin of aplastic anemia and periodic hematopoiesis. Blood **51**(5), 941–956 (1978)
20. Mackey, M.C.: Cell kinetic status of haematopoietic stem cells. Cell Prolif. **34**, 71–83 (2001)
21. Mackey, M.C., Dormer, P.: Continuous maturation of proliferating erythroid precursors. Cell Tissue Kinet. **15**, 381–392 (1982)
22. Mahaffy, J.M., Bélair, J., Mackey, M.C.: Hematopoietic model with moving boundary condition and state dependent delay. J. Theor. Biol. **190**, 135–146 (1998)
23. Orr, J.S., Kirk, J., Gray, K.G., Anderson, J.R.: A study of the interdependence of red cell and bone marrow stem cell populations. Br. J. Haematol. **15**, 23–24 (1968)
24. Othmer, H.G., Adler, F.R., Lewis, M.A., Dalton, J.C.: The Art of Mathematical Modeling: Case Studies in Ecology, Physiology and Biofluids. Prentice Hall, New York (1997)
25. Pitchford, S.C., Furze, R.C., Jones, C.P., Wengner, A.M., Rankin, S.M.: Differential mobilization of subsets of progenitor cells from the bone marrow. Cell Stem Cell **4**, 62–72 (2009)

26. Santillan, M., Mahaffy, J.M., Bélair, J., Mackey, M.C.: Regulation of platelet production: the normal response to perturbation and cyclical platelet disease. J. Theor. Biol. **206**, 585–603 (2000)
27. Smith, J.A., Martin, L.: Do cells cycle? Proc. Natl. Acad. Sci. USA **70**, 1263–1267 (1973)
28. Swinburune, J., Mackey, M.C.: Cyclical thrombocytopenia: characterization by spectral analysis and a review. J. Theor. Med. **2**, 81–91 (2000)
29. Wazewska-Czyzewska, M., Lasota, A.: Mathematical problems of the dynamics of the red blood cell system. Mathematyka Stosowana **6**, 23–40 (1976)
30. Yanabu, M., Nomura, S., Fukuroi, T., Kawakatsu, T., Kido, H., Yamaguchi, K., Suzuki, M., Kokawa, T., Yasunaga, K.: Periodic production of antiplatelet autoantibody directed against GP IIIa in cyclic thrombocytopenia. Acta Haematol. **89**, 155–159 (1993)

Chapter 4
Random Noninstantaneous Impulsive Models for Studying Periodic Evolution Processes in Pharmacotherapy

JinRong Wang, Michal Fečkan, and Yong Zhou

Abstract In this chapter we offer a new class of impulsive models for studying the dynamics of periodic evolution processes in pharmacotherapy, which is given by random, noninstantaneous, impulsive, nonautonomous periodic evolution equations. This type of impulsive equation can describe the injection of drugs in the bloodstream, and the consequent absorption of them in the body is a random, periodic, gradual, and continuous process. Sufficient conditions on the existence of periodic and subharmonic solutions are established, as are other related results such as their globally asymptotic stability. The dynamical properties are also derived for the whole system, leading to the theory of fractals. Finally, examples are given to illustrate our theoretical results.

4.1 Introduction

It has been recognized that a periodically varying environment plays an important role in many biological and ecological dynamic systems. The periodic evolutionary process of many nonautonomous biological and ecological dynamical models

J. Wang
Department of Mathematics, Guizhou University, Guiyang, Guizhou 550025, People's Republic of China
e-mail: sci.jrwang@gzu.edu.cn

M. Fečkan (✉)
Department of Mathematical Analysis and Numerical Mathematics, Comenius University in Bratislava, Mlynská dolina, 842 48 Bratislava, Slovakia

Mathematical Institute of Slovak Academy of Sciences, Štefánikova 49, 814 73 Bratislava, Slovakia
e-mail: Michal.Feckan@fmph.uniba.sk

Y. Zhou
Department of Mathematics, Xiangtan University, Xiangtan, Hunan 411105, People's Republic of China
e-mail: yzhou@xtu.edu.cn

A.C.J. Luo, H. Merdan (eds.), *Mathematical Modeling and Applications in Nonlinear Dynamics*, Nonlinear Systems and Complexity 14,
DOI 10.1007/978-3-319-26630-5_4

whose motions depend on abrupt changes in their states is best described by differential equations with instantaneous periodic impulses.

Hernández and O'Regan [12] and Pierri et al. [26] recently studied a new class of evolution equations with noninstantaneous impulses to describe certain dynamics of evolution processes in pharmacotherapy. Taking into account the hemodynamic equilibrium of a person in a random, periodically varying environment, the injection of drugs in the bloodstream and their consequent absorption in the body are periodic, gradual, and continuous processes. So this situation should be regarded as a continuous impulsive and periodic action, which starts at an arbitrary fixed point and stays active during one periodic time interval. This motivated us to study a model described by nonautonomous periodic evolution equations with either deterministic or random, noninstantaneous, periodic impulses.

Periodic solutions of nonlinear evolution equations have been studied extensively; we refer to the excellent monographs [7, 27] and the contributions [5, 13–15, 17–22, 28, 30] of nonlinear evolution equations and the references therein. More complex discontinuous systems are investigated in [2–4].

The main objective of this chapter is looking for periodic solutions and subharmonic solutions of the following nonautonomous periodic evolution equations with random noninstantaneous impulses:

$$\begin{cases} u'(t) + A(t)u(t) = q_i(t, u(t)), \ t \in (s_i, t_{i+1}), \ i = 0, 1, 2, \cdots, \infty, \\ u(t_i^+) = g_i^j(t_i, u(t_i^-)), \ i = 1, 2, \cdots, \infty, \ j = 1, 2, \cdots, n_i, \\ u(t) = g_i^j(t, u(t_i^-)), \ t \in (t_i, s_i], \ i = 1, 2, \cdots, \infty, \ j = 1, 2, \cdots, n_i, \end{cases} \tag{4.1}$$

where $A(t) : D(A(t)) \to X, t \geq 0$, is a family of T-periodic, linear unbounded operators on a Banach space X, which can generate a strongly continuous evolutionary process $\{U(t,s), t \geq s \geq 0\}$. The fixed points s_i and t_i satisfy $0 = s_0 < t_1 \leq s_1 \leq t_2 < \cdots < t_m \leq s_m \leq t_{m+1} < \cdots$ with $\lim_{i\to\infty} t_i = \infty$ and $t_{i+m} = t_i + T$, $s_{i+m} = s_i + T$, and $m \in \mathbb{N}$ denotes the number of impulsive points between 0 and T. Moreover, $g_i^j : [t_i, s_i] \times X \to X$ are T-periodic continuous functions for all $i = 1, 2, \cdots, \infty$ with $g_{i+m}^j = g_i^j$, $j = 1, 2, \cdots, n_i$, $n_{i+m} = n_i$, and g_i^j appears in (4.1) with a given probability $p_{ij} > 0$. Hence $\sum_{j=1}^{n_i} p_{ij} = 1$ and $p_{(i+m)j} = p_{ij}$ for all i. We establish a lower bound for the probability of the existence of periodic and subharmonic solutions of (4.1) under addition conditions (see Theorems 5 and 6). In the final theoretical section, we also study in more detail the dynamic properties of the random system (4.1) (see Theorem 7). In particular, we study

$$\begin{cases} u'(t) - \mu(t)\mathbb{A}u(t) = 0, \ t \in (s_i, t_{i+1}), \ i = 0, 1, 2, \cdots, \infty, \\ u(t_i^+) = \zeta_j a(t_i^+) + b(t_i^+)Bu(t_i^-), \ i = 1, 2, \cdots, \infty, \ j = 1, 2, \\ u(t) = \zeta_j a(t) + b(t)Bu(t_i^-), \ t \in (t_i, s_i], \ i = 1, 2, \cdots, \infty, \ j = 1, 2, \end{cases} \tag{4.2}$$

where $\mathbb{A}$ is the infinitesimal generator of a C_0-semigroup $\{S(t), t \geq 0\}$ in X, $a : \mathbb{R} \to X$, $\mu : \mathbb{R} \to (0, \infty)$, and $b : \mathbb{R} \to \mathbb{R}$ are T-periodic continuous functions, $s_{i+1} = s_i + T$, $t_{i+1} = t_i + T$, so $m = 1$, and $\zeta_j \in \mathbb{R}$ appears in (4.2) with given probabilities $p_j > 0$. Hence $p_1 + p_2 = 1$. Furthermore, $B \in L(X)$ and μ is Lipschitz continuous. We present a method in Theorem 8 that (4.2) can have a random attractor/fractal, either the random Cantor set in $\mathbb{R}$ or the random Sierpinski triangle in $\mathbb{R}^2$ [8, p. 335], for instance, or in any finite-dimensional space. Then we derive a lower bound for the probability that (4.1) has a globally asymptotically stable periodic mild solution (see Theorem 11). Note that results of Sects. 4.2 and 4.3 are preparatory achievements for obtaining our main results of this chapter, presented in Sects. 4.4 and 4.5. In particular, we start this chapter by looking for periodic solutions of the following nonautonomous periodic evolution equations with deterministic noninstantaneous impulses:

$$\begin{cases} u'(t) + A(t)u(t) = q_i(t, u(t)), \ t \in (s_i, t_{i+1}), \ i = 0, 1, 2, \cdots, \infty, \\ u(t_i^+) = g_i(t_i, u(t_i^-)), \ i = 1, 2, \cdots, \infty, \\ u(t) = g_i(t, u(t_i^-)), \ t \in (t_i, s_i], \ i = 1, 2, \cdots, \infty, \end{cases} \tag{4.3}$$

where $q_i : [s_i, t_{i+1}] \times X \to X$ is a T-periodic continuous function for all $i = 0, 1, 2, \cdots, \infty$ with $q_{i+m} = q_i$, and $g_i : [t_i, s_i] \times X \to X$ is another T-periodic continuous function for all $i = 1, 2, \cdots, \infty$ with $g_{i+m} = g_i$. To achieve our aim, we have to show the continuity and compactness of the corresponding Poincaré map $P : X \to X$ of (4.3) given by (4.7). Then we establish new sufficient conditions on the existence of periodic mild solutions when PC-mild solutions are ultimate bounded (see Theorems 2 and 3). Furthermore, a global asymptotic stability result of periodic solutions is presented in Theorem 4. These results are applied in Sects. 4.4 and 4.5 for the random cases (4.1) and (4.2). The final section, Sect. 4.6, is devoted to concrete examples to illustrate the theory.

This chapter is a continuation of our recent related papers [11, 29]. However, we note that it seems that we are the first to study the above evolution equations with random noninstantaneous impulses, which is the main novelty of this work. It is interesting that fractals are studied similarly in [16] as in our chapter for an economic, random, discrete-time, two-sector optimal growth model in which the production of the homogeneous consumption good uses Cobb–Douglas technology.

4.2 Preliminaries

Let $J = [0, T]$. Denote $C(J, X)$ by the Banach space of all continuous functions from J into X with the norm $\|x\|_C := \sup\{\|x(t)\| : t \in J\}$ for $u \in C(J, X)$. We consider the Banach space $PC(J, X) := \{x : J \to X : x \in C((t_k, t_{k+1}], X), \ k = 0, 1, \cdots, m,$ and there exist $x(t_k^-)$ and $x(t_k^+)$, $k = 1, \cdots, m$, with $x(t_k^-) = x(t_k)\}$ endowed with the Chebyshev PC-norm $\|x\|_{PC} := \sup\{\|x(t)\| : t \in J\}$. Similarly, we can define the Banach space $PC([0, \infty), X)$ endowed with the PC-norm $\|x\|_{PC} := \sup\{\|x(t)\| : t \in [0, \infty)\}$.

Let $\{A(t), T \geq t \geq 0\}$ be a family of closed, densely defined, linear unbounded operators acting on X and assume that this family ascribes to the following three standard conditions [1]:

[A_1]: The domain $D(A(t)) := D$ is independent of t and is dense in X.
[A_2]: For $t \geq 0$, the resolvent $R(\lambda, A(t)) = (\lambda I - A(t))^{-1}$ exists for all λ with $\Re\lambda \leq 0$, and there is a constant M independent of λ and t such that

$$\|R(\lambda, A(t))\| \leq M(1 + |\lambda|)^{-1} \text{ for } \Re\lambda \leq 0.$$

[A_3]: There exist constants $L > 0$ and $0 < \alpha \leq 1$ such that

$$\left\|\big(A(t) - A(s)\big)A^{-1}(\tau)\right\| \leq L|t - s|^{\alpha} \text{ for } t, s, \tau \in J.$$

Lemma 1. *(see [1, p. 159]) Assume that [A_1]–[A_3] are satisfied. Then*

$$\begin{cases} x'(t) + A(t)x(t) = 0, \ t \in (0, T], \\ x(0) = \bar{x}, \end{cases} \tag{4.4}$$

has a unique evolution system $\{U(t, s) : 0 \leq s \leq t \leq T\}$ in X satisfying the following properties:

(i) There exists $M > 0$ such that $\sup_{0 \leq s \leq t \leq T} \|U(t, s)\| \leq M$.
(ii) $U(t, r)U(r, s) = U(t, s)$ for $0 \leq s \leq r \leq t \leq T$.
(iii) $U(\cdot, \cdot)x \in C(\Delta, X)$ for $x \in X$, $\Delta = \{(t, \theta) \in [0, T] \times [0, T] : 0 \leq s \leq t \leq T\}$.
(iv) For $0 \leq s < t \leq T$, $U(t, s)$: $X \longrightarrow D$ and $t \longrightarrow U(t, s)$ is strongly differentiable in X. The derivative $\sup_{0 \leq s \leq t \leq T} \|\frac{\partial}{\partial t} U(t, s)\| \leq M$ and it is strongly continuous on $0 \leq s < t \leq T$. Moreover,

$$\frac{\partial}{\partial t} U(t, s) = -A(t)U(t, s) \text{ for } 0 \leq s < t \leq T,$$

$$\left\|\frac{\partial}{\partial t} U(t, s)\right\| = \|A(t)U(t, s)\| \leq \frac{C}{t - s},$$

$$\left\|A(t)U(t, s)A(s)^{-1}\right\| \leq C \text{ for } 0 \leq s \leq t \leq T.$$

(v) For every $v \in D$ and $t \in (0, T]$, $U(t, s)v$ is differentiable with respect to s on $0 \leq s \leq t \leq T$:

$$\frac{\partial}{\partial \theta} U(t, s)v = U(t, s)A(s)v.$$

For each $\bar{x} \in X$, (4.4) also has a unique classical solution $x \in C^1(J, X)$ given by $x(t) = U(t, 0)\bar{x}, t \in J$.

The preceding lemma implies that (4.4) is well posed. For the concept of well-posedness and further details, we refer to [10, Theorem 2.6], [23], and [25, Theorem 6.1].

Moreover, we impose the following additional assumption:

[A_4]: $A(t)$ is T-periodic in t; that is, $A(t+T) = A(t)$ for $t \geq 0$.

Lemma 2. *Assume that [A_1]–[A_4] are satisfied. Then evolution system* $\{U(t,s) : 0 \leq s \leq t \leq T\}$ *satisfies the following property:*

(vi) $U(t+T, s+T) = U(t,s)$ *for* $0 \leq s \leq t \leq T$.

Now we can introduce the following standard definition.

Definition 1. A function $u \in PC(J,X)$ is called a mild solution of Cauchy problem

$$\begin{cases} u'(t) + A(t)u(t) = q_i(t,u(t)), \ t \in (s_i, t_{i+1}), \ i = 0,1,2,\cdots,m, \\ u(t_i^+) = g_i(t_i, u(t_i^-)), \ i = 1,2,\cdots,m, \\ u(t) = g_i(t, u(t_i^-)), \ t \in (t_i, s_i], \ i = 1,2,\cdots,m, \\ u(0) = \bar{x} \end{cases} \tag{4.5}$$

if u satisfies

$$\begin{cases} u(t) = g_i(t, u(t_i^-)), \ t \in (t_i, s_i], \ i = 1,2,\cdots,m; \\ u(t_i^+) = g_i(t_i, u(t_i^-)), \ i = 1,2,\cdots,m; \\ u(t) = U(t,0)\bar{x} + \int_0^t U(t,s)q_i(s,u(s))ds, \ t \in [0,t_1]; \\ u(t) = U(t,s_i)g_i(s_i, u(t_i^-)) + \int_{s_i}^t U(t,s)q_i(s,u(s))ds, \\ \qquad t \in [s_i, t_{i+1}], \ i = 1,2,\cdots,m. \end{cases} \tag{4.6}$$

In what follows, we introduce the following assumptions and establish an existence result for (4.5).

[Q_1]: $q_i : [s_i, t_{i+1}] \times X \to X$ is continuous for all $i = 0,1,2,\cdots,m$ and for any $u, v \in X$ satisfying $\|u\|, \|v\| \leq \rho$, there exists a positive constant $L_{q_i}(\rho) > 0$ such that

$$\|q_i(t,u) - q_i(t,v)\| \leq L_{q_i}(\rho)\|u - v\|.$$

[Q_2]: There exist constants $m_{q_i} \geq 0$ and $M_{q_i} \geq 0$, $i = 0,1,2,\cdots,m$, such that

$$\|q_i(t,u)\| \leq m_{q_i} + M_{q_i}\|u\| \text{ for all } u \in X.$$

[G_1]: $g_i : [t_i, s_i] \times X \to X$ is continuous for all $i = 1,2,\cdots,m$ and for any $u, v \in X$ satisfying $\|u\|, \|v\| \leq \rho$, there exists a positive constant $L_{g_i}(\rho) > 0$ such that

$$\|g_i(t,u) - g_i(t,v)\| \leq L_{g_i}(\rho)\|u - v\|.$$

[G_2]: There exist constants $m_{g_i} \geq 0$ and $M_{g_i} \geq 0$, $i = 0, 1, 2, \cdots, m$, such that

$$\|g_i(t, u)\| \leq m_{g_i} + M_{g_i}\|u\| \text{ for each } t \in [t_i, s_i] \text{ and all } u \in X.$$

By adopting a similar procedure as in [11, Theorem 2.1], [29, Theorem 2.2] and using the standard method [1, p. 169, Theorem 5.3.3] via the Banach contraction principle, we can obtain the following existence and uniqueness results.

Theorem 1. *Let the assumptions [A_1]–[A_3], [Q_1]–[Q_2], and [G_1]–[G_2] be satisfied. Then (4.5) has a unique mild solution.*

4.3 Existence Results on Periodic Solutions for the Determined Case

In this section, we present the existence of periodic solutions of Eq. (4.3). In addition to [A_1]–[A_3], [Q_1]–[Q_2], and [G_1]–[G_2], we need the following periodicity conditions on q_i and g_i:

[Q_3]: $q_i(t, u)$ is a T-periodic in t; that is, $q_i(t+T, u) = q_i(t, u)$, $t \in [s_i, t_{i+1}]$; hence $q_{i+m} = q_i$.

[G_3]: $g_i(t, u)$ is a T-periodic in t; that is, $g_i(t+T, u) = g_i(t, u)$, $t \in [s_i, t_{i+1}]$; hence $g_{i+m} = g_i$.

Definition 2. A function $u \in PC([0, \infty), X)$ is said to be a T-periodic PC-mild solution of Eq. (4.3) if it is a PC-mild solution of (4.5) corresponding to some $\bar{x}$ and $u(t + T) = u(t)$ for $t \geq 0$.

By adopting a similar procedure as in Fečkan et al. [11, Lemma 2.2] and using Lemma 2(iv), [Q_3], and [G_3], we have the following result.

Lemma 3. *Equation (4.3) has a T-periodic PC-mild solution if and only if the Poincaré operator P has a fixed point where $P : X \to X$ is given by*

$$\begin{aligned} P(\bar{x}) &= u(T, \bar{x}) \\ &= g_m(T, u(t_m^-, \bar{x})) \\ &= g_m\left(T, U(t, s_{m-1})g_{m-1}(s_{m-1}, u(t_{m-1}^-, \bar{x})) + \int_{s_{m-1}}^t U(t, s)q_i(s, u(s, \bar{x}))ds\right). \end{aligned} \tag{4.7}$$

Now we show basic properties of P.

Lemma 4. *$P : X \to X$ defined in (4.7) is a continuous and compact operator.*

Proof. Note that P is a composition of the maps given by

$$\begin{aligned} P_0(x) &= u(t_1, x),\ u(t,x) = U(t,0)x + \int_0^t U(t,s)q_i(s, u(s,x))ds,\ t \in [0, t_1]; \\ G_i(x) &= g_i(s_i, x),\ i = 1, 2, \cdots, m; \\ P_i(x) &= u(t_{i+1}, x),\ u(t,x) = U(t, s_i)x \\ &\quad + \int_{s_i}^t U(t,s)q_i(s, u(s,x))ds,\ t \in [s_i, t_{i+1}],\ i = 1, 2, \cdots, m. \end{aligned}$$

Thus, we can rewrite

$$P = G_m \circ P_{m-1} \circ \cdots \circ G_1 \circ P_0. \tag{4.8}$$

By following the same procedure as in Fečkan et al. [11], we know that each P_i and G_i are locally Lipschitz, so they are continuous, and thus P is continuous as well. Next, by following the proof of Park et al. [24, Theorem 3.1] and using the fact that the embedding $X_\eta \to X,\ \eta \in (0,1)$ is compact, we can prove that each P_i is compact, and so P is compact as well. The proof is finished. □

To proceed, we introduce the following definitions [9, 24].

Definition 3. We say that PC-mild solutions of (4.5) are locally bounded if for each $B_1 > 0$ and $k_0 > 0$, there is a $B_2 > 0$ such that $\|\bar{x}\| \leq B_1$ implies $\|u(t, \bar{x})\| \leq B_2$ for $0 \leq t \leq k_0$.

Definition 4. We say that PC-mild solutions of (4.5) are ultimate bounded if there is a bound $B > 0$ such for each $B_3 > 0$, there is a $k > 0$ such that $\|\bar{x}\| \leq B_3$ and $t \geq k$ imply $\|u(t, \bar{x})\| \leq B$.

Now we show the local boundedness of the solutions.

Lemma 5. *Under the above assumptions, PC-mild solutions of (4.5) are locally bounded.*

Proof. Using Lemma 1(i) and [G$_2$], we have

$$\|G_i(x)\| \leq m_{g_i} + M_{g_i}\|x\|,\ x \in X,\ i = 1, 2, \cdots, m, \tag{4.9}$$

while using [Q$_2$], we derive

$$\begin{aligned} \|u(t,x)\| \leq M\|x\| \ &+ M \int_{s_i}^t \left(m_{q_i} + M_{q_i}\|u(s,x)\|\right) ds, \\ &t \in [s_i, t_{i+1}],\ i = 1, 2, \cdots, m-1,\ x \in X. \end{aligned}$$

By Gronwall's inequality [6, p. 5] we have

$$\|u(t,x)\| \leq M(\|x\| + m_{q_i}(t - s_i))e^{MM_{q_i}(t-s_i)},$$

which implies that

$$\|P_i(x)\| = \|u(t_{i+1}, x)\| \leq m_{q_i}(t_{i+1} - s_i)e^{MM_{q_i}(t_{i+1}-s_i)} + Me^{MM_{q_i}(t_{i+1}-s_i)}\|x\|. \quad (4.10)$$

Now linking (4.9) with (4.10), we have

$$\|(G_i \circ P_{i-1})(x)\| \leq a + b\|x\|, \; x \in X, \; i = 1, 2, \cdots, m, \quad (4.11)$$

where

$$a = \max_{i=1,2,\cdots,m} \left\{ m_{g_i} + M_{g_i} m_{q_{i-1}}(t_i - s_{i-1}) e^{MM_{q_{i-1}}(t_i - s_{i-1})} \right\},$$

$$b = M \max_{i=1,2,\cdots,m} \left\{ M_{g_i} e^{MM_{q_{i-1}}(t_i - s_{i-1})} \right\}.$$

Note that from (4.11) and by repeating the similar process again and again, we can derive by (4.8) that

$$\|P(x)\| = \|(G_m \circ P_{m-1} \circ \cdots \circ G_1 \circ P_0)(x)\| \leq B + C\|x\|, \quad (4.12)$$

where

$$C = M^m \prod_{i=1}^{m} M_{g_i} e^{M \sum_{i=1}^{m} M_{q_{i-1}}(t_i - s_{i-1})}, \quad B = a(1 + b + \cdots + b^{m-1}). \quad (4.13)$$

Clearly (4.12) gives the desired result. □

Now we are ready to present the main results of this section.

Theorem 2. *Assume that [A_1]–[A_4], [Q_1]–[Q_3], and [G_1]–[G_3] are satisfied. If the solutions of Eq. (4.3) are ultimate bounded, then Eq. (4.3) has at least a T-periodic PC-mild solution.*

Proof. Using Lemma 5, we can directly follow the proof of Park et al. [24, Theorem 3.2] to get our result via Horn's fixed-point theorem. Thus we do not go into details. □

To proceed in this section, we present another existence result via the well-known Schauder's fixed-point theorem.

Theorem 3. *Assume that [A_1]–[A_4], [Q_1]–[Q_3], and [G_1]–[G_3] are satisfied. If $C < 1$, as defined in (4.13), then Eq. (4.3) has at least a T-periodic PC-mild solution.*

Proof. Take $\rho \geq \frac{B}{1-C}$, and define

$$W := \{x \in X : \|x\| \leq \rho\} \subset X.$$

By Lemma 4, $P : W \to W$ is continuous and compact. Hence Schauder's fixed-point theorem gives the result. □

Remark 1. As a matter of fact, the condition $C < 1$ implies that the solutions of Eq. (4.3) are ultimate bounded. So Theorem 3 also follows from Theorem 2.

Remark 2. Results of Wang and Fečkan [29] can be directly applied for Ulam's type of stability problems for Eq. (4.3), and so we refer the reader to that paper [29] for more details.

To end this section, we suppose that the functions L_{q_i} in $[Q_1]$ and L_{g_i} in $[G_1]$ are constants, and so q_i and g_i are globally Lipschitz continuous with constants L_{q_i} and L_{g_i}, respectively. Moreover, we suppose the asymptotic stability of $U(\cdot,\cdot)$:

$[A_5]$ $\|U(t,s)\| \le Me^{-\omega(t-s)}$ for any $t \ge s \ge 0$ and some positive constants $M \ge 1$ and $\omega > 0$.

Then following the above procedure, we see that each P_i is globally Lipschitz continuous with a constant $Me^{(ML_{q_i}-\omega)(t_{i+1}-s_i)}$. In fact, using our assumptions, we can obtain

$$\|u(t,\bar{x}) - v(t,\bar{y})\| \le Me^{-\omega(t-s_i)}\|\bar{x}-\bar{y}\| + ML_{q_i}\int_{s_i}^{t} e^{-\omega(t-s)}\|u(s,\bar{x}) - v(s,\bar{y})\|ds,$$

for $t \in [s_i, t_{i+1}]$, $i = 0, 1, \cdots, m$. By virtue of Gronwall's inequality, we obtain

$$\|u(t,\bar{x}) - v(t,\bar{y})\| \le Me^{(ML_{q_i}-\omega)(t-s_i)}\|\bar{x}-\bar{y}\| \text{ for all } t \in [s_i, t_{i+1}],\ i = 0, 1, \cdots, m,$$

which implies that

$$\|P_j(\bar{x}) - P_j(\bar{y})\| \le Me^{(ML_{q_i}-\omega)(t_{i+1}-s_i)}\|\bar{x}-\bar{y}\|.$$

Then each $G_i \circ P_{i-1}$ is globally Lipschitz continuous with a constant $ML_{g_i}e^{(ML_{q_{i-1}}-\omega)(t_i-s_{i-1})}$. Consequently, P is globally Lipschitz continuous with a constant

$$L_p := M^m \prod_{i=1}^{m} L_{g_i}e^{(ML_{q_{i-1}}-\omega)(t_i-s_{i-1})}. \tag{4.14}$$

Thanks to the well-known Banach's fixed-point theorem, we have the following result.

Theorem 4. *Assume that $[A_1]$–$[A_5]$, $[Q_1]$, $[Q_3]$, $[G_1]$, and $[G_3]$ are satisfied. If $L_p < 1$ as defined in (4.14), then Eq. (4.3) has a unique T-periodic PC-mild solution which is globally asymptotically stable.*

Proof. By our assumptions, P is globally contractive. By Banach's fixed-point theorem, it has a unique fixed point which is a global attractor. The proof is finished. □

4.4 Existence Results on Periodic and Subharmonic Solutions for Random Case

In this section, we seek periodic solutions of (4.1). First, we set

$$\mathscr{P} := \prod_{i=1}^{m} \{1, 2, \cdots, n_i\},$$

and for any $\chi \in \mathscr{P}$, $\chi = (j_1, j_2 \cdots, j_m)$, $j_i \in \{1, 2, \cdots, n_i\}$, we define its probability as

$$\eta(\chi) := \prod_{i=1}^{m} p_{ij_i}.$$

Note that $\sum_{\chi \in \mathscr{P}} \eta(\chi) = 1$; that is, $\eta : \mathscr{P} \to [0, \infty)$ is really a discrete probability measure on $\mathscr{P}$.

Furthermore, for any $\chi \in \mathscr{P}$, $\chi = (j_1, j_2 \cdots, j_m), j_i \in \{1, 2, \cdots, n_i\}$, we consider

$$\begin{cases} u'(t) + A(t)u(t) = q_i(t, u(t)), \ t \in (s_i, t_{i+1}), \ i = 0, 1, 2, \cdots, m-1, \\ u(t_i^+) = g_i^{j_i}(t_i, u(t_i^-)), \ i = 1, 2, \cdots, m, \\ u(t) = g_i^{j_i}(t, u(t_i^-)), \ t \in (t_i, s_i], \ i = 1, 2, \cdots, m. \end{cases} \tag{4.15}$$

Now we extend [G$_1$] and [G$_2$], so there are corresponding constants $M_{g_i^j}$ with appropriate properties. Then as in (4.7), we can give the corresponding Poincaré map P_χ of (4.15). Furthermore, as in (4.13), we have the constant

$$C(\chi) = \prod_{i=1}^{m} M_{g_i^{j_i}} e^{\sum_{i=1}^{m} M_{g_{i-1}}(t_i - s_{i-1})}. \tag{4.16}$$

Theorem 3 implies that if $C(\chi) < 1$, then Eq. (4.15) has at least a T-periodic PC-mild solution. Note that the probability of P_χ in (4.1) is just $\eta(\chi)$. Summarizing, we arrive at the following result.

Theorem 5. *The probability that* (4.1) *has a T-periodic mild solution is greater than or equal to*

$$\sum_{\prod_{i=1}^{m} M_{g_i^{j_i}} e^{\sum_{i=1}^{m} M_{g_{i-1}}(t_i - s_{i-1})} < 1} \ \prod_{i=1}^{m} p_{ij_i}.$$

Proof. Indeed, the probability that (4.1) has a T-periodic mild solution is greater than or equal to

$$\eta(C^{-1}([0,1))) = \sum_{C(\chi)<1} \eta(\chi) = \sum_{\prod_{i=1}^{m} M_{g_i^{j_i}} e^{\sum_{i=1}^{m} M_{g_{i-1}}(t_i - s_{i-1})} < 1} \prod_{i=1}^{m} p_{ij_i},$$

where C is considered to be $C : \mathscr{P} \to [0, \infty)$. □

Certainly, we can extend the preceding method for studying nT-subharmonic mild solutions of (4.1) as follows. We take $\mathscr{P}^n$ with the corresponding probability measure η_n, and for any $\chi = (\chi_1, \cdots, \chi_n) \in \mathscr{P}^n$, we set $P_\chi = P_{\chi_n} \circ \cdots \circ P_{\chi_1}$. Note that $\eta_n(\chi) = \eta(\chi_1) \cdots \eta(\chi_n)$. The fixed points of P_χ determine nT-subharmonic mild solutions of (4.1). Now we have constants $C_n(\chi) = C(\chi_1) \cdots C(\chi_n)$, so $C_n : \mathscr{P}^n \to [0, \infty)$. So we have the next result.

Theorem 6. *The probability that* (4.1) *has a nT-subharmonic mild solution is greater than or equal to*

$$\sum_{C(\chi_1)\cdots C(\chi_n)<1} \eta(\chi_1) \cdots \eta(\chi_n).$$

Proof. Indeed, the probability that (4.1) has a nT-subharmonic mild solution is greater than or equal to

$$\eta(C_n^{-1}([0,1))) = \sum_{C_n(\chi)<1} \eta_n(\chi) = \sum_{C(\chi_1)\cdots C(\chi_n)<1} \eta(\chi_1) \cdots \eta(\chi_n).$$

The proof is completed. □

4.5 Dynamics of Random Systems

In this section, we study in more detail the dynamics of the random system (4.1) by assuming

[D$_1$]: $C(\chi) < 1$ for all $\chi \in \mathscr{P}$.

Then any P_χ has a fixed point. Moreover, there are $0 < C_M < 1$ and $B_M > 0$, so that

$$\|P_\chi(x)\| \le B_M + C_M \|x\|, \quad \forall\, x \in X,\ \forall\, \chi \in \mathscr{P}. \tag{4.17}$$

So for any $\chi \in \mathscr{P}^n$, we have

$$\begin{aligned} \|P_{\chi_n} \circ \cdots \circ P_{\chi_1}(x)\| &\le B_M(1 + C_M + \cdots + C_M^{n-1}) + C_M^n \|x\| \\ &\le \frac{B_M}{1 - C_M} + C_M^n \|x\|. \end{aligned}$$

This means that any random iteration $\{P_{\chi_n} \circ \cdots \circ P_{\chi_1}(x)\}_{n=1}^{\infty}$ for $(\chi_1, \chi_2, \cdots) \in \mathscr{P}^{\infty}$ enters after a finite iteration in the ball

$$\mathfrak{B} := \left\{ x \in X : \|x\| \leq \frac{2B_M}{1 - C_M} \right\}.$$

Consequently, the study of random iterations can be restricted on $\mathfrak{B}$. Then by [Q$_1$], [G$_1$] and using Gronwall's inequality as we did earlier, we see that each G_i and P_i, $i = 1, \cdots, m$, are globally Lipschitz on $\mathfrak{B}$. Thus all P_χ, $\chi \in \mathscr{P}$ are globally Lipschitz on $\mathfrak{B}$ as well, with constants $L(\chi)$, respectively (see details on (4.14)). Now we suppose

[D$_2$]: $L(\chi) < 1$ for all $\chi \in \mathscr{P}$.

Then, of course, [D$_2$] implies [D$_1$] on $\mathfrak{B}$, since $C(\chi) \leq L(\chi)$ on $\mathfrak{B}$. By our assumptions, the set

$$Y := \overline{\bigcup_{\chi \in \mathscr{P}} P_\chi(\mathfrak{B})}$$

is a compact subset of X. Moreover, (4.17) gives $Y \subset \mathfrak{B}$, so then $P_\chi(Y) \subset P_\chi(\mathfrak{B}) \subset Y$ for each $\chi \in \mathscr{P}$. Summarizing, the study of random iterations can be restricted on the compact set Y. Then we consider the compact metric space $(Y, \|\cdot\|)$ and let $(H(Y), h)$ be the corresponding space of nonempty compact subsets of Y with the Hausdorff metric h [8]. Since by [D$_2$], all $P_\chi : Y \to Y$ are contractions, they create an iterated function system (IFS) [8, Definition 7.1, p. 80]. Next, we define the transformation $W : H(Y) \to H(Y)$ by

$$W(Z) := \bigcup_{\chi \in \mathscr{P}} P_\chi(Z).$$

It is well known [8, Theorem 7.1, p. 81] that W is a contraction with a constant $L_Y = \max_{\chi \in \mathscr{P}} L(\chi) < 1$. So by Banach's fixed-point theorem, W has a unique fixed point F_W, called the attractor or fractal of the IFS. Furthermore, since each P_χ has probability $\eta(\chi)$, we are dealing with an IFS with probabilities (IFSP) according to [8, Definition 1.1, p. 330]. By following [8, p. 349], we consider the set $P(Y)$ of normalized Borel measures on Y with the Hutchinson metric d_H. Then $(P(X), d_H)$ is a compact metric space. Next, our IFSP defines a Markov operator $M : P(Y) \to P(Y)$ by

$$M(\nu) := \sum_{\chi \in \mathscr{P}} \eta(\chi) \nu \circ P_\chi^{-1}.$$

We know that M is a contraction on Y with a constant L_Y, so it has a unique fixed point $\mu \in P(Y)$; that is, $M(\mu) = \mu$. This μ is called the invariant measure of the IFSP. The relationship between F_W and μ is given by the fact that the support of

μ is F_W; namely, μ "lives" in F_W. Furthermore, using the Carathéodory theorem [8, p. 342], η_n can be uniquely extended to η_∞ on $\mathscr{P}^\infty$. Now for any $x_0 \in Y$ and $\chi \in \mathscr{P}^\infty$, we consider the random iteration $\{x_0, x_1, \cdots\}$ for $x_n := P_{\chi_n} \circ \cdots \circ P_{\chi_1}(x_0)$. Then for a Borel set $Z \subset Y$, we define

$$N(Z, n) := \text{number of points in } \{x_0, x_1, \cdots\} \cap \mathfrak{B}.$$

Finally, we are ready to formulate the basic result on random iterations, a consequence of Elton's theorem [8, p. 364]:

Theorem 7. *Assuming [D_2], for almost all $\chi \in \mathscr{P}^\infty$ with respect to η_∞, the following holds:*

$$\eta(Z) = \lim_{n\to\infty} \frac{N(Z, n)}{n+1}$$

for any Borel set $Z \subset Y$ and $x_0 \in Y$.

Hence the dynamics of almost all random iterations are concentrated on F_W. This means that the dynamics of the random impulsive system (4.1) is determined by its dynamics on the compact fractal F_W. The fractal dimension of F_W can be estimated by Barnsley [8, Theorem 2.3, p. 183] for special affine cases. Also, F_W can be approximated with respect to the Hausdorff metric h in concrete examples by using the iterations $I_n := W^n(\{0\})$ for n sufficiently large. Note that I_n are finite sets and

$$h(F_W, I_n) \le \frac{L_Y^n}{1 - L_Y} h(I_1, I_0) \le \frac{2B_M L_Y^n}{(1 - L_Y)(1 - C_M)}.$$

To be more concrete, we consider Eq. (4.2) and suppose

[B_1]: The domain $D(\mathbb{A})$ is dense in X.
[B_2]: The resolvent $R(\lambda, \mathbb{A}) = (\lambda I - \mathbb{A})^{-1}$ exists for all λ with $\Re\lambda \ge 0$, and there is a constant M independent of λ such that

$$\|R(\lambda, \mathbb{A})\| \le M(1 + |\lambda|)^{-1} \text{ for } \Re\lambda \ge 0.$$

Then $A(t) = -\mu(t)\mathbb{A}$ satisfies assumptions [A_1]–[A_4] and by Lemma 1, we see

$$U(t, s) = S\left(\int_s^t \mu(\tau)d\tau\right), \quad t \ge s \ge 0$$

for the C_0-semigroup $\{S(t),\ t \ge 0\}$ of $\mathbb{A}$. Now $\mathscr{P} = \{1, 2\}$ and then

$$P_j = \zeta_j a(T) + b(T)BS\left(\int_0^{t_1^-} \mu(\tau)d\tau\right), \quad j = 1, 2. \tag{4.18}$$

To further simplify our consideration, we assume

[B$_3$]: $X = X_1 \oplus X_2$, $0 < \dim X_1 < \infty$ with $D(\mathbb{A}) = D(\mathbb{A}_1) \oplus D(\mathbb{A}_2)$, $D(\mathbb{A}_i) \subset X_i$ and $\mathbb{A} : D(\mathbb{A}_i) \to X_i$, so $\mathbb{A}_i := \mathbb{A}/X_i$, $i = 1, 2$.
[B$_4$]: $B : X_i \to X_i$, $i = 1, 2$.

Using the splitting $x = x_1 + x_2$, $x_i \in X_i$, $i = 1, 2$, we have $B = (B_1, B_2)$, $S = (S_1, S_2)$, $B_i, S_i : X_i \to X_i$, $i = 1, 2$. Hence (4.18) splits into

$$P_j = \left(\zeta_j a(T) + b(T)B_1S_1\left(\int_0^{t_1^-} \mu(\tau)d\tau\right), b(T)B_2S_2\left(\int_0^{t_1^-} \mu(\tau)d\tau\right)\right), \quad j = 1, 2. \tag{4.19}$$

Assuming

$$M_i|b(T)|\|B_i\|e^{\omega_i\left(\int_0^{t_1^-} \mu(\tau)d\tau\right)} < 1, \quad i = 1, 2, \tag{4.20}$$

when

$$\|S_i(t)\| \leq M_i e^{\omega_i t}, \quad t \geq 0$$

for some $M_i \geq 1$ and $\omega_i \in \mathbb{R}$, we can apply the above theory to (4.19). Then the corresponding $F_W \subset X_1$ is generated by the restricted IFSP

$$\left\{\zeta_1 a(T) + b(T)B_1S_1\left(\int_0^{t_1^-} \mu(\tau)d\tau\right), \zeta_2 a(T) + b(T)B_1S_1\left(\int_0^{t_1^-} \mu(\tau)d\tau\right)\right\} \tag{4.21}$$

on X_1. Summarizing, we have the following result.

Theorem 8. *Under assumptions [B$_1$]–[B$_4$] and* (4.20)*, the random dynamics of Eq.* (4.2) *is restricted on finite-dimensional X_1 with an IFSP represented by* (4.21)*, where the corresponding attractor F_W is located in X_1.*

Consequently, we can construct problem (4.2) for which F_W can be either the random Cantor set in $\mathbb{R}$ or the random Sierpinski triangle in $\mathbb{R}^2$ [8, p. 335], for instance. Furthermore, denoting

$$E := b(T)B_1S_1\left(\int_0^{t_1^-} \mu(\tau)d\tau\right), \quad b_i = \zeta_i a(T),\ i = 1, 2, \tag{4.22}$$

IFSP (4.21) has the form $\{w_i, w_2\}$ with $w_i(x_1) = b_i + Ex_1$, $i = 1, 2$, $x_1 \in X_1$. Let $\delta > 0$ and $x_1 \in B_\delta(b_1) := \{x_1 \in X_1 : \|x_1 - b_1\| \leq \delta\}$; then we compute

$$\|w_1(x_1) - b_1\| = \|Ex_1\| \leq \|E\|(\|b_1\| + \delta). \tag{4.23}$$

By (4.20), $\|E\| < 1$, so if $\frac{\|b_1\|\|E\|}{1-\|E\|} \leq \delta$, then (4.23) gives $\|w_1(x_1) - b_1\| \leq \delta$; that is, $w_1(B_\delta(b_1)) \subset B_\delta(b_1)$. Similarly, if $\frac{\|b_2\|\|E\|}{1-\|E\|} \leq \delta$, then $w_1(B_\delta(b_2)) \subset B_\delta(b_1)$. Consequently, for

$$\delta = \max\{\|b_1\|, \|b_2\|\}\frac{\|E\|}{1-\|E\|}, \tag{4.24}$$

it holds that

$$w_i(B_\delta(b_j)) \subset B_\delta(b_i), \quad i,j = 1,2,$$

which implies

$$w_1(B_\delta(b_1) \cup B_\delta(b_2)) \subset B_\delta(b_1) \cup B_\delta(b_2).$$

So we arrive at

$$F_W \subset B_\delta(b_1) \cup B_\delta(b_2), \quad w_i(F_W)) \subset B_\delta(b_i), \quad i = 1,2. \tag{4.25}$$

Clearly $F_W \cap B_\delta(b_i) \neq \emptyset$, $i = 1,2$. Next, if

$$\delta < \frac{\|b_1 - b_2\|}{2}, \tag{4.26}$$

then $B_\delta(b_1) \cap B_\delta(b_2) = \emptyset$. So by (4.25), we get $w_1(F_W) \cap w_2(F_W) = \emptyset$; that is, [8, Theorem 2.2, p. 125] gives that IFSP (4.21) is totally disconnected. Then [8, Theorem 5.1, p. 147, Theorem 8.1, p. 167] can be applied to get the chaotic dynamics of IFSP (4.21). Combining (4.22), (4.24), and (4.26), we get

$$\|E\| < \frac{\|b_1 - b_2\|}{2\max\{\|b_1\|, \|b_2\|\} + \|b_1 - b_2\|} = \frac{|\zeta_1 - \zeta_2|}{2\max\{|\zeta_1|, |\zeta_2|\} + |\zeta_1 - \zeta_2|}, \tag{4.27}$$

and summarizing, we arrive at

Theorem 9. *Under the assumptions of Theorem 8 and*

$$M_1|b(T)|\|B_1\|e^{\omega_1\left(\int_0^{t_1^-} \mu(\tau)d\tau\right)} < \frac{|\zeta_1 - \zeta_2|}{2\max\{|\zeta_1|, |\zeta_2|\} + |\zeta_1 - \zeta_2|},$$

the dynamics of IFSP on attractor F_W is chaotic.

Now let us consider the case when E is a similitude, namely, $\|Ex_1\| = \|E\|\|x_1\|$ for any $x_1 \in X_1$. Then [8, Theorem 4.3, p. 199] implies the following:

Theorem 10. *Assume [B_1]–[B_4],* (4.20) *with $i = 2$. Moreover, E is a similitude satisfying* (4.27). *Then the dynamics of IFSP on attractor F_W is chaotic, and the Hausdorff–Besicovitch and fractal dimensions of F_W are equal to $-\frac{\ln 2}{\ln \|E\|}$.*

We end this section with the following result.

Theorem 11. *If we assume* [A_5] *and that all mappings are globally Lipschitz continuous, then the probability that* (4.1) *has a globally asymptotically stable T-periodic mild solution is greater than or equal to*

$$\sum_{\prod_{i=1}^{m} L_{g_i^{j_i}} e^{(ML_{q_{i-1}}-\omega)(t_i-s_{i-1})}<1} \prod_{i=1}^{m} p_{ij_i}. \tag{4.28}$$

A similar result can be stated for globally asymptotically stable subharmonic mild solutions.

Proof. Under the assumptions, each P_χ is globally Lipschitz continuous with a constant $L(\chi)$. Then, similar to what we saw earlier, the probability that (4.1) has a globally asymptotically stable T-periodic mild solution is greater than or equal to

$$\eta(L^{-1}([0,1))) = \sum_{L(\chi)<1} \eta(\chi) = \sum_{\prod_{i=1}^{m} L_{g_i^{j_i}} e^{(ML_{q_{i-1}}-\omega)(t_i-s_{i-1})}<1} \prod_{i=1}^{m} p_{ij_i},$$

where L is considered to be $L : \mathscr{P} \to [0,\infty)$. The proof is finished. □

4.6 Examples

This section is devoted to concrete examples for illustration of the above general theory.

Example 1. First, we consider the following impulsive nonautonomous periodic problem:

$$\begin{cases} \frac{\partial}{\partial t}x(t,y) = (1+\sin^2 t)\frac{\partial^2}{\partial y^2}x(t,y) + \frac{2\gamma}{\pi}\arctan|x(t,y)|, \\ \qquad\qquad \gamma \in \mathbb{R},\ y \in (0,1),\ t \in (0,\frac{\pi}{2}), \\ x(\frac{\pi}{2}^+, y) = -\frac{|x(\frac{\pi}{2}^-,y)|}{\zeta},\ \zeta \in \mathbb{R}, \\ x(t,y) = \cos t + \frac{\cos(\frac{t}{2}+\frac{3\pi}{4})|x(\frac{\pi}{2}^-,y)|}{\zeta},\ t \in (\frac{\pi}{2},\pi],\ y \in (0,1), \\ x(t,0) = x(t,1) = 0,\ t \in [0,\pi], \\ x(0,y) = x(\pi,y),\ y \in (0,1). \end{cases} \tag{4.29}$$

In order to rewrite (4.29) in the abstract form, we set $X = L^2([0,1])$ endowed with the usual norm and choose $0 = t_0 = s_0, t_1 = \frac{\pi}{2}, s_1 = T = \pi$, and $J = [0,\pi]$. Define $\mathbb{A}x = \frac{\partial^2}{\partial y^2}x$ for $x \in D(\mathbb{A})$ with $D(\mathbb{A}) = \left\{x \in X : \frac{\partial x}{\partial y}, \frac{\partial^2 x}{\partial y^2} \in X,\ x(0) = x(1) = 0\right\}$. Set $A(t) = -\mu(t)\mathbb{A}$, where $\mu(t) := 1 + \sin^2 t$, $t \geq 0$. Then assumption [A_1] clearly holds. Next, for all λ with $\Re\lambda \leq 0$, we have

$$(\lambda I - A(t))x(\xi) = \frac{2}{\pi}\sum_{n=1}^{\infty}(\lambda - n^2\mu(t))\sin n\xi\left(\int_0^{\pi} x(s)\sin nsds\right), \ \xi \in [0,1]$$

for any $x \in D(A)$. Since $|\lambda - n^2\mu(t)|^2 \geq |\lambda|^2 + n^4 \geq 1$, we derive

$$(\lambda I - A(t))^{-1}x(\xi) = \frac{\pi}{2}\sum_{n=1}^{\infty}\frac{1}{\lambda - n^2\mu(t)}\sin n\xi\left(\int_0^{\pi} x(s)\sin nsds\right), \ \xi \in [0,1]$$

for any $x \in X$. So $R(\lambda, A(t))$ exists. Moreover, we have

$$\begin{aligned}\|R(\lambda, A(t))x\|^2 &= \sum_{n=1}^{\infty}\frac{1}{|\lambda - n^2\mu(t)|^2}\left|\int_0^{\pi} x(s)\sin nsds\right|^2 \\ &\leq \sum_{n=1}^{\infty}\frac{1}{|\lambda|^2 + n^4}\left|\int_0^{\pi} x(s)\sin nsds\right|^2 \leq \frac{\pi}{2}\sum_{n=1}^{\infty}\frac{1}{|\lambda|^2 + n^4}\|x\|^2 \\ &\leq \frac{\pi}{2}\|x\|^2\int_0^{\infty}\frac{dz}{|\lambda|^2 + z^4} = \frac{\pi^2}{4\sqrt{2}|\lambda|^{3/2}}\|x\|^2 \\ &\leq \frac{4\pi^2}{3\sqrt{6}(|\lambda|+1)^2}\|x\|^2\end{aligned}$$

for $|\lambda| \geq 1$, since $\frac{(|\lambda|+1)^2}{|\lambda|^{3/2}} \leq \frac{16}{3\sqrt{3}}$ for $|\lambda| \geq 1$. Consequently, we have

$$\|R(\lambda, A(t))\| \leq \frac{2\pi}{\sqrt{3}\sqrt[4]{6}(|\lambda|+1)}$$

for $|\lambda| \geq 1$, $\Re\lambda \leq 0$, and $t \geq 0$. Moreover, the above computation shows that

$$\|R(\lambda, A(t))\| \leq \sqrt{\frac{\pi}{2}\sum_{n=1}^{\infty}\frac{1}{n^4}} = \sqrt{\frac{\pi^5}{180}}$$

for $|\lambda| \leq 1$, $\Re\lambda \leq 0$, and $t \geq 0$. Summarizing, assumption [A_2] is verified. Next, using

$$\begin{aligned}(A(t) - A(s))A^{-1}(\tau)x(\xi) &= \frac{\mu(s) - \mu(t)}{\mu(\tau)}\sum_{n=1}^{\infty}\sin n\xi\left(\int_0^{\pi} x(s)\sin nsds\right) \\ &= \frac{\pi}{2}\frac{\mu(s) - \mu(t)}{\mu(\tau)}x(\xi)\end{aligned}$$

for any $x \in X$, we get

$$\begin{aligned} \|(A(t)-A(s))A^{-1}(\tau)\| &= \frac{\pi}{2}\left|\frac{\mu(s)-\mu(t)}{\mu(\tau)}\right| \le \frac{\pi}{2}|\mu(s)-\mu(t)| \\ &= \frac{\pi}{2}|\sin^2 s - \sin^2 t| \le \pi|t-s|. \end{aligned}$$

Hence assumption [A$_3$] is verified as well. So the family $\{A(t), t \ge 0\}$ is well posed (see also [10, Example 2.9b]). Moreover, $\mu(t+\pi) = \mu(t)$, so $\mathbb{A}(t+\pi) = \mathbb{A}(t)$ for $t \ge 0$. Thus, $\{A(t), t \ge 0\}$ is π-periodic and assumption [A$_4$] is satisfied.

On the other hand, $\mathbb{A}$ is the infinitesimal generator of a C_0-semigroup $\{S(t),\ t \ge 0\}$ in X which is given by Zabczyk [31, Example 1.3]:

$$(S(t))x(\xi) = \frac{2}{\pi}\sum_{n=1}^{\infty} e^{-tn^2}\sin n\xi\left(\int_0^{\pi} x(s)\sin nsds\right),\ \xi \in [0,1],\ t \ge 0.$$

So the associated π-periodic evolution family of $\{A(t), t \ge 0\}$ is given by

$$U(t,s)x = S\left(\int_s^t \mu(\tau)d\tau\right)x,\ x \in X,\ t \ge 0.$$

Since $\int_s^t \mu(\tau)d\tau \ge t-s$ for any $t \ge s \ge 0$, we have $\|U(t,s)x\| \le e^{-(t-s)}\|x\| \le \|x\|$ for all $t \ge s \ge 0$. Thus, $M = 1$ in part (i) of Lemma 1.

Now denote

$$q_1(t,x)(y) = \frac{2\gamma}{\pi}\arctan|x(y)|, \quad g_1(t,x)(y) = \cos t + \frac{\cos(\frac{t}{2}+\frac{3\pi}{4})|x(y)|}{\zeta}.$$

Obviously, f and g_1 are π-periodic functions in t and satisfy global Lipschitz conditions, respectively, and growth conditions

$$\|q_1(\cdot,x) - q_1(\cdot,z)\| \le |\gamma|\frac{2}{\pi}\|x-z\|, \quad \|q_1(\cdot,x)\| \le |\gamma|,$$

$$\|g_1(\cdot,x) - g_1(\cdot,z)\| \le \frac{1}{|\zeta|}\|x-z\|, \quad \|g_1(\cdot,x)\| \le 1 + \frac{1}{|\zeta|}\|x\|.$$

- By applying Theorem 2, we obtain the following: If PC-mild solutions of (4.29) are ultimate bounded, then the problem (4.29) has at least one π-periodic PC-mild solution.
- By applying Theorem 3, we obtain the following: If $|\zeta| > 1$ ($\Longrightarrow$ ultimate boundedness), then the problem (4.29) has at least one π-periodic PC-mild solution.

- By applying Theorem 4, we obtain the following: If $|\zeta| > e^{|\gamma|-\frac{\pi}{2}}$, then the problem (4.29) has a unique globally asymptotically stable π-periodic PC-mild solution.

Example 2. Problem (4.29) can be easily extended to the random case when $\zeta \in \{\zeta_1, \cdots, \zeta_n\}$ with probability $p_i > 0$ for given ζ_i, $i = 1, 2, \cdots, n$. Then this random problem has a π-periodic PC-mild solution with a probability greater than or equal to $\sum_{|\zeta_i|>1} p_i$. More generally, this random problem has a $n\pi$-subharmonic PC-mild solution with a probability greater than or equal to $\sum_{|\zeta_{i_1}\cdots\zeta_{i_n}|>1} p_{i_1}\cdots p_{i_n}$. Furthermore, this random problem has a globally asymptotically stable π-periodic PC-mild solution with a probability greater than or equal to $\sum_{|\zeta_i|>e^{|\gamma|-\frac{\pi}{2}}} p_i$.

Example 3. The final example is as follows:

$$\begin{cases} \frac{\partial}{\partial t}x(t,y) = (1+\sin^2 t)\frac{\partial^2}{\partial y^2}x(t,y), \ y \in (0,1), \ t \in (0,\frac{\pi}{2}), \\ x(\frac{\pi}{2}^+, y) = -x(\frac{\pi}{2}^-, y), \\ x(t,y) = \zeta_j \cos t \sin y - \frac{\sqrt{2}e^{\frac{3\pi}{4}}}{3}\cos(\frac{t}{2} + \frac{3\pi}{4})x(\frac{\pi}{2}^-, y), \\ \qquad\qquad t \in (\frac{\pi}{2}, \pi], \ y \in (0,1), \\ x(t,0) = x(t,1) = 0, \ t \in [0,\pi], \\ x(0,y) = x(\pi, y), \ y \in (0,1), \end{cases} \tag{4.30}$$

when ζ_j has a probability $p_j \in (0,1)$, $j = 1, 2$. (4.30) has a form of (4.2) with the above X, $\mathbb{A}$ and

$$a(t) = \cos t \sin y, \quad b(t) = -\frac{\sqrt{2}e^{\frac{3\pi}{4}}}{3}\cos\left(\frac{t}{2} + \frac{3\pi}{4}\right), \quad B = I.$$

We take a splitting

$$X = X_1 \oplus X_1^{\perp}, \quad X_1 := \text{span}[\sin y].$$

Then

$$\left(S_1\left(\int_0^{t_1} \mu(\tau)d\tau\right)\right)x_1(\xi) = \frac{2}{\pi}e^{-\int_0^{t_1}\mu(\tau)d\tau} \sin \xi \left(\int_0^{\pi} x_1(s)\sin s ds\right) = c_1 e^{-\frac{3\pi}{4}} \sin \xi$$

for $x_1(\xi) = c_1 \sin \xi$, $c_1 \in \mathbb{R}$. Similarly,

$$\left(S_2\left(\int_0^{t_1} \mu(\tau)d\tau\right)\right)x_2(\xi) = \frac{2}{\pi}\sum_{n=2}^{\infty} e^{-\frac{3\pi}{4}n^2} \sin n\xi \left(\int_0^{\pi} x(s)\sin ns ds\right)$$

for $x_2 \in X_1^{\perp}$, which gives

$$\left\| S_1\left(\int_0^{t_1} \mu(\tau)d\tau\right)\right\| = e^{-\frac{3\pi}{4}}, \quad \left\| S_2\left(\int_0^{t_1} \mu(\tau)d\tau\right)\right\| = e^{-3\pi}. \tag{4.31}$$

Using $b(T) = \frac{e^{\frac{3\pi}{4}}}{3}$, we satisfy the inequalities of (4.20):

$$\frac{e^{\frac{3\pi}{4}}}{3}e^{-\frac{3\pi}{4}} = \frac{1}{3} < 1, \quad \frac{e^{\frac{3\pi}{4}}}{3}e^{-3\pi} = \frac{e^{\frac{-9\pi}{4}}}{3} < 0.0003.$$

Hence, Theorem 8 can be applied. Furthermore, mappings of (4.21) have the form

$$c_1 \to -\zeta_j + \frac{c_1}{3}$$

when $x_1(\xi) = c_1 \sin\xi \in X_1$. Hence, taking $\zeta_1 = 0$ and $\zeta_2 = -\frac{2}{3}$, mappings of (4.21) have the form

$$c_1 \to \left\{\frac{c_1}{3}, \frac{2}{3} + \frac{c_1}{3}\right\}$$

whose F_W is just the Cantor set [8, p. 81]. Hence, the random dynamics of (4.30) for $\zeta_1 = 0$ and $\zeta_2 = -\frac{2}{3}$ is concentrated on the Cantor set of span[$\sin y$]. We get something similar for general ζ_j.

Acknowledgements The first author acknowledges the support of the National Natural Science Foundation of China (11201091) and the Outstanding Scientific and Technological Innovation Talent Award of the Education Department of Guizhou Province [2014]240. The second author acknowledges the support of grants VEGA-MS 1/0071/14 and VEGA-SAV 2/0029/13. The third author acknowledges the support of the National Natural Science Foundation of China (11271309), the Specialized Research Fund for the Doctoral Program of Higher Education (20114301110001), and Key Projects of Hunan Provincial Natural Science Foundation of China (12JJ2001).

References

1. Ahmed, N.U.: Semigroup Theory with Applications to Systems and Control. Pitman Research Notes in Mathematics Series, vol. 246. Longman Scientific and Technical, Harlow (1991)
2. Akhmet, M.U.: Principles of Discontinuous Dynamical Systems. Springer, New York (2010)
3. Akhmet, M.U.: Nonlinear Hybrid Continuous/Discrete-Time Models. Atlantis Press, Paris (2011)
4. Akhmet, M.U., Alzabut, J., Zafer, A.: On uniform asymptotic stability of linear impulsive delay differential equations. Dyn. Contin. Discrete Impulsive Syst. Ser. A **15**, 621–631 (2008)
5. Amann, H.: Periodic solutions of semi-linear parabolic equations. In: Cesari, L., Kannan, R., Weinberger, H.F. (eds.) Nonlinear Analysis, a Collection of Papers in Honor of Erich Roth. Academic Press, New York, 1–29 (1978)

6. Bainov, D.D., Simeonov, P.S.: Integral Inequalities and Applications. Kluwer Academic Publishers, Dordrecht (1992)
7. Bainov, D.D., Simeonov, P.S.: Impulsive differential equations: periodic solutions and applications. Pitman Monographs and Surveys in Pure and Applied Mathematics, vol. 66. Longman Scientific and Technical Harlow (1993)
8. Barnsley, M.F.: Fractals Everywhere, 2nd ed. Morgan Kaufmann, San Diego (1993)
9. Buşe, C., Lassoued, D., Nguyen, T.L., Saierli, O.: Exponential stability and uniform boundedness of solutions for nonautonomous periodic abstract Cauchy problems. An evolution semigroup approach. Integr. Equ. Oper. Theory **74**, 345–362 (2012)
10. Daners, D., Medina, P.K.: Abstract Evolution Equations, Periodic Problems and Applications. Pitman Research Notes in Mathematics Series, vol. 279. Longman Scientific and Technical, Harlow (1992)
11. Fečkan, M., Wang, J., Zhou, Y.: Existence of periodic solutions for nonlinear evolution equations with non-instantaneous impulses. Nonauton. Dyn. Syst. **1**, 93–101 (2014)
12. Hernández, E., O'Regan, D.: On a new class of abstract impulsive differential equations. Proc. Am. Math. Soc. **141**, 1641–1649 (2013)
13. Kokocki, P.: Existence and asymptotic stability of periodic solution for evolution equations with delays. J. Math. Anal. Appl. **392**, 55–74 (2012)
14. Kokocki, P.: Connecting orbits for nonlinear differential equations at resonance. J. Differ. Equ. **255**, 1554–1575 (2013)
15. Kokocki, P.: The averaging principle and periodic solutions for nonlinear evolution equations at resonance. Nonlinear Anal. Theory Methods Appl. **85**, 253–278 (2013)
16. La Torre, D., Marsiglio, S., Privileggi, F.: Fractals and self-similarity in economics: the case of a stochastic two-sector growth model. Image Anal. Stereol. **30**, 143–151 (2011)
17. Li, Y.: Existence and asymptotic stability of periodic solution for evolution equations with delays. J. Funct. Anal. **261**, 1309–1324 (2011)
18. Liu, J.H.: Bounded and periodic solutions of differential equations in Banach space. Appl. Math. Comput. **65**, 141–150 (1994)
19. Liu, J.H.: Bounded and periodic solutions of semilinear evolution equations. Dyn. Syst. Appl. **4**, 341–350 (1995)
20. Liu, J.H.: Bounded and periodic solutions of finite delay evolution equations. Nonlinear Anal. Theory Methods Appl. **34**, 101–111 (1998)
21. Liu, J.H., Naito, T., Minh, N.V.: Bounded and periodic solutions of infinite delay evolution equations. J. Math. Anal. Appl. **286**, 705–712 (2003)
22. Liu, Z.: Anti-periodic solutions to nonlinear evolution equations. J. Funct. Anal. **258**, 2026–2033 (2010)
23. Nguyenm, T.L.: On nonautonomous functional differential equations. J. Math. Anal. Appl. **239**, 158–174 (1999)
24. Park, J.Y., Kwun, Y.C., Jeong, J.M.: Existence of periodic solutions for delay evolution integrodifferential equations. Math. Comput. Model. **40**, 597–603 (2004)
25. Pazy, A.: Semigroup of Linear Operators and Applications to Partial Differential Equations. Springer, New York (1983)
26. Pierri, M., O'Regan, D., Rolnik, V.: Existence of solutions for semi-linear abstract differential equations with not instantaneous impulses. Appl. Math. Comput. **219**, 6743–6749 (2013)
27. Samoilenko, A.M., Perestyuk, N.A.: Impulsive Differential Equations, vol. 14 of World Scientific Series on Nonlinear Science. Series A: Monographs and Treatises, World Scientific, Singapore (1995)
28. Sattayatham, P., Tangmanee, S., Wei, W.: On periodic solutions of nonlinear evolution equations in Banach spaces. J. Math. Anal. Appl. **276**, 98–108 (2002)
29. Wang, J., Fečkan, M., A general class of impulsive evolution equations. Top. Meth. Nonl. Anal. (2015) DOI: http://dx.doi.org/10.12775/TMNA.2015.072
30. Xiang, X., Ahmed, N.U.: Existence of periodic solutions of semilinear evolution equations with time lags. Nonlinear Anal. Theory Methods Appl. **18**, 1063–1070 (1992)
31. Zabczyk, J.: Mathematical Control Theory: An Introduction to Systems and Control. Birkhäuser, Basel (1992)

Chapter 5
Boundedness of Solutions to a Certain System of Differential Equations with Multiple Delays

Cemil Tunç

Abstract In this chapter, we consider a system of differential equations of second order with multiple delays. Based on the Lyapunov–Krasovskii functional approach, we investigate the boundedness of solutions. The obtained results essentially complement and improve some known results in the literature.

5.1 Introduction

In recent years, the theory of delay differential equations (DDEs) with retarded arguments has provided a natural framework for the mathematical modeling of many real-world phenomena related to the engineering technique fields, mechanics, models of economic dynamics, optimal control problems, physics, chemistry, life sciences, medicine, atomic energy, information sciences, nerve conduction theory, the slowing down of neutrons in nuclear reactors, the description of traveling waves in a spatial lattice, among others. See, in particular, the books of Bellman and Cooke [1], Erneux [2], Kolmanovskii and Myshkis [3], Smith [4], and Wu et al. [5] for more applications of differential equations of retarded type. The concept of delay is related to the memory of systems, where past events influence the current behavior, and which could be useful for decision making. Further, finding solutions of DDEs is difficult and many times cannot be obtained in closed form. In the absence of a closed form, a viable alternative is studying the qualitative behavior of solutions. In this case, for nonlinear systems without and with delay, the Lyapunov function and Lyapunov–Krasovskii functional approach, respectively, provide a way to analyze the qualitative behavior of solutions (stability, instability, boundedness, asymptotic behaviors, global existence, etc.) of a system without explicitly solving the differential equations. However, since the method requires an auxiliary function or functional, which is not easy to find, it remains an open problem in the literature at this time. When we look at the related literature, the qualitative properties of solutions to second-order DDEs have been intensively discussed and are still being

C. Tunç (✉)
Department of Mathematics, Faculty of Sciences, Yüzüncü Yıl University, 65080 Van, Turkey
e-mail: cemtunc@yahoo.com

A.C.J. Luo, H. Merdan (eds.), *Mathematical Modeling and Applications in Nonlinear Dynamics*, Nonlinear Systems and Complexity 14,
DOI 10.1007/978-3-319-26630-5_5

investigated. We refer the reader to the papers or books of Ahmad and Rama Mohana Rao [6], Anh et al. [7], Barnett [8], Burton [9], Burton and Zhang [10], Caldeira-Saraiva [11], Cantarelli [12], Èl'sgol'ts and Norkin [13], Gao and Zhao [14], Hale [15], Hara and Yoneyama [16, 17], Heidel [18], Huang and Yu [19], Jitsuro and Yusuke [20], Kato [21, 22], Krasovskiĭ [23], LaSalle and Lefschetz [24], Li [25], Liu and Huang [26, 27], Liu and Xu [28], Liu [29], Long and Zhang [30], Luk [31], Lyapunov [32], Malyseva [33], Muresan [34], Sugie [35], Sugie and Amano [36], Sugie et al. [37], Tunç [38–45], Tunç and Tunç [46], Yang [47], Ye et al. [48], Yu and Xiao [49], Yoshizawa [50], Zhang [51, 52], Zhang and Yan [53], Zhou and Jiang [54], Zhou and Liu [55], Zhou and Xiang [56], Wei and Huang [57], and Wiandt [58].

At the same time, very recently Omeike et al. [59] considered the following second-order nonlinear system of differential equations of the form

$$X'' + F\left(X, X'\right) X' + H(X) = P\left(t, X, X'\right). \tag{5.1}$$

Omeike et al. [59] proved two new results dealing with the boundedness of solutions of Eq. (5.1). In their work, the authors extended some known results, in the literature, on the boundedness of certain second-order nonlinear scalar differential equations to a system of second-order differential equations, Eq. (5.1). However, to the best of our knowledge of the literature, there is no work based on the results of Omeike et al. [59] to discuss the boundedness of solutions of certain systems of second-order DDEs.

In this chapter, we study the boundedness of solutions to the retarded system of differential equations with multiple constant delays,

$$X'' + F\left(X, X'\right) X' + \sum_{i=1}^{n} H_i\left(X\left(t - \tau_i\right)\right) = P\left(t, X, X'\right), \tag{5.2}$$

where $t \in \Re^+$, $\Re^+ = [0, \infty)$, $X \in \Re^n$, τ_i are positive constants with $t - \tau_i \geq 0$, F is a continuous $n \times n-$ symmetric matrix function, $H_i : \Re^n \to \Re^n$ and $P : \Re^+ \times \Re^n \times \Re^n \to \Re^n$ are continuous, and H_i are also differentiable with $H_i(0) = 0$. The existence and uniqueness of the solutions of Eq. (5.2) are assumed (see Èl'sgol'ts and Norkin [13]).We can write Eq. (5.2) in the differential system form

$$X' = Y,$$

$$\begin{aligned} Y' &= -F\left(X, Y\right) Y - \sum_{i=1}^{n} H_i(X) \\ &+ \sum_{i=1}^{n} \int_{t-\tau_i}^{t} J_{H_i}\left(X(s)\right) Y(s) ds + P\left(t, X, Y\right) \end{aligned} \tag{5.3}$$

or

$$X' = Y,$$
$$Y' = -F(X,Y)\,Y - \sum_{i=1}^{n} H_i\Big(X(t-\tau_i)\Big) + P(t,X,Y), \tag{5.4}$$

which were obtained by setting $X' = Y$ from Eq. (5.2), and $X(t)$ and $Y(t)$ are respectively abbreviated as X and Y throughout the chapter.

Throughout the chapter, the Jacobian matrices of $H_1(X), \ldots, H_n(X)$ will be given by

$$J_{H_1}(X) = \left(\frac{\partial h_{1i}}{\partial x_j}\right), \ldots, J_{H_n}(X) = \left(\frac{\partial h_{ni}}{\partial x_j}\right), \quad (i, j = 1, 2, \ldots, n),$$

where $(x_1, \ldots, x_n)$ and $(h_{1i}), \ldots, (h_{ni})$ are the components of X and H_i, respectively. Moreover, we assume that the given Jacobian matrices exist and are continuous.

The symbol $\langle X, Y\rangle$ corresponding to any pair X, Y in $\Re^n$ stands for the usual scalar product $\sum_{i=1}^{n} x_i y_i$, that is, $\langle X, Y\rangle = \sum_{i=1}^{n} x_i y_i$; thus $\langle X, X\rangle = \|X\|^2$, and $\lambda_i(A)$ are the eigenvalues of the real symmetric $n \times n$ matrix A. The matrix A is said to be negative definite when $\langle AX, X\rangle \leq 0$ for all nonzero X in $\Re^n$. Finally, by $\operatorname{sgn} X$, we mean $(\operatorname{sgn} x_1, \operatorname{sgn} x_2, \ldots, \operatorname{sgn} x_n)$ and $\|\operatorname{sgn} X\| = \sqrt{n}$.

Motivated by the work in Omeike et al. [59], in this chapter we will improve and extend the results in [59] to DDE (5.2). This work is also a first attempt to obtain certain sufficient conditions on the ultimate boundedness of solutions of a vector Lienard equation with multiple delays; and it is a contribution to the subject in the literature and may be useful for researchers' work on the qualitative behaviors of solutions.

We need the following preliminary result.

Lemma 1. (Bellman [60]) Let A be a real symmetric $n \times n$ matrix and

$$\bar{a} \geq \lambda_i(A) \geq a > 0, \quad (i = 1, 2, \ldots, n),$$

where $\bar{a}$ and a are constants.

Then

$$\bar{a}\|X\|^2 \geq \langle AX, X\rangle \geq a\|X\|^2$$

and

$$\overline{a}^2\|X\|^2 \geq \langle AX, AX\rangle \geq a^2\|X\|^2.$$

5.2 The Main Results

In this section, we introduce the main results.

Theorem 1. We assume that there exist some positive constants ρ, β, δ_f, Δ_f, α_i, β_i and τ such that the following conditions hold in Eq. (5.2):

(i) $\rho - \beta > 0$, the matrix F is symmetric,
$0 < \delta_f \leq \lambda_i\left(F\left(X, Y\right)\right) \leq \Delta_f$ for all X, Y,

(ii) $H_i(0) = 0$, $H_i(X) \neq 0$, $(X \neq 0)$, $J_{H_i}(X)$ are symmetric, $\alpha_i \leq \lambda_i\left(J_{H_i}(X)\right) \leq \beta_i$ for all X, $\langle H_i(X), X\rangle \to +\infty$ as $\|X\| \to \infty$ or $\langle H_i(X), X\rangle \to -\infty$ as $\|X\| \to \infty$,

(iii)

$$\lim_{\|X\|\to\infty}\Big\{\langle \alpha H_1\left(X\left(t-\tau_1\right)\right), \operatorname{sgn} X\rangle + \cdots + \langle \alpha H_n\left(X\left(t-\tau_n\right)\right), \operatorname{sgn} X\rangle - 2\gamma\Delta_f\Big\} > 2\gamma\beta,$$

where

$$\alpha = \operatorname{sgn}\langle H_i\left(X\left(t-\tau_i\right)\right), \operatorname{sgn} X\rangle \quad \text{and} \quad \gamma = \sqrt{n},$$

(iv) $\|P\left(t, X, Y\right)\| \leq \beta$ for all t, X and Y.
If

$$\tau < \frac{\delta_f}{\sqrt{n}\sum_{i=1}^{n}\beta_i},$$

then there exists a positive constant D, whose magnitude depends only on the constants ρ, β, δ_f, Δ_f, α_i, β_i as well as on $F(X, Y)$, $J_{H_1}(X), \ldots, J_{H_n}(X)$ and $P(t, X, Y)$ such that every solution X of Eq. (5.2) ultimately satisfies

$$\|X\| \leq D, \ \left\|X'\right\| \leq D.$$

Proof We define a Lyapunov–Krasovskii functional $V = V_1 + V_2 = V_1\left(X_t, Y_t\right) + V_2\left(X, Y\right)$ by

$$V_1 = \sum_{i=1}^{n} \int_0^1 \langle H_i(\sigma X), X \rangle \, d\sigma + \frac{1}{2} \langle Y, Y \rangle + \sum_{i=1}^{n} \mu_i \int_{-\tau_i}^{0} \int_{t+s}^{t} \|Y(\theta)\|^2 d\theta ds \tag{5.5}$$

and

$$V_2 = \begin{cases} \alpha \langle Y, \operatorname{sgn} X \rangle & \text{if } \|Y\| \le \|X\|, \\ \langle X, \operatorname{sgn} Y \rangle & \text{if } \|X\| \le \|Y\|, \end{cases}$$

where s is a real variable such that the integrals $\int_{-\tau_i}^{0} \int_{t+s}^{t} \|Y(\theta)\|^2 d\theta ds$ are nonnegative and μ_i are some positive constants to be determined later in the proof.

It is clear that $V_1(0,0) = 0$. Using the estimates $H_i(0) = 0$, $\frac{\partial}{\partial \sigma} H_i(\sigma X) = J_{H_i}(\sigma X) X$, $\lambda_i(J_{H_i}(X)) \ge \alpha_i$, $(i = 1, 2, \ldots, n)$, we obtain

$$H_i(X) = \int_0^1 J_{H_i}(\sigma X) X d\sigma$$

so that

$$\begin{aligned} \int_0^1 \langle H_1(\sigma X), X \rangle \, d\sigma &= \int_0^1 \int_0^1 \langle \sigma_1 J_{H_1}(\sigma_1 \sigma_2 X) X, X \rangle \, d\sigma_2 d\sigma_1 \\ &\ge \int_0^1 \int_0^1 \langle \sigma_1 \alpha_1 X, X \rangle \, d\sigma_2 d\sigma_1 \ge \tfrac{1}{2} \alpha_1 \|X\|^2, \end{aligned}$$

$$\begin{aligned} \int_0^1 \langle H_2(\sigma X), X \rangle \, d\sigma &= \int_0^1 \int_0^1 \langle \sigma_1 J_{H_2}(\sigma_1 \sigma_2 X) X, X \rangle \, d\sigma_2 d\sigma_1 \\ &\ge \int_0^1 \int_0^1 \langle \sigma_1 \alpha_2 X, X \rangle \, d\sigma_2 d\sigma_1 \ge \tfrac{1}{2} \alpha_2 \|X\|^2, \end{aligned}$$

$$\vdots$$

$$\begin{aligned} \int_0^1 \langle H_n(\sigma X), X \rangle \, d\sigma &= \int_0^1 \int_0^1 \langle \sigma_1 J_{H_n}(\sigma_1 \sigma_2 X) X, X \rangle \, d\sigma_2 d\sigma_1 \\ &\ge \int_0^1 \int_0^1 \langle \sigma_1 \alpha_n X, X \rangle \, d\sigma_2 d\sigma_1 \ge \tfrac{1}{2} \alpha_n \|X\|^2. \end{aligned}$$

Then it follows from V_1 that

$$V_1 \geq \frac{1}{2}\left(\sum_{i=1}^{n}\alpha_i\right)\|X\|^2 + \frac{1}{2}\|Y\|^2 + \sum_{i=1}^{n}\mu_i\int_{-\tau_i}^{0}\int_{t+s}^{t}\|Y(\theta)\|^2 d\theta ds$$

$$\geq D_1\left(\|X\|^2 + \|Y\|^2\right) + \sum_{i=1}^{n}\mu_i\int_{-\tau_i}^{0}\int_{t+s}^{t}\|Y(\theta)\|^2 d\theta ds$$

$$\geq D_1\left(\|X\|^2 + \|Y\|^2\right),$$

where $D_1 = \min\left\{\frac{1}{2}\left(\sum_{i=1}^{n}\alpha_i\right), \frac{1}{2}\right\}$.

Further, it follows from the definition of V_2 that $|V_2| \leq \delta\,\|Y\|$, where δ is a positive constant. Hence, we can conclude

$$V \geq D_1\left(\|X\|^2 + \|Y\|^2\right) - \delta\,\|Y\|.$$

Then it is clear that the right-hand side of the last estimate tends to $+\infty$ when

$$\|X\|^2 + \|Y\|^2 \to +\infty.$$

Using a basic calculation, by the time derivatives of V_1 and V_2 along the solutions of (5.3), we have

$$\dot{V}_1 = -\langle F(X,Y)\,Y, Y\rangle + < \sum_{i=1}^{n}\int_{t-\tau_i}^{t} J_{H_i}(X(s))\,Y(s)ds, Y >$$
$$+ \langle Y, P(t,X,Y)\rangle + < \sum_{i=1}^{n}\left(\mu_i\tau_i\right)Y, Y > - \sum_{i=1}^{n}\mu_i\int_{t-\tau_i}^{t}\|Y(\theta)\|^2 d\theta$$

and

$$\dot{V}_2 = \begin{cases} -\alpha\Big\{\langle F(X,Y)\,Y, \operatorname{sgn} X\rangle + < \sum_{i=1}^{n} H_i\left(X(t-\tau_i)\right), \operatorname{sgn} X > \\ -\Big\langle P(t,X,Y), \operatorname{sgn} X\Big\}\Big\rangle \quad \text{if } \|Y\| \leq \|X\|, \\ \langle Y, \operatorname{sgn} Y\rangle \quad \text{if } \|X\| \leq \|Y\|. \end{cases}$$

In view of Lemma 1, the assumptions $\lambda_i\left(F(X,Y)\right) \geq \delta_f$, $\lambda_i\left(J_{H_i}(X)\right) \leq \beta_i$ and the estimate $2\,|a|\,|b| \leq a^2 + b^2$ (with a and b real numbers) combined with the classical Cauchy–Schwartz inequality, it follows that

$$
\begin{aligned}
&-\langle F(X,Y)\,Y,Y\rangle \le -\delta_f\|Y\|^2,\\
&< \int_{t-\tau_i}^{t} J_{H_i}(X(s))\,Y(s)ds, Y > \le \|Y\| \left\| \int_{t-\tau_i}^{t} J_{H_i}(X(s))\,Y(s)ds \right\|\\
&\le \sqrt{n}\beta_i\,\|Y\| \left\| \int_{t-\tau_i}^{t} Y(s) \right\| ds\\
&\le \sqrt{n}\beta_i\,\|Y\| \int_{t-\tau_i}^{t} \|Y(s)\|ds\\
&\le \frac{1}{2}\sqrt{n}\beta_i \int_{t-\tau_i}^{t} \left(\|Y(t)\|^2 + \|Y(s)\|^2\right)ds\\
&\le \frac{1}{2}\sqrt{n}\beta_i\tau_i\|Y\|^2 + \frac{1}{2}\sqrt{n}\beta_i \int_{t-\tau_i}^{t} \|Y(s)\|^2 ds
\end{aligned}
$$

so that

$$
\begin{aligned}
\dot{V}_1 \le &-\delta_f\|Y\|^2 + \left(\sum_{i=1}^{n}\mu_i\tau_i\right)\|Y\|^2 + \tfrac{1}{2}\left(\sqrt{n}\sum_{i=1}^{n}\beta_i\tau_i\right)\|Y\|^2\\
&-\sum_{i=1}^{n}\left(\mu_i - \frac{1}{2}\sqrt{n}\beta_i\right)\int_{t-\tau_i}^{t}\|Y(s)\|^2 ds + \langle Y, P(t,X,Y)\rangle .
\end{aligned}
$$

Let $\mu_i = \dfrac{1}{2}\sqrt{n}\beta_i$, $\tau = \max \tau_i$, and $\overline{\delta} = \sqrt{n}\sum_{i=1}^{n}\beta_i$. Then it is clear that

$$
\begin{aligned}
\dot{V}_1 &\le -\left\{\delta_f - \sqrt{n}\sum_{i=1}^{n}\left(\beta_i\tau_i\right)\right\}\|Y\|^2 + \langle Y, P(t,X,Y)\rangle\\
&\le -\left(\delta_f - \overline{\delta}\tau\right)\|Y\|^2 + \langle Y, P(t,X,Y)\rangle .
\end{aligned}
$$

If $\tau < \frac{\delta_f}{\overline{\delta}}$, then we can obtain, for some positive constant ρ, that

$$
\dot{V}_1 \le -\rho\|Y\|^2 + \langle Y, P(t,X,Y)\rangle .
$$

In view of the last estimates for $\dot{V}_1$ and $\dot{V}_2$, if $\|Y\| \le \|X\|$, then

$$
\begin{aligned}
\dot{V} \le &-\rho\|Y\|^2 + \langle Y, P(t,X,Y)\rangle - \alpha\,\langle F(X,Y)\,Y, \operatorname{sgn} X\rangle\\
&-\alpha < \sum_{i=1}^{n} H_i\left(X\left(t-\tau_i\right)\right), \operatorname{sgn} X > +\alpha\,\langle P(t,X,Y), \operatorname{sgn} X\rangle
\end{aligned}
$$

and if $\|X\| \leq \|Y\|$, then

$$\dot{V} \leq -\rho\|Y\|^2 + \langle Y, P(t, X, Y)\rangle + \langle Y, \operatorname{sgn} Y\rangle .$$

Hence, it follows that

$$\begin{aligned} \dot{V} \leq -\rho\|Y\|^2 + |\langle F(X, Y)\, Y, \operatorname{sgn} X\rangle| - \alpha < \sum_{i=1}^{n} H_i(X(t-\tau_i)), \operatorname{sgn} X > \\ + (\|Y\| + \gamma)\ \|P(t, X, Y)\| \quad \text{if } \|Y\| \leq \|X\| \end{aligned} \tag{5.6}$$

and

$$\dot{V} \leq -\rho\|Y\|^2 + (\|P(t, X, Y)\| + \gamma)\,\|Y\| \quad \text{if } \|X\| \leq \|Y\| . \tag{5.7}$$

The assumption

$$\lim_{\|X\|\to\infty} \Big\{\alpha\,\langle H_1(X(t-\tau_1)), \operatorname{sgn} X\rangle + \cdots + \alpha\,\langle H_n(X(t-\tau_n)), \operatorname{sgn} X\rangle \\ - 2\gamma\Delta_f\Big\} > 2\gamma\beta$$

with $\alpha = (\operatorname{sgn}\langle H_i(X(t-\tau_i)), \operatorname{sgn} X\rangle)$, and $\gamma = \sqrt{n}$, β are positive constants, implies the existence of finite constants $\alpha_0 > 0$ and $D_2 > 0$ such that $\|X\| \geq \alpha_0$.

By assumption (iii) of Theorem 1, $\|X\| \geq \alpha_0$ implies that

$$\begin{aligned} \alpha\,\langle H_1(X(t-\tau_1)), \operatorname{sgn} X\rangle + \cdots + \alpha\,\langle H_n(X(t-\tau_i)), \operatorname{sgn} X\rangle \\ - 2\gamma\Delta_f - 2\gamma\beta \geq D_2. \end{aligned} \tag{5.8}$$

Let

$$\alpha_1 = \max\{1, \alpha_0, \mu\}$$

with $\mu = \rho^{-1}(\beta + \gamma)$.

We claim, for some finite positive constant $D_3 > 0$, that

$$\dot{V} \leq -D_3 \quad \text{if } \|X\| \geq \alpha_1 .$$

In fact, if $\|Y\| \leq \|X\|$, then it is clear that $\dot{V}$ satisfies (5.6), and if $\|Y\| \geq 1$, then, by the assumptions $0 < \delta_f \leq \lambda_i(F(X, Y)) \leq \Delta_f$, $\|P(t, X, Y)\| \leq \beta$, and $\rho - \beta > 0$, we have

$$\begin{aligned}\dot{V} &\leq -\alpha < \sum_{i=1}^{n} H_i\left(X\left(t-\tau_i\right)\right), \operatorname{sgn} X > -\|Y\|\left(\rho\|Y\|-\gamma \Delta_f\right)+\beta(\|Y\|+\gamma) \\ &\leq -\alpha < \sum_{i=1}^{n} H_i\left(X\left(t-\tau_i\right)\right), \operatorname{sgn} X > -\left(\rho\|Y\|-\gamma \Delta_f\right)+\beta(\|Y\|+\gamma) \\ &= -(\rho-\beta)\|Y\|-\alpha < \sum_{i=1}^{n} H_i\left(X\left(t-\tau_i\right)\right), \operatorname{sgn} X > +\gamma\left(\beta+\Delta_f\right) \\ &\leq -\alpha < \sum_{i=1}^{n} H_i\left(X\left(t-\tau_i\right)\right), \operatorname{sgn} X > +2\gamma\left(\beta+\Delta_f\right).\end{aligned} \tag{5.9}$$

By noting assumption (iii) of Theorem 1,

$$\lim_{\|X\| \to \infty}\Big\{\alpha\left\langle H_1\left(X\left(t-\tau_1\right)\right), \operatorname{sgn} X\right\rangle+\cdots+\alpha\left\langle H_n\left(X\left(t-\tau_n\right)\right), \operatorname{sgn} X\right\rangle - 2\gamma \Delta_f\Big\}>2\gamma\beta$$

it can be followed from (5.9), for some positive constant D_2, that

$$\dot{V} \leq -D_2 \text{ if } \|X\| \geq \alpha_1. \tag{5.10}$$

Next, we suppose that $\|Y\| \leq 1$. Then

$$\begin{aligned}\dot{V} &\leq -\rho\|Y\|^2+\gamma \Delta_f\|Y\|-\alpha < \sum_{i=1}^{n} H_i\left(X\left(t-\tau_i\right)\right), \operatorname{sgn} X > +\beta(\|Y\|+\gamma) \\ &\leq -\alpha < \sum_{i=1}^{n} H_i\left(X\left(t-\tau_i\right)\right), \operatorname{sgn} X > +\gamma \Delta_f+\beta(1+\gamma) \\ &\leq -\alpha < \sum_{i=1}^{n} H_i\left(X\left(t-\tau_i\right)\right), \operatorname{sgn} X > +2\gamma\left(\beta+\Delta_f\right).\end{aligned}$$

Hence, it can be concluded that estimates (5.8) and (5.10) still hold in this case.

We now consider estimate (5.7) when $\|X\| \leq \|Y\|$. If $\|X\| \geq \alpha_1$ with $\alpha_1 = \max\{1, \alpha_0, \mu\}$, then $\|Y\| \geq \alpha_1$. Hence,

$$\dot{V} \leq -\rho\|Y\|^2+(\beta+\gamma)\|Y\|=-\|Y\|\{\rho\|Y\|-(\beta+\gamma)\} \leq -1$$

if $\|Y\| \geq \mu=\rho^{-1}(\beta+\gamma)$.

This means that if $\|X\| \geq \alpha_1$, then $\dot{V} \leq -1$. In view of the last estimate and (5.10), $\dot{V} \leq -D_2$, if $\|X\| \geq \alpha_1$, it follows that

$$\dot{V} \leq -D_3,$$

where $D_3 = \max\{1, D_2\}$.

To conclude the end of the proof, we suppose, on the contrary, that $\|X\| \leq \alpha_1$. Let $\|Y\| \geq \alpha_1$. Then $\|Y\| \geq \|X\|$. Hence,

$$\dot{V} \leq -\rho\|Y\|^2 + (\beta+\gamma)\|Y\| = -\|Y\|\{\rho\|Y\| - (\beta+\gamma)\} \leq -1$$

if $\|Y\| \geq \rho^{-1}(\beta+\gamma)$.

Then, in view of the last estimate and $\dot{V} \leq -D_2$ if $\|X\| \geq \alpha_1$, it follows that

$$\dot{V} \leq -D_3,$$

where $D_3 = \max\{1, D_2\}$. Therefore, we can conclude that

$$\dot{V} \leq -D_3 \quad \text{if } \|X\|^2 + \|Y\|^2 \geq 2\alpha_1.$$

The proof of Theorem 1 is complete. □

Our second main result is the following theorem.

Theorem 2. Let all assumptions of Theorem 1 hold, except (iii), and assume that

(i) $$\lim_{\|X\|\to\infty} \Big\{\alpha\langle H_1(X(t-\tau_1)), \operatorname{sgn} X\rangle + \cdots + \alpha\langle H_n(X(t-\tau_n)), \operatorname{sgn} X\rangle - 2\gamma\Delta_f\Big\} > 2\gamma\beta*,$$

where

$$\beta* = \max\left\{\frac{\gamma}{8}(\Delta_f+\beta)^2(\rho-\beta)^{-1}, \beta\right\},$$

and

(ii) $\|P(t,X,Y)\| \leq \beta\|Y\|$ for all $t, X, Y \in \Re^n$.

If

$$\tau < \frac{\rho}{\sqrt{n}\sum_{i=1}^{n}\beta_i},$$

then there exists a positive constant D, whose magnitude depends only on the constants ρ, β, δ_f, Δ_f, α_i, β_i as well as on $F(X,Y)$, $J_{H_1}(X), \ldots, J_{H_n}(X)$, and $P(t,X,Y)$ such that every solution X of Eq. (5.2) ultimately satisfies

$$\|X\| \le D, \ \|X'\| \le D.$$

Proof The main tool to prove Theorem 1 is the Lyapunov–Krasovskii functional $V = V_1 + V_2$ used in Theorem 1. We use the same procedure as that used in the proof of Theorem 1. The proof of Theorem 2 is immediate when we show the following estimates:

$$V \to \infty \quad \text{as } \|X\|^2 + \|Y\|^2 \to \infty \tag{5.11}$$

and

$$\dot{V} \le -D_0 \quad \text{if } \|X\|^2 + \|Y\|^2 \ge D_1. \tag{5.12}$$

The verification of (5.11) can be easily checked from the discussion made in the proof of Theorem 1. Therefore, we omit the details to verify (5.11), $V \to \infty$ as $\|X\|^2 + \|Y\|^2 \to \infty$. To verify estimate (5.12), we benefit from estimates (5.6) and (5.7), which are still valid in this case. In fact, in view of assumption (i) of Theorem 2, that is, the definition of an infinite limit, we can say that there are positive constants α_0 and D_4 such that $\|X\| \ge \alpha_0$ implies the existence of the following estimate:

$$\begin{aligned} &\alpha \langle H_1 (X(t-\tau_1)), \operatorname{sgn} X \rangle + \cdots + \alpha \langle H_n (X(t-\tau_n)), \operatorname{sgn} X \rangle \\ &- 2\gamma\Delta_f - 2\gamma\beta * \ge D_4. \end{aligned} \tag{5.13}$$

We also assume that there exists a positive constant ξ_0 such that $\|Y\| \ge \xi_0$ implies

$$-(\rho - \beta)\|Y\|^2 + \|Y\| \le -1.$$

Let

$$\alpha_1 = \max\{1, \alpha_0, \delta_0\}. \tag{5.14}$$

We claim, for some finite positive constant $D_5 > 0$, that

$$\dot{V} \le -D_5 \quad \text{if } \|X\| \ge \alpha_1.$$

To conclude the preceding claim; we consider the following two cases:
$\|Y\| \le \|X\|$ and $\|X\| \le \|Y\|$, separately.

In fact, if $\|Y\| \le \|X\|$ and $\|Y\| \ge 1$, then in view of (5.6) and the assumptions of Theorem 2, it follows that

$$\begin{aligned}\dot{V} &\leq -\rho\|Y\|^2 + \gamma\Delta_f \|Y\| - \alpha < \sum_{i=1}^{n} H_i\left(X\left(t-\tau_i\right)\right), \operatorname{sgn} X > \\ &\quad + \beta \|Y\| \left(\|Y\| + \gamma\right) \\ &\leq -\alpha < \sum_{i=1}^{n} H_i\left(X\left(t-\tau_i\right)\right), \operatorname{sgn} X > -(\rho-\beta)\left[\|Y\| - \tfrac{\gamma(\Delta_f+\beta)}{2(\rho-\beta)}\right]^2 \\ &\quad + \tfrac{\gamma^2(\Delta_f+\beta)^2}{4(\rho-\beta)} \\ &\leq -\alpha < \sum_{i=1}^{n} H_i\left(X\left(t-\tau_i\right)\right), \operatorname{sgn} X > + \tfrac{\gamma^2(\Delta_f+\beta)^2}{4(\rho-\beta)} \\ &\leq -\alpha < \sum_{i=1}^{n} H_i\left(X\left(t-\tau_i\right)\right), \operatorname{sgn} X > +2\gamma\Delta_f + \tfrac{\gamma^2(\Delta_f+\beta)^2}{4(\rho-\beta)}.\end{aligned}$$

Next, we suppose that $\|Y\| \leq 1$. Then

$$\begin{aligned}\dot{V} &\leq -\rho\|Y\|^2 + \gamma\Delta_f \|Y\| - \alpha < \sum_{i=1}^{n} H_i\left(X\left(t-\tau_i\right)\right), \operatorname{sgn} X > \\ &\quad + \beta \|Y\| \left(\|Y\| + \gamma\right) \\ &= -\alpha < \sum_{i=1}^{n} H_i\left(X\left(t-\tau_i\right)\right), \operatorname{sgn} X > \\ &\quad + \gamma\left(\Delta_f + \beta\right)\|Y\| \\ &\leq -\alpha < \sum_{i=1}^{n} H_i\left(X\left(t-\tau_i\right)\right), \operatorname{sgn} X > +2\gamma\left(\beta + \Delta_f\right).\end{aligned}$$

Hence, in view of (5.13), assumption (i) of Theorem 2, and (5.14), it follows in either case that if $\|X\| \geq \alpha_1$, then

$$\dot{V} \leq -D_5,\ D_5 > 0, \tag{5.15}$$

hold. Thus, we have

$$\dot{V} \leq -\left(\rho-\beta\right)\|Y\|^2 + \gamma\|Y\| \quad \text{for } \|Y\| \geq \gamma(\rho\text{-}\beta)^{-1}.$$

On the contrary, we now suppose that $\|X\| \leq \alpha_1$ and assume $\|Y\| \geq \alpha_1$. In this case, it is clear that $\|Y\| \geq \|X\|$. Then in view of $\|Y\| \geq \alpha_1$, we get

$$\dot{V} \leq -\Big\{(\rho-\beta)\|Y\| - \gamma\Big)\Big\}\|Y\| \leq -1$$

for $\|Y\| \geq \gamma(\rho-\beta)^{-1}$.

The last estimate and (5.15) together imply that

$$\dot{V} \leq -D_5 \quad \text{if } \|X\|^2 + \|Y\|^2 \geq 2\alpha_1,$$

which verifies

$$\dot{V} \leq -D_0 \quad \text{if } \|X\|^2 + \|Y\|^2 \geq D_1.$$

The proof of Theorem 2 is complete. □

5.3 Conclusion

We have considered a system of second-order differential equations with multiple delays. By using the Lyapunov–Krasovskii functional approach, we proved two new theorems on the boundedness of solutions to the considered system. The obtained results complement and improve the recent results obtained by Omeike et al. [59].

References

1. Bellman, R., Cooke, K.L.: Modern Elementary Differential Equations. Reprint of the 1971 second edition. Dover Publications, New York (1995)
2. Erneux, T.: Applied Delay Differential Equations. Surveys and Tutorials in the Applied Mathematical Sciences, vol. 3. Springer, New York (2009)
3. Kolmanovskii, V., Myshkis, A.: Introduction to the Theory and Applications of Functional Differential Equations. Kluwer Academic, Dordrecht (1999)
4. Smith, H.: An Introduction to Delay Differential Equations with Applications to the Life Sciences. Texts in Applied Mathematics, vol. 57. Springer, New York (2011)
5. Wu, M., He, Y., She, J.-H.: Stability Analysis and Robust Control of Time-Delay Systems. Science Press Beijing/Springer, Beijing/Berlin (2010)
6. Ahmad, S., Rama Mohana Rao, M.: Theory of Ordinary Differential Equations. With Applications in Biology and Engineering. Affiliated East–West Press Pvt. Ltd., New Delhi (1999)
7. Anh, T.T., Hien, L.V., Phat, V.N.: Stability analysis for linear non-autonomous systems with continuously distributed multiple time-varying delays and applications. Acta Math. Vietnam. **36**(2), 129–144 (2011)
8. Barnett, S.A.: New formulation of the Liénard–Chipart stability criterion. Proc. Cambridge Philos. Soc. **70**, 269–274 (1971)
9. Burton, T.A.: Stability and Periodic Solutions of Ordinary and Functional Differential Equations. Academic, Orlando, FL (1985)
10. Burton, T.A., Zhang, B.: Boundedness, periodicity, and convergence of solutions in a retarded Liénard equation. Ann. Mat. Pure Appl. **4**(165), 351–368 (1993)
11. Caldeira-Saraiva, F.: The boundedness of solutions of a Liénard equation arising in the theory of ship rolling. IMA J. Appl. Math. **36**(2), 129–139 (1986)
12. Cantarelli, G.: On the stability of the origin of a non autonomous Liénard equation. Boll. Un. Mat. Ital. A **7**(10), 563–573 (1996)
13. Èl'sgol'ts, L.È., Norkin, S.B.: Introduction to the theory and application of differential equations with deviating arguments. Translated from the Russian by John L. Casti. Mathematics in Science and Engineering, vol. 105. Academic Press, New York (1973)
14. Gao, S.Z., Zhao, L.Q.: Global asymptotic stability of generalized Liénard equation. Chin. Sci. Bull. **40**(2), 105–109 (1995)

15. Hale, J.: Sufficient conditions for stability and instability of autonomous functional-differential equations. J. Differ. Equ. **1**, 452–482 (1965)
16. Hara, T., Yoneyama, T.: On the global center of generalized Liénard equation and its application to stability problems. Funkcial. Ekvac. **28**(2), 171–192 (1985)
17. Hara, T., Yoneyama, T.: On the global center of generalized Liénard equation and its application to stability problems. Funkcial. Ekvac. **31**(2), 221–225 (1988)
18. Heidel, J.W.: Global asymptotic stability of a generalized Liénard equation. SIAM J. Appl. Math. **19**(3), 629–636 (1970)
19. Huang, L.H., Yu, J.S.: On boundedness of solutions of generalized Liénard's system and its application. Ann. Differ. Equ. **9**(3), 311–318 (1993)
20. Jitsuro, S., Yusuke, A.: Global asymptotic stability of non-autonomous systems of Lienard type. J. Math. Anal. Appl. **289**(2), 673–690 (2004)
21. Kato, J.: On a boundedness condition for solutions of a generalized Liénard equation. J. Differ. Equ. **65**(2), 269–286 (1986)
22. Kato, J.: A simple boundedness theorem for a Liénard equation with damping. Ann. Polon. Math. **51**, 183–188 (1990)
23. Krasovskiĭ, N.N.: Stability of Motion. Applications of Lyapunov's Second Method to Differential Systems and Equations with Delay. Stanford University Press, Stanford, CA (1963)
24. LaSalle, J., Lefschetz, S.: Stability by Liapunov's Direct Method, with Applications. Mathematics in Science and Engineering, vol. 4. Academic, New York/London (1961)
25. Li, H.Q.: Necessary and sufficient conditions for complete stability of the zero solution of the Liénard equation. Acta Math. Sin. **31**(2), 209–214 (1988)
26. Liu, B., Huang, L.: Boundedness of solutions for a class of retarded Liénard equation. J. Math. Anal. Appl. **286**(2), 422–434 (2003)
27. Liu, B., Huang, L.: Boundedness of solutions for a class of Liénard equations with a deviating argument. Appl. Math. Lett. **21**(2), 109–112 (2008)
28. Liu, C.J., Xu, S.L.: Boundedness of solutions of Liénard equations. J. Qingdao Univ. Nat. Sci. Ed. **11**(3), 12–16 (1998)
29. Liu, Z.R.: Conditions for the global stability of the Liénard equation. Acta Math. Sin. **38**(5), 614–620 (1995)
30. Long, W., Zhang, H.X.: Boundedness of solutions to a retarded Liénard equation. Electron. J. Qual. Theory Differ. Equ. **24**, 9 pp (2010)
31. Luk, W.S.: Some results concerning the boundedness of solutions of Lienard equations with delay. SIAM J. Appl. Math. **30**(4), 768–774 (1976)
32. Lyapunov, A.M.: Stability of Motion. Academic Press, London (1966)
33. Malyseva, I.A.: Boundedness of solutions of a Liénard differential equation. Differetial'niye Uravneniya **15**(8), 1420–1426 (1979)
34. Muresan, M.: Boundedness of solutions for Liénard type equations. Mathematica **40**(63) (2), 243–257 (1998)
35. Sugie, J.: On the boundedness of solutions of the generalized Liénard equation without the signum condition. Nonlinear Anal. **11**(12), 1391–1397 (1987)
36. Sugie, J., Amano, Y.: Global asymptotic stability of non-autonomous systems of Liénard type. J. Math. Anal. Appl. **289**(2), 673–690 (2004)
37. Sugie, J., Chen, D.L., Matsunaga, H.: On global asymptotic stability of systems of Liénard type. J. Math. Anal. Appl. **219**(1), 140–164 (1998)
38. Tunç, C.: Some new stability and boundedness results of solutions of Liénard type equations with deviating argument. Nonlinear Anal. Hybrid Syst. **4**(1), 85–91 (2010)
39. Tunç, C.: A note on boundedness of solutions to a class of non-autonomous differential equations of second order. Appl. Anal. Discret. Math. **4**(2), 361–372 (2010)
40. Tunç, C.: New stability and boundedness results of Liénard type equations with multiple deviating arguments. Izv. Nats. Akad. Nauk Armenii Mat. **45**(4), 47–56 (2010)
41. Tunç, C.: Boundedness results for solutions of certain nonlinear differential equations of second order. J. Indones. Math. Soc. **16**(2), 115–128 (2010)

42. Tunç, C.: On the stability and boundedness of solutions of a class of Liénard equations with multiple deviating arguments. Vietnam J. Math. **39**(2), 177–190 (2011)
43. Tunç, C.: Uniformly stability and boundedness of solutions of second order nonlinear delay differential equations. Appl. Comput. Math. **10**(3), 449–462 (2011)
44. Tunç, C.: On the uniform boundedness of solutions of Liénard type equations with multiple deviating arguments. Carpathian J. Math. **27**(2), 269–275 (2011)
45. Tunç, C.: Stability and uniform boundedness results for non-autonomous Liénard-type equations with a variable deviating argument. Acta Math. Vietnam **37**(3), 311–326 (2012)
46. Tunç, C., Tunç, E.: On the asymptotic behavior of solutions of certain second-order differential equations. J. Franklin Inst. **344**(5), 391–398 (2007)
47. Yang, Q.G.: Boundedness and global asymptotic behavior of solutions to the Liénard equation. J. Syst. Sci. Math. Sci. **19**(2), 211–216 (1999)
48. Ye, G.R., Ding, S., Wu, X.L.: Uniform boundedness of solutions for a class of Liénard equations. Electron. J. Differ. Equ. **97**, 5 pp (2009)
49. Yu, Y., Xiao, B.: Boundedness of solutions for a class of Liénard equation with multiple deviating arguments. Vietnam J. Math. **37**(1), 35–41 (2009)
50. Yoshizawa, T.: Stability Theory by Lyapunov's Second Method. Publications of the Mathematical Society of Japan, vol. 9. The Mathematical Society of Japan, Tokyo (1966)
51. Zhang, B.: On the retarded Liénard equation. Proc. Amer. Math. Soc. **115**(3), 779–785 (1992)
52. Zhang, B.: Boundedness and stability of solutions of the retarded Liénard equation with negative damping. Nonlinear Anal. **20**(3), 303–313 (1993)
53. Zhang, X.S., Yan, W.P.: Boundedness and asymptotic stability for a delay Liénard equation. Math. Pract. Theory **30**(4), 453–458 (2000)
54. Zhou, X., Jiang, W.: Stability and boundedness of retarded Liénard–type equation. Chin. Quart. J. Math. **18**(1), 7–12 (2003)
55. Zhou, J., Liu, Z.R.: The global asymptotic behavior of solutions for a nonautonomous generalized Liénard system. J. Math. Res. Exposition **21**(3), 410–414 (2001)
56. Zhou, J., Xiang, L.: On the stability and boundedness of solutions for the retarded Liénard-type equation. Ann. Differ. Equ. **15**(4), 460–465 (1999)
57. Wei, J., Huang, Q.: Global existence of periodic solutions of Liénard equations with finite delay. Dyn. Contin. Discrete Impuls. Syst. **6**(4), 603–614 (1999)
58. Wiandt, T.: On the boundedness of solutions of the vector Liénard equation. Dyn. Syst. Appl. **7**(1), 141–143 (1998)
59. Omeike, M.O., Oyetunde, O.O., Olutimo, A.L.: New result on the ultimate boundedness of solutions of certain third-order vector differential equations. Acta Univ. Palack. Olomuc. Fac. Rerum Natur. Math. **53**(1), 107–115 (2014)
60. Bellman, R.: Introduction to Matrix Analysis. Reprint of the second (1970) edition. With a foreword by Gene Golub. Classics in Applied Mathematics, vol. 19. Society for Industrial and Applied Mathematics (SIAM), Philadelphia (1997)

Chapter 6
Delay Effects on the Dynamics of the Lengyel–Epstein Reaction-Diffusion Model

Hüseyin Merdan and Şeyma Kayan

Abstract We investigate bifurcations of the Lengyel–Epstein reaction-diffusion model involving time delay under the Neumann boundary conditions. We first give stability and Hopf bifurcation analysis of the ordinary differential equation (ODE) models, including delay associated with this model. Later, we extend this analysis to the partial differential equation (PDE) model. We determine conditions on parameters of both models to have Hopf bifurcations. Bifurcation analysis for both models show that Hopf bifurcations occur by regarding the delay parameter as a bifurcation parameter. Using the normal form theory and the center manifold reduction for partial functional differential equations, we also determine the direction of the Hopf bifurcations and the stability of bifurcating periodic solutions for the PDE model. Finally, we perform some numerical simulations to support analytical results obtained for the ODE models.

6.1 Introduction

Representing the events by using mathematical terms to produce a better understanding of the world around us as well as to find solutions to technical problems is called mathematical modeling. The model is initially kept as simple as possible. After adding new terms and variables to the model at later stages, it becomes more realistic. The obtained models are often used to understand the dynamics of the structure of the system that changes with respect to time. However, mostly this is

---

H. Merdan (✉)
Department of Mathematics, TOBB University of Economics and Technology, Söğütözü Cad. No 43. 06560, Ankara, Turkey
e-mail: merdan@etu.edu.tr

Ş. Kayan
Çankaya University, Eskişehir Yolu 29.km, Yukarıyurtçu Mahallesi Mimar Sinan Caddesi No 4. 06790, Etimesgut, Ankara, Turkey
e-mail: sbilazeroglu@cankaya.edu.tr

A.C.J. Luo, H. Merdan (eds.), *Mathematical Modeling and Applications in Nonlinear Dynamics*, Nonlinear Systems and Complexity 14,
DOI 10.1007/978-3-319-26630-5_6

not enough to understand the whole story. For example, some problems in real life usually depend on time, but they may also depend on various independent variables such as location or age. In an assemblage of particles (for example, cells, bacteria, chemicals, animals), each particle usually moves around in a random way. The particles spread out as a result of this irregular motion by individual particles. When this microscopic irregular movement results in some macroscopic or gross regular motion of the group, we can think of it as a diffusion process. Therefore, a model that contains a diffusion process can be improved by adding a spatial variable to the model. In 1952, Turing [34] proposed a reaction-diffusion system that is an activator-inhibitor mechanism as a model for the chemical basis of morphogenesis. Such systems have been widely studied since about 1970 (see, for example, [10–12, 17, 18, 23–25, 29–33, 38, 39] and references therein).

On the other hand, in many applications it is assumed that the system under consideration is governed by a principle of causality; that is, the future state of the system is independent of the past and is determined solely by the present. One should keep in mind that this is only a first approximation to the true situation. A more realistic model must include some of the past history of the system [4, 5, 21]. To reflect the dynamical behavior of models that depend on the past history of the system, it is often necessary to take the time-delay effect into account in forming a biologically meaningful mathematical model. Exploring the dynamical behaviors of models involving time delays has attracted very much interest in chemistry, mathematical biology, medicine, ecology, population dynamics, neural networks, economics, and other fields (see, for example, [1, 2, 4, 6, 9, 19, 20, 26, 27, 36, 37, 40] and references therein).

It is well known that studies of dynamical systems include not only a discussion of stability, attractivity, and persistence, but also many dynamical behaviors, such as periodic phenomenon, bifurcation, and chaos [4, 5, 13, 14, 21, 22, 28, 30]. In delay differential equations (DDEs), periodic solutions can arise through the (local) Hopf bifurcation. Several methods for analyzing the nature of Hopf bifurcations have been described in the literature. Integral averaging has been used by Chow and Mallet–Paret, the Fredholm alternative has been used by Iooss and Joseph, the implicit function theorem has been used by Hale and Lunel, multiscale expansion has been used by Nayfeh et al., and center manifold projection has been used by Hassard et al. and Stépán and Kalmár–Nagy [3, 4, 16, 22, and see the references therein]. Center manifold theory is one of the rigorous mathematical tools to study bifurcations of DDEs [15].

In this chapter we consider the following reaction-diffusion model under the Neumann boundary conditions:

$$\begin{cases} \frac{\partial u(x,t)}{\partial t} = d_1 \frac{\partial^2 u(x,t)}{\partial x^2} + a - u(x,t) - \frac{4u(x,t)v(x,t)}{1+u^2(x,t)}, \text{ for } x \in \Omega,\ t > 0, \\ \frac{\partial v(x,t)}{\partial t} = d_2 \frac{\partial^2 v(x,t)}{\partial x^2} + \sigma b \left(u(x,t) - \frac{u(x,t)v(x,t)}{1+u^2(x,t)} \right), \text{ for } x \in \Omega,\ t > 0, \\ \frac{\partial u}{\partial \overrightarrow{n}} = \frac{\partial v}{\partial \overrightarrow{n}} = 0, \text{ for } x \in \partial\Omega,\ t > 0, \\ u(x,0) = u_0(x),\ v(x,0) = v_0(x), \text{ for } x \in \Omega, \end{cases} \tag{6.1}$$

where Ω is a bounded domain in $\mathbb{R}^n$ with sufficient and smooth boundary $\partial\Omega$, $\overrightarrow{n}$ is the unit outer normal to $\partial\Omega$, and $u_0,\ v_0 \in \mathbf{C}^2(\Omega) \cap \mathbf{C}(\overline{\Omega})$. System (6.1) is known as the Lengyel–Epstein reaction-diffusion model based on the chlorite–iodide–malonic acid chemical (CIMA) reaction (see De Kepper et al. [10], Lengyel and Epstein [23, 24], and Epstein and Pojman [12] and the references therein). In the model, $u(x,t)$ and $v(x,t)$ denote the chemical concentration of the activator iodide and the inhibitor chlorite, respectively, at time $t > 0$ and a spatial point $x \in \Omega$. The parameters a and b are positive numbers related to concentration, and $\sigma > 0$ is a rescaling parameter. Here, the positive constants d_1 and d_2 are diffusion coefficients of the activator and the inhibitor, respectively.

In the past two decades, numerous mathematical investigations have been conducted for the Lengyel–Epstein system (see, for example, [11, 17, 18, 23, 24, 29, 31, 32, 38, 39]). For example, Yi et al. [38] and Rovinsky and Menzinger [32] derived the conditions on the parameters at which the spatial homogeneous equilibrium solution and the spatial homogeneous periodic solution became Turing unstable, and they performed a Hopf bifurcation analysis for both ODE and PDE models. Du and Wang [11] investigated the existence of multiple spatially nonhomogeneous periodic solutions when all parameters of the system are spatially homogeneous in the one-dimensional case. Ni and Tang [31] obtained a priori bound for solutions, the nonexistence of nonconstant steady states for a small effective diffusion rate, and the existence of nonconstant steady states for a large effective diffusion rate which partially verify Turing stability for the CIMA reaction. Jang et al. [17] studied the limiting behavior of the steady-state solutions by using a shadow system approach and the global bifurcation of the nonconstant equilibriums for the one-dimensional case.

The ODE model associated with system (6.1) is

$$\begin{cases} \frac{du(t)}{dt} = a - u(t) - \frac{4u(t)v(t)}{1+u^2(t)} \\ \frac{dv(t)}{dt} = \sigma b \left(u(t) - \frac{u(t)v(t)}{1+u^2(t)} \right). \end{cases} \tag{6.2}$$

In this chapter we study the delay effect on both ODE and PDE models given by systems (6.2) and (6.1), respectively. We include a delay term in both models to discuss their dynamical behavior from a mathematical point of view, and we give a detailed Hopf bifurcation analysis for the models incorporating delay. We determine conditions on parameters of the system to have Hopf bifurcations. Furthermore, we determine some properties of the Hopf bifurcation for the PDE model by applying the normal form theory and the center manifold reduction for partial functional differential equations [15, 35].

The chapter is organized as follows. Section 6.1 involves the Hopf bifurcation analysis of system (6.2) involving a discrete delay. In Section 6.2 we give a detailed Hopf bifurcation analysis of system (6.1) with a discrete delay. We investigate the existence of the periodic solution and determine the direction of this solution by using Poincaré's normal form and the center manifold reduction for partial functional differential equations. Section 6.3 presents some numerical simulations for the ODE model (6.2) with delay to support the theoretical results. Finally, the chapter ends with some concluding remarks.

6.2 The ODE Model with Delay

We want to make system (6.2) more realistic by adding a discrete delay term into it, which is obtained by reducing the original CIMA reaction model. Recently, Çelik and Merdan [9] studied the bifurcation analysis of the following model:

$$\begin{cases} \frac{du(t)}{dt} = a - u(t) - \frac{4uv(t-\tau)}{1+u^2(t)} \\ \frac{dv(t)}{dt} = \sigma b\left(u(t) - \frac{u(t)v(t-\tau)}{1+u^2(t)}\right). \end{cases} \tag{6.3}$$

They added a discrete delay term to the chemical concentration of the inhibitor $v(t)$, gave a detailed Hopf bifurcation analysis by choosing the delay parameter τ as a bifurcation parameter, and locally studied the asymptotic behavior of the equilibrium point of (6.3). Their work illustrates that Hopf bifurcation occurs when the delay parameter passes through a sequence of critical values, namely, $(\tau_n)_{n=1}^{\infty}$. They also analyzed the direction of the Hopf bifurcation and the stability of bifurcating periodic solution through the normal form theory and the center manifold reduction for functional differential equations. Theoretical results are supported by some numerical simulations.

In this section we study the dynamics of the following models that include delay:

$$\begin{cases} \frac{du}{dt} = a - u(t) - \frac{4u(t-\tau)v(t)}{1+u^2(t)} \\ \frac{dv}{dt} = \sigma b\left(u(t) - \frac{u(t-\tau)v(t)}{1+u^2(t)}\right) \end{cases} \tag{6.4}$$

and

$$\begin{cases} \frac{du}{dt} = a - u(t-\tau) - \frac{4u(t-\tau)v(t)}{1+u^2(t-\tau)} \\ \frac{dv}{dt} = \sigma b\left(u(t-\tau) - \frac{u(t-\tau)v(t)}{1+u^2(t-\tau)}\right). \end{cases} \tag{6.5}$$

In these models, $u(t)$ and $v(t)$ denote the chemical concentration of the activator iodide and the inhibitor chlorite, respectively, at time $t > 0$. The parameters a and b are positive parameters related to the concentration, $\sigma > 0$ is a rescaling parameter, and τ is the delay parameter.

The aim of this section is to investigate the impact of delay on the chemical concentration of the activator $u(t)$ on the dynamics of the system from two different aspects. First, in Eq. (6.4) we add a discrete delay term to $u(t)$, which interacts with $v(t)$, in order to analyze the effect of delay on the interactions of the two variables. Second, in Eq. (6.5) we add a discrete delay term not only to $u(t)$ [which interacts with $v(t)$] but also to all variables $u(t)$ in the model to analyze how the delay in the chemical concentration of the activator $u(t)$ effects the dynamics of the system from a mathematical point of view. Now, we first give stability analysis of a positive equilibrium point, and then the existence of Hopf bifurcation will be examined for Eqs. (6.4) and (6.5) separately. Notice that both Eqs. (6.4) and (6.5) have a unique equilibrium point $(u^*, v^*) = (\alpha, 1 + \alpha^2)$, where $\alpha = \frac{a}{5}$.

6.2.1 Analysis of System (6.4)

Let us begin with Eq. (6.4). By shifting the positive equilibrium point (u^*, v^*) of Eq. (6.4) to the origin, and then linearizing the new system around (0, 0), we get the following system:

$$\begin{cases} \frac{du}{dt} = \left(\frac{7\alpha^2-1}{1+\alpha^2}\right) u(t) - \frac{4\alpha}{1+\alpha^2} v(t) - 4u(t-\tau) + \text{h.o.t.} \\ \frac{dv}{dt} = \sigma b \left(\frac{1+3\alpha^2}{1+\alpha^2}\right) u(t) - \frac{\sigma b \alpha}{1+\alpha^2} v(t) - \sigma b u(t-\tau) + \text{h.o.t.}, \end{cases} \tag{6.6}$$

where h.o.t. represents the higher-order terms.

The characteristic equation associated with the linearization of Eq. (6.6) is

$$\lambda^2 + (r+m)\lambda + 4e^{-\lambda\tau}\lambda + 5m = 0, \tag{6.7}$$

where

$$r = \frac{1-7\alpha^2}{1+\alpha^2} \text{ and } m = \frac{\sigma b \alpha}{1+\alpha^2}.$$

Theorem 1. *If* $|r+m| < 4$, *then there is a positive integer* s *such that the equilibrium point* $(0,0)$ *is stable when* $\tau \in [0, \tau_{1,0}) \cup (\tau_{2,0}, \tau_{1,1}) \cup \cdots \cup (\tau_{2,s-1}, \tau_{1,s})$ *and unstable when* $\tau \in (\tau_{1,0}, \tau_{2,0}) \cup (\tau_{1,1}, \tau_{2,1}) \cdots \cup (\tau_{1,s-1}, \tau_{2,s-1}) \cup (\tau_{1,s}, \infty)$. *Moreover, system (6.6) undergoes a Hopf bifurcation at the equilibrium point* $(0,0)$ *when* $\tau_{n,k} = \frac{1}{\omega_n}\left\{\arccos\left(-\frac{r+m}{4}\right)\right\} + \frac{2k\pi}{\omega_n}$, $n = 1, 2$, *and* $k = 0, 1, 2, \cdots$.

Proof. The proof will be given in two cases.

Case 1. Let us first assume that $\tau = 0$. In this case, Eq. (6.7) turns into

$$\lambda^2 + (r+m+4)\lambda + 5m = 0. \tag{6.8}$$

The roots of this equation are $\lambda_{1,2} = \frac{-(r+m+4) \pm \sqrt{(r+m+4)^2 - 20m}}{2}$. Since $m > 0$, real parts of $\lambda_{1,2}$ are negative if and only if $(r+m+4) > 0$. Eventually, if $(r+m) > -4$, then (u^*, v^*) is stable when $\tau = 0$.

Case 2. Now, let us take $\tau > 0$. Assume $\lambda = i\omega$, $\omega > 0$, is a solution of Eq. (6.7). Substituting $\lambda = i\omega$ into the characteristic equation (6.7) and separating the real and imaginary parts yield

$$\begin{cases} -\omega^2 + 4\omega \sin \omega\tau + 5m = 0, \\ (r+m)\omega + 4\omega \cos \omega\tau = 0. \end{cases} \tag{6.9}$$

From (6.9) we get

$$\omega^4 + \left((r+m)^2 - 10m - 16\right)\omega^2 + 25m^2 = 0. \tag{6.10}$$

It follows from (6.10) that

$$\omega^2 = \frac{-\left((r+m)^2-10m-16\right) \pm \sqrt{\left((r+m)^2-10m-16\right)^2-100m^2}}{2}. \quad (6.11)$$

Let $A = \left((r+m)^2 - 10m - 16\right)$ and $B = 25m^2$. If $A < 0$ and $A^2 - 4B > 0$, then we get $\omega > 0$. Assume that $A < 0$; then $A^2 - 4B > 0$ if and only if $(r+m)^2 < 16$. This condition also guarantees our assumption, $A < 0$. Hence, if $(r+m)^2 < 16$, then we have two different $\omega > 0$ values from (6.11), namely,

$$\omega_1 = \sqrt{\frac{-A + \sqrt{A^2 - 4B}}{2}}, \quad (6.12)$$

$$\omega_2 = \sqrt{\frac{-A - \sqrt{A^2 - 4B}}{2}}. \quad (6.13)$$

By substituting ω_1 and ω_2 into Eq. (6.9), we can calculate τ as follows:

$$\tau_{n,k} = \frac{1}{\omega_n}\left\{\arccos\left(-\frac{r+m}{4}\right)\right\} + \frac{2k\pi}{\omega_n}, n=1,2 \text{ and } k = 0, 1, 2, \cdots. \quad (6.14)$$

Next, we check the transversality condition. Let

$$F(\omega) = \omega^4 + \left((r+m)^2 - 10m - 16\right)\omega^2 + 25m^2.$$

Then one has

$$\frac{dF}{d\omega} = 2\omega\left(2\omega^2 + (r+m)^2 - 10m - 16\right),$$

so that

$$\left.\frac{dF}{d\omega}\right|_{\omega_1} = 2\omega_1\sqrt{A^2 - 4B} > 0 \text{ and } \left.\frac{dF}{d\omega}\right|_{\omega_2} = -2\omega_2\sqrt{A^2 - 4B} < 0. \quad (6.15)$$

Therefore, from Cooke and Driessche's article [7], the transversality condition holds for both $\omega = \omega_n$, $n = 1, 2$. Furthermore, again by the same article, due to the fact that $\left.\frac{dF}{d\omega}\right|_{\omega_1} > 0$ and $\left.\frac{dF}{d\omega}\right|_{\omega_2} < 0$ from (6.15), there is a positive integer s such that the arrangement of τ is $0 < \tau_{1,0} < \tau_{2,0} < \tau_{1,1} < \tau_{2,1} < \cdots < \tau_{2,s-1} < \tau_{1,s}$. As a result, the sign of the real parts of the eigenvalues λ_n is negative (for $n = 1, 2$) when $\tau \in [0, \tau_{1,0}) \cup (\tau_{2,0}, \tau_{1,1}) \cup \cdots \cup (\tau_{2,s-1}, \tau_{1,s})$, and positive when $\tau \in (\tau_{1,0}, \tau_{2,0}) \cup (\tau_{1,1}, \tau_{2,1}) \cdots \cup (\tau_{1,s-1}, \tau_{2,s-1}) \cup (\tau_{1,s}, \infty)$. In addition, by the Hopf bifurcation theorem (see [15]), system (6.6) undergoes a Hopf bifurcation at the equilibrium point $(0,0)$ for eigenvalues $\lambda_n = i\omega_n$, $n = 1, 2$, and possesses a family of real-valued periodic solutions at $\tau = \tau_n$, where $\tau_{n,k} = \frac{1}{\omega_n}\left\{\arccos\left(-\frac{r+m}{4}\right)\right\} + \frac{2k\pi}{\omega_n}$, $n = 1, 2$ and $k = 0, 1, 2, \cdots$. This completes the proof.

6.2.2 *Analysis of System (6.5)*

Once again, let us shift the positive equilibrium point (u^*, v^*) of Eq. (6.5) to the origin to obtain the linearized system that follows:

$$\begin{cases} \frac{du}{dt} = -\frac{4\alpha}{1+\alpha^2}v(t) + \frac{3\alpha^2-5}{1+\alpha^2}u(t-\tau) + \text{h.o.t.}, \\ \frac{dv}{dt} = -\frac{\sigma b\alpha}{1+\alpha^2}v(t) + \frac{2\sigma b\alpha^2}{1+\alpha^2}u(t-\tau) + \text{h.o.t.}, \end{cases} \tag{6.16}$$

where h.o.t. represents the higher-order terms.

The characteristic equation associated with Eq. (6.16) is

$$\lambda^2 + m\lambda - p\lambda e^{-\lambda\tau} + 5me^{-\lambda\tau} = 0, \tag{6.17}$$

where

$$p = \frac{3\alpha^2 - 5}{1+\alpha^2} \text{ and } m = \frac{\sigma b\alpha}{1+\alpha^2}.$$

Theorem 2. *If* $(m-p) > 0$*, then the equilibrium point* $(0,0)$ *is stable for* $\tau < \tau_{0,0}$ *and unstable for* $\tau > \tau_{0,0}$*. Moreover, system (6.16) undergoes a Hopf bifurcation at the equilibrium point* $(0,0)$ *when* $\tau_{0,0} = \frac{1}{\omega_0}\left\{\arccos\left(\frac{m\omega^2(p+5)}{25m^2+p^2\omega^2}\right)\right\}$.

Proof. As with the proof of Theorem 1, the proof will be given in two cases, as follows:

Case 1. Assume that $\tau = 0$ in Eq. (6.17). Then it turns into

$$\lambda^2 + (m-p)\lambda + 5m = 0.$$

The roots of this equation are determined by $\lambda_{1,2} = \frac{-(m-p)\mp\sqrt{(m-p)^2-20m}}{2}$. Because $m > 0$, the real parts of $\lambda_{1,2}$ are negative if and only if $(m-p) > 0$. Ultimately, if $(m-p) > 0$, then (u^*, v^*) is stable when $\tau = 0$.

Case 2. Suppose now that $\tau > 0$. Let us take $\lambda = i\omega$, $\omega > 0$, as a solution of (6.17). Substituting $\lambda = i\omega$ into the characteristic equation (6.17) and separating the real and imaginary parts, we obtain

$$\begin{cases} -\omega^2 + 5m\cos\omega\tau - p\omega\sin\omega\tau = 0, \\ m\omega - 5m\sin\omega\tau - p\omega\cos\omega\tau = 0. \end{cases} \tag{6.18}$$

From (6.18) we get

$$\omega^4 + \left(m^2 - p^2\right)\omega^2 - 25m^2 = 0. \tag{6.19}$$

It follows from (6.19) that

$$\omega^2 = \frac{-\left(m^2 - p^2\right) \pm \sqrt{\left(m^2 - p^2\right)^2 + 100m^2}}{2}. \tag{6.20}$$

Equation (6.20) gives only one positive ω, namely,

$$\omega_0 = \sqrt{\frac{-\left(m^2 - p^2\right) + \sqrt{\left(m^2 - p^2\right)^2 + 100m^2}}{2}}. \tag{6.21}$$

By putting ω_0 into Eq. (6.18), we can calculate τ as follows:

$$\tau_{0,k} = \frac{1}{\omega_0}\left\{\arccos\left(\frac{m\omega^2(p+5)}{25m^2 + p^2\omega^2}\right)\right\} + \frac{2k\pi}{\omega_0}, k = 0, 1, 2, \cdots. \tag{6.22}$$

Utilizing results in [8], one may conclude that if there is a simple pure imaginary root of Eq. (6.17), it means that a stable equilibrium point never stays stable forever. In other words, if it is stable for $\tau = 0$, then it is unstable after the smallest value of τ for which an imaginary root exists. Because of this, the equilibrium point of system (6.16) is stable for $\tau < \tau_{0,0}$ and unstable for $\tau > \tau_{0,0}$ under the condition $(m - p) > 0$ as long as the transversality condition holds.

Now, let us check the transversality condition. Let

$$F(\omega) = \omega^4 + \left(m^2 - p^2\right)\omega^2 - 25m^2.$$

Then we obtain its derivative as follows:

$$\frac{dF}{d\omega} = 2\omega\left(2\omega^2 + \left(m^2 - p^2\right)\right)$$

so that

$$\left.\frac{dF}{d\omega}\right|_{\omega_0} = 2\omega_0\sqrt{\left(m^2 - p^2\right)^2 + 100m^2} > 0. \tag{6.23}$$

Equation (6.23) shows $\left.\frac{dF}{d\omega}\right|_{\omega_0} > 0$. Because of this, from Cooke and Driessche's article [7] the transversality condition holds for $\omega = \omega_0$. Consequently, by the Hopf bifurcation theorem in [15], system (6.16) undergoes a Hopf bifurcation at $(0,0)$ when $\omega = \omega_0$, and possesses a family of real-valued periodic solutions at $\tau = \tau_{0,0}$, where $\tau_{0,k} = \frac{1}{\omega_n}\left\{\arccos\left(-\frac{r+m}{4}\right)\right\} + \frac{2k\pi}{\omega_n}$, $k = 0, 1, 2, \cdots$.

Note: Some properties of the Hopf bifurcations such as direction of bifurcation, stability of periodic solutions, and so forth can be calculated by following steps in [9].

6.3 The PDE Model with Delay

In this section, we investigate bifurcations of the Lengyel–Epstein reaction-diffusion model involving time delay under the Neumann boundary conditions. We first determine conditions on parameters to have a Hopf bifurcation. Then we determine some properties of the Hopf bifurcation. Finally, we complete this section with a bifurcation analysis of a spatially homogeneous Lengyel–Epstein system with delay.

6.3.1 Occurrence of Hopf Bifurcation

In this section, we study the Hopf bifurcation of system (6.1) with delay on a spatial domain. For simplicity, we chose the spatial domain as $\Omega = (0, \pi) \subset \mathbb{R}$, but all calculations can be extended to higher dimensions. In this case, the one-dimensional delayed Lengyel–Epstein reaction-diffusion model can be written as follows:

$$\begin{cases} \frac{\partial u(x,t)}{\partial t} = d_1 \frac{\partial^2 u(x,t)}{\partial x^2} + a - u(x,t) - 4\frac{u(x,t)v(x,t-\tau)}{1+u^2(x,t)}, \text{ for } x \in (0,\pi), \ t > 0, \\ \frac{\partial v(x,t)}{\partial t} = d_2 \frac{\partial^2 v(x,t)}{\partial x^2} + \sigma b\left(u(x,t) - \frac{u(x,t)v(x,t-\tau)}{1+u^2(x,t)}\right), \text{ for } x \in (0,\pi), \ t > 0, \\ \left.\frac{\partial u}{\partial x}\right|_{x=0,\pi} = \left.\frac{\partial v}{\partial x}\right|_{x=0,\pi} = 0, \text{ for } t > 0, \\ u(x,0) = u_0(x), \ v(x,0) = v_0(x), \text{ for } x \in (0,\pi). \end{cases} \tag{6.24}$$

System (6.24) has a unique equilibrium point $(u^*, v^*) = (\alpha, 1+\alpha^2)$, where $\alpha = \frac{a}{5}$. First, we get the following system by shifting the equilibrium point (u^*, v^*) to the origin:

$$\begin{cases} u_t(x,t) = d_1 u_{xx}(x,t) + \left(\frac{3\alpha^2-5}{1+\alpha^2}\right)u(x,t) + \left(\frac{-4\alpha}{1+\alpha^2}\right)v(x,t-\tau) + f(u,v,\tau), \\ v_t(x,t) = d_2 v_{xx}(x,t) + \left(\frac{2\sigma b\alpha^2}{1+\alpha^2}\right)u(x,t) + \left(\frac{-\sigma b\alpha}{1+\alpha^2}\right)v(x,t-\tau) + g(u,v,\tau), \\ \left.\frac{\partial u}{\partial x}\right|_{x=0,\pi} = \left.\frac{\partial v}{\partial x}\right|_{x=0,\pi} = 0, \\ u(x,0) = u_0(x) - u^*, \ v(x,0) = v_0(x) - v^*, \text{ for } x \in (0,\pi), \end{cases} \tag{6.25}$$

where

$$f(u, v, \tau) := \frac{4\alpha(3-\alpha^2)}{(1+\alpha^2)^2}u^2 + \frac{4(\alpha^2-1)}{(1+\alpha^2)^2}uv + \text{h.o.t.}, \tag{6.26}$$

$$g(u, v, \tau) := \frac{\sigma b}{4} f(u, v, \tau), \tag{6.27}$$

in which the term h.o.t. denotes the higher-order terms. Second, we find the characteristic equation of system (6.25). Let the linear operator Δ be defined by Δ :=diag$\left\{\frac{\partial^2}{\partial x^2}, \frac{\partial^2}{\partial x^2}\right\}$, and $U(t) := (u(t), v(t))^T = (u(\cdot, t), v(\cdot, t))^T$. With this notation, system (6.25) can be rewritten as an abstract ODE in the Banach space $\mathbf{C} = \mathbf{C}([-\tau, 0], \mathbf{X})$, where

$$\mathbf{X} = \left\{(u, v) : u, v \in \mathbf{W}^{2,2}(0, \pi);\ \frac{\mathrm{d}u}{\mathrm{d}x} = \frac{\mathrm{d}v}{\mathrm{d}x} = 0,\ x = 0, \pi\right\},$$

as follows:

$$\frac{\mathrm{d}}{\mathrm{d}t}U(t) = d\Delta U(t) + L(U_t) + \text{h.o.t.}, \tag{6.28}$$

where $d = (d_1, d_2)^T$, $U_t(\theta) = U(t+\theta)$, $-\tau \le \theta \le 0$, $L : \mathbf{C} \to \mathbf{X}$. Here, L is defined by

$$L(\varphi) = \begin{pmatrix} \frac{3\alpha^2-5}{1+\alpha^2}\varphi_1(0) & \frac{-4\alpha}{1+\alpha^2}\varphi_2(-\tau) \\ \frac{2\sigma b\alpha^2}{1+\alpha^2}\varphi_1(0) & \frac{-\sigma b\alpha}{1+\alpha^2}\varphi_2(-\tau) \end{pmatrix}_{2x2}$$

for $\varphi(\theta) = U_t(\theta)$, $\varphi = (\varphi_1,\varphi_2)^T \in \mathbf{C}$. The characteristic equation of (6.28) is equivalent to

$$\lambda y - d\Delta y - L(e^{\lambda} y) = 0, \tag{6.29}$$

where $y \in dom(\Delta)$ and $y \neq 0$, $dom(\Delta) \subset \mathbf{X}$. From properties of the Laplacian operator defined on a bounded domain, the operator Δ has eigenvalues $-n^2$, $n \in \mathbb{N}_0 = \{0, 1, 2, \ldots\}$. The corresponding eigenfunctions for each n are

$$\beta_n^1 = \begin{pmatrix} \gamma_n \\ 0 \end{pmatrix}, \beta_n^2 = \begin{pmatrix} 0 \\ \gamma_n \end{pmatrix}, \ \gamma_n = \cos(nx).$$

It is easy to see that $\left(\beta_n^1, \beta_n^2\right)_{n=0}^{\infty}$ forms a basis for the phase space $\mathbf{X}$. Therefore, any arbitrary y in $\mathbf{X}$ can be written as a Fourier series in the following form:

$$y = \sum_{n=0}^{\infty} Y_n^T \begin{pmatrix} \beta_n^1 \\ \beta_n^2 \end{pmatrix}, \; Y_n^T = \begin{pmatrix} < y, \beta_n^1 > \\ < y, \beta_n^2 > \end{pmatrix}. \tag{6.30}$$

One can show that

$$L(\varphi^T \begin{pmatrix} \beta_n^1 \\ \beta_n^2 \end{pmatrix}) = L(\varphi)^T \begin{pmatrix} \beta_n^1 \\ \beta_n^2 \end{pmatrix}, \; n \in \mathbb{N}_0. \tag{6.31}$$

From (6.30) and (6.31), (6.29) is equivalent to

$$\sum_{n=0}^{\infty} Y_n^T \left[(\lambda I_2 + dn^2 I_2) - \begin{pmatrix} \frac{3\alpha^2-5}{1+\alpha^2} & \frac{-4\alpha}{1+\alpha^2} e^{-\lambda\tau} \\ \frac{2\sigma b\alpha^2}{1+\alpha^2} & \frac{-\sigma b\alpha}{1+\alpha^2} e^{-\lambda\tau} \end{pmatrix} \right] \begin{pmatrix} \beta_n^1 \\ \beta_n^2 \end{pmatrix} = 0, \tag{6.32}$$

where I_2 is the 2×2 identity matrix here. Notice that the sum in (6.32) is zero if and only if the determinant of the matrix in brackets is zero; that is, $\det(\lambda I_2 - J) = 0$, where

$$J = \begin{pmatrix} -d_1 n^2 + \frac{3\alpha^2-5}{1+\alpha^2} & -\frac{4\alpha}{1+\alpha^2} e^{-\lambda\tau} \\ \frac{2\sigma b\alpha^2}{1+\alpha^2} & -d_2 n^2 - \frac{\sigma b\alpha}{1+\alpha^2} e^{-\lambda\tau} \end{pmatrix}.$$

Hence, we conclude that the characteristic equation of system (6.25) is

$$\lambda^2 + A\lambda + Be^{-\lambda\tau} + C\lambda e^{-\lambda\tau} + D = 0, \tag{6.33}$$

where

$$A = (d_1 + d_2)\, n^2 - m, \; B = 5k + kd_1 n^2, \; C = k, \; D = d_1 d_2 n^4 - md_2 n^2, \tag{6.34}$$

in which

$$m = \frac{3\alpha^2 - 5}{1 + \alpha^2}, \; k = \frac{\sigma b\alpha}{1 + \alpha^2}, n \in \mathbb{N}_0 = \{0, 1, 2, \ldots\}.$$

We also conclude that system (6.25) is equivalent to the following system of DDEs:

$$\frac{du}{dt} = \left(-d_1 n^2 + \frac{3\alpha^2 - 5}{1 + \alpha^2} \right) u(t) + \left(\frac{-4\alpha}{1 + \alpha^2} \right) v(t - \tau) + f(u, v, \tau),$$
$$\frac{dv}{dt} = \left(\frac{2\sigma b\alpha^2}{1 + \alpha^2} \right) u(t) - d_2 n^2 v(t) + \left(\frac{-\sigma b\alpha}{1 + \alpha^2} \right) v(t - \tau) + g(u, v, \tau),$$

where f and g are defined by (6.26) and (6.27), respectively. Now, we can apply the general Hopf bifurcation theorem (see [15]) to this system. We will state the main theorem of this work after the next two lemmas.

Lemma 1. *The characteristic equation (6.33) has a pair of pure imaginary roots* $\lambda = \pm i\omega,\ \omega > 0$ *if any of these two conditions hold:*

1. $(X_n^2 - 4Y_n = 0)$ *and* $(X_n < 0)$,
2. $(X_n = 0$ *and* $Y_n < 0)$ *or*
3. $(X_n > 0$ *and* $Y_n < 0)$ *or*
4. $(X_n < 0$ *and* $Y_n < 0)$ *or*
5. $(X_n < 0$ *and* $Y_n = 0)$ *or*
6. $(X_n^2 - 4Y_n > 0)$ *and* $(X_n < 0$ *and* $Y_n > 0)$,

where $X_n = (A^2 - C^2 - 2D)$ *and* $Y_n = \left(D^2 - B^2\right)$.

Proof. Assume that $\lambda = i\omega$, $\omega \in \mathbb{R}$ and $\omega > 0$, is a solution of (6.33). First, substituting it into the characteristic equation (6.33) and then separating its real and imaginary parts by utilizing Euler's formula give us the following two equations in ω:

$$\omega^2 - D = B\cos(\omega\tau) + C\omega\sin(\omega\tau),$$
$$A\omega = B\sin(\omega\tau) - C\omega\cos(\omega\tau).$$

Second, by squaring each side of these equations and then adding them, one can obtain the following equation:

$$\omega^4 + (A^2 - C^2 - 2D)\omega^2 + D^2 - B^2 = 0.$$

Its roots are given by

$$\omega^2 = \frac{-(A^2 - C^2 - 2D) \pm \sqrt{(A^2 - C^2 - 2D)^2 - 4\left(D^2 - B^2\right)}}{2}. \tag{6.35}$$

Since $X_n = (A^2 - C^2 - 2D)$ and $Y_n = \left(D^2 - B^2\right)$, we can now write (6.35) as follows:

$$\omega^2 = \frac{-X_n \pm \sqrt{X_n^2 - 4Y_n}}{2}. \tag{6.36}$$

Notice that for each $n \in \mathbb{N}_0 = \{0, 1, 2, \ldots\}$, (6.36) gives a different ω since A, B, and D depend on n [see (6.34)]. Therefore, for each $n \in \mathbb{N}_0$, let us denote it by ω_n; that is, $\omega_n^2 := \omega^2$. Our goal is to get a strictly positive real ω_n. Analyzing the quantity in the radical in (6.36) yields the following results:

1. If $X_n^2 - 4Y_n < 0$, then $\omega_n^2 \in \mathbb{C}$ so that there is no real root.
2. If $X_n^2 - 4Y_n = 0$, then $\omega_{n1,2} = \pm\sqrt{\frac{(-X_n)}{2}}$. Thus

   a. $X_n > 0 \Longrightarrow \omega_{n1,2} \in \mathbb{C}$,
   b. $X_n = 0 \Longrightarrow \omega_{n1,2} = 0$, and

c. $X_n < 0 \Longrightarrow$ there is only one positive real root, namely, $\omega_n = \sqrt{\frac{(-X_n)}{2}}$, where $n \in \mathbb{N}_0$.

3. If $X_n^2 - 4Y_n > 0$, then

a. $X_n = 0$ and $Y_n < 0 \Longrightarrow$ there is only one positive real root $\omega_n = \sqrt[4]{-Y_n}$, where $n \in \mathbb{N}_0$,
b. $X_n > 0$ and $Y_n > 0 \Longrightarrow X_n > \sqrt{X_n^2 - 4Y_n} \Longrightarrow \omega_n^2 < 0$ so that there is no real root,
c. $X_n > 0$ and $Y_n = 0 \Longrightarrow \omega_n^2 = \frac{-X_n \pm \sqrt{X_n^2}}{2} \Longrightarrow \omega_n^2 = -X_n$ or $\omega_n^2 = 0$ so that there is no positive real root,
d. $X_n > 0$ and $Y_n < 0 \Longrightarrow$ there is only one positive real root, namely, $\omega_n = \sqrt{\frac{-X_n + \sqrt{X_n^2 - 4Y_n}}{2}}$, where $n \in \mathbb{N}_0$,
e. $X_n < 0$ and $Y_n < 0 \Longrightarrow$ there is only one positive real root, which is $\omega_n = \sqrt{\frac{-X_n + \sqrt{X_n^2 - 4Y_n}}{2}}$, where $n \in \mathbb{N}_0$,
f. $X_n < 0$ and $Y_n = 0 \Longrightarrow \omega_n^2 = \frac{-X_n \pm \sqrt{X_n^2}}{2} \Longrightarrow \omega_n^2 = -X_n$ or $\omega_n^2 = 0 \Longrightarrow$ there is only one positive real root, namely, $\omega_n^2 = -X_n$, where $n \in \mathbb{N}$,
g. $X_n < 0$ and $Y_n > 0 \Longrightarrow$ there are two positive real roots, which are $\omega_{n_1} = \sqrt{\frac{-X_n + \sqrt{X_n^2 - 4Y_n}}{2}}$ and $\omega_{n_2} = \sqrt{\frac{-X_n - \sqrt{X_n^2 - 4Y_n}}{2}}$, where $n \in \mathbb{N}_0$.

We conclude from this analysis that there exists only one positive real ω_n for 2c, 3a, 3d, 3e, and 3f while there exist two different positive real ω_n-values for 3g. This completes the proof.

Lemma 1 basically underlines that the characteristic equation (6.33) has a pair of complex conjugate eigenvalues of the form $\lambda(\tau) = \gamma(\tau) \pm i\omega(\tau)$, and there are some critical values, namely, τ_n, of the bifurcation parameter τ at which $\gamma(\tau_n) = 0$ and $\omega(\tau_n) = \omega_n$ for each $n \in \mathbb{N}_0$. Next we determine these critical values τ_n. To do this we substitute $\lambda(\tau_n) = i\omega(\tau_n) = i\omega_n$ into (6.33), separate the real and imaginary parts utilizing Euler's formula, and obtain the following two equations in ω_n and τ_n :

$$\begin{aligned} \omega_n^2 - D &= B\cos(\omega_n \tau_n) + C\omega_n \sin(\omega_n \tau_n), \\ A\omega_n &= B\sin(\omega_n \tau_n) - C\omega_n \cos(\omega_n \tau_n). \end{aligned}$$

Solving these equations for τ_n, one has the following:

$$\tau_n = \frac{1}{\omega_n} \arctan\left(\frac{C\omega_n^3 + (AB - CD)\omega_n}{(B - AC)\omega_n^2 - BD} \right).$$

On the other hand, since $\tan x$ is a periodic function with period π, the critical values have the following form for each n and k:

$$\tau_{n,k} = \frac{1}{\omega_n} \arctan\left(\frac{C{\omega_n}^3 + (AB - CD)\omega_n}{(B - AC)\omega_n^2 - BD}\right) + \frac{k\pi}{\omega_n},$$

where $n, k \in \mathbb{N}_0$. Note that $\gamma(\tau_n) = \gamma(\tau_{n,k}) = 0$ and $\omega(\tau_n) = \omega(\tau_{n,k}) = \omega_n$. Note also that for each $n^* \in \mathbb{N}_0$ we uniquely determine $\tau_{n^*,k}$ such that $\lambda(\tau_{n^*,k}) = \mathrm{i}\omega_{n^*}$. This underlines that all other roots of the characteristic equation (6.33) have nonzero real parts at $\tau = \tau_{n^*,k}$.

We now check whether the transversality condition holds. The following lemma gives the required conditions under which it holds.

Lemma 2. *The transversality condition holds, that is,*

$$\left.\frac{\mathrm{d}(\mathrm{Re}\,\lambda)}{\mathrm{d}\tau}\right|_{\tau=\tau_{n,k}} \neq 0,$$

where $n, k \in \mathbb{N}_0$, if one of the following conditions is satisfied:

1. *$X_n = 0$ and $Y_n < 0$,*
2. *$X_n > 0$ and $Y_n < 0$,*
3. *$X_n < 0$ and $Y_n = 0$,*
4. *$X_n < 0$ and $Y_n < 0$,*
5. *$X_n < 0$ and $Y_n > 0$ and $\sqrt{X_n^2 - 4Y} > 0$,*

where $X_n = (A^2 - C^2 - 2D)$ and $Y_n = \left(D^2 - B^2\right)$.

Proof. Differentiating the characteristic equation (6.33) with respect to τ, we get the following equation:

$$\frac{\mathrm{d}\lambda}{\mathrm{d}\tau} = \frac{B\lambda + C\lambda^2}{2\lambda \mathrm{e}^{\lambda\tau} + A\mathrm{e}^{\lambda\tau} - B\tau + C - C\lambda\tau}.$$

Substituting $\tau = \tau_{n,k}$ into the preceding equation yields

$$\left.\frac{\mathrm{d}\lambda}{\mathrm{d}\tau}\right|_{\tau=\tau_{n,k}} = \frac{B\mathrm{i}\omega_n + C\,(\mathrm{i}\omega_n)^2}{2\mathrm{i}\omega_n \mathrm{e}^{\mathrm{i}\omega_n\tau_{n,k}} + A\mathrm{e}^{\mathrm{i}\omega_n\tau_{n,k}} - B\tau_{n,k} + C - C\mathrm{i}\omega_n\tau_{n,k}}. \tag{6.37}$$

Since

$$\left.\frac{\mathrm{d}\lambda}{\mathrm{d}\tau}\right|_{\tau=\tau_{n,k}} = \left.\frac{\mathrm{d}\gamma}{\mathrm{d}\tau}\right|_{\tau=\tau_{n,k}} + i\left.\frac{\mathrm{d}\omega}{\mathrm{d}\tau}\right|_{\tau=\tau_{n,k}},$$

we can find the equation of $\left.\frac{\mathrm{d}\gamma}{\mathrm{d}\tau}\right|_{\tau=\tau_{n,k}}$ explicitly from (6.37). Notice that

$$\mathrm{Re}\left(\left.\frac{\mathrm{d}\lambda}{\mathrm{d}\tau}\right|_{\tau=\tau_{n,k}}\right) \neq 0 \Longleftrightarrow \mathrm{Re}\left(\left.\frac{\mathrm{d}\lambda}{\mathrm{d}\tau}\right|_{\tau=\tau_{n,k}}\right)^{-1} \neq 0.$$

From (6.37) we obtain

$$\mathrm{Re}\left(\frac{\mathrm{d}\lambda}{\mathrm{d}\tau}\bigg|_{\tau=\tau_{n,k}}\right)^{-1} = \frac{2\omega_n^2 + (A^2 - C^2 - 2D)}{B^2 + C^2\omega_n^2} = \frac{2\omega_n^2 + X_n}{B^2 + C^2\omega_n^2}. \tag{6.38}$$

Then by substituting ω_n-values which are obtained from 2c, 3a, 3d–3g in the proof of Lemma 1 into (6.38), we can check whether the transversality condition holds. We conclude that $\frac{\mathrm{d}\gamma}{\mathrm{d}\tau}\Big|_{\tau=\tau_{n,k}} \neq 0$ in all cases 3a, 3d–3g, but $\frac{d\gamma}{d\tau}\Big|_{\tau=\tau_{n,k}} = 0$ in case 2c as follows:

1. If $X_n^2 - 4Y_n = 0$, $X_n < 0$, then $\omega_n = \sqrt{-\frac{X_n}{2}}$ and $\mathrm{Re}\left(\frac{\mathrm{d}\lambda}{\mathrm{d}\tau}\Big|_{\tau=\tau_{n,k}}\right) = 0$.
2. If $X_n = 0$, $Y_n < 0$, then $\omega_n = \sqrt[4]{-Y_n}$ and $\mathrm{Re}\left(\frac{\mathrm{d}\lambda}{\mathrm{d}\tau}\Big|_{\tau=\tau_{n,k}}\right) = \frac{2\sqrt{-Y_n}}{B^2+C^2\omega_n^2} > 0$.
3. If $X_n > 0$, $Y_n < 0$, then $\omega_n = \sqrt{\frac{-X_n+\sqrt{X_n^2-4Y_n}}{2}}$ and $\mathrm{Re}\left(\frac{\mathrm{d}\lambda}{\mathrm{d}\tau}\Big|_{\tau=\tau_{n,k}}\right) = \frac{\sqrt{X_n^2-4Y_n}}{B^2+C^2\omega_n^2} > 0$.
4. If $X_n < 0$, $Y_n < 0$, then $\omega_n = \sqrt{\frac{-X_n+\sqrt{X_n^2-4Y_n}}{2}}$ and $\mathrm{Re}\left(\frac{\mathrm{d}\lambda}{\mathrm{d}\tau}\Big|_{\tau=\tau_{n,k}}\right) = \frac{\sqrt{X_n^2-4Y_n}}{B^2+C^2\omega_n^2} > 0$.
5. If $X_n < 0$, $Y_n = 0$, then $\omega_n = -X_n$ and $\mathrm{Re}\left(\frac{\mathrm{d}\lambda}{\mathrm{d}\tau}\Big|_{\tau=\tau_{n,k}}\right) = -\frac{X_n}{B^2+C^2\omega_n^2} > 0$.
6. If $X_n < 0$, $Y_n > 0$, and $X_n^2 - 4Y_n > 0$, then

   a. for $\omega_n = \sqrt{\frac{-X_n+\sqrt{X_n^2-4Y_n}}{2}}$, $\mathrm{Re}\left(\frac{\mathrm{d}\lambda}{\mathrm{d}\tau}\Big|_{\tau=\tau_{n,k}}\right) = \frac{\sqrt{X_n^2-4Y_n}}{B^2+C^2\omega_n^2} > 0$,

   b. for $\omega_n = \sqrt{\frac{-X_n-\sqrt{X_n^2-4Y_n}}{2}}$, $\mathrm{Re}\left(\frac{\mathrm{d}\lambda}{D\tau}\Big|_{\tau=\tau_{n,k}}\right) = -\frac{\sqrt{X_n^2-4Y_n}}{B^2+C^2\omega_n^2} < 0$.

This completes the proof.

Thus, using the Hopf bifurcation theorem [15] together with Lemmas 1 and 2 one will be able to show that for each of these cases obtained above, system (6.25) undergoes a Hopf bifurcation at (u^*, v^*) as τ passes through $\tau_{n,k}$ $(n, k \in \mathbb{N}_0)$, and possesses a family of real-valued periodic solutions at these values. These results are summarized in the following theorem.

Theorem 3. *Let* $X_n = (A^2 - C^2 - 2D)$, $Y_n = \left(D^2 - B^2\right)$. *If one of the following conditions holds:*

1. $X_n = 0$ *and* $Y_n < 0$,
2. $X_n > 0$ *and* $Y_n < 0$,
3. $X_n < 0$ *and* $Y_n = 0$,
4. $X_n < 0$ *and* $Y_n < 0$,
5. $X_n < 0$ *and* $Y_n > 0$ *and* $X_n^2 - 4Y_n > 0$,

then system (6.24) undergoes a Hopf bifurcation at (u^*, v^*) as τ passes through $\tau_{n,k}$ and possesses a family of real-valued periodic solutions when $\lambda(\tau)$ crosses the imaginary axis at $\tau = \tau_{n,k}$.

6.3.2 Direction and Stability of the Hopf Bifurcation

In this section, we determine some of the properties of Hopf bifurcation by applying the normal form theory and the center manifold reduction for partial functional differential equations.

Remember that the system whose equilibrium is shifted to the origin is

$$\begin{cases} u_t(x,t) = d_1 u_{xx}(x,t) + \left(\frac{3\alpha^2-5}{1+\alpha^2}\right) u(x,t) + \left(\frac{-4\alpha}{1+\alpha^2}\right) v(x,t-\tau) + f(u,v,\tau), \\ v_t(x,t) = d_2 v_{xx}(x,t) + \left(\frac{2\sigma b\alpha^2}{1+\alpha^2}\right) u(x,t) + \left(\frac{-\sigma b\alpha}{1+\alpha^2}\right) v(x,t-\tau) + g(u,v,\tau), \end{cases} \tag{6.39}$$

where the functions f and g have the forms in (6.26) and (6.27), respectively. In order to determine the direction and stability of the Hopf bifurcation, we consider the following system, which is equivalent to (6.39):

$$\begin{cases} \frac{du}{dt} = \left(-d_1 n^2 + \frac{3\alpha^2-5}{1+\alpha^2}\right) u(t) + \left(\frac{-4\alpha}{1+\alpha^2}\right) v(t-\tau) + f(u,v,\tau), \\ \frac{dv}{dt} = \frac{2\sigma b\alpha^2}{1+\alpha^2} u(t) - d_2 n^2 v(t) + \left(\frac{-\sigma b\alpha}{1+\alpha^2}\right) v(t-\tau) + g(u,v,\tau), \end{cases} \tag{6.40}$$

where $u(t) = u(.,t)$ and $v(t) = v(.,t)$, so we can continue our analysis with system (6.40). Let $\phi(\theta) = \begin{pmatrix} \phi_1(\theta) \\ \phi_2(\theta) \end{pmatrix} \in \mathbf{C}^1[-\tau, 0]$ and $L_n : \mathbf{C}^1[-\tau, 0] \to \mathbb{R}^2$. Now we define L_n and F as follows:

$$L_n(\phi(\theta)) = \begin{bmatrix} \frac{3\alpha^2-5}{1+\alpha^2} - d_1 n^2 & 0 \\ \frac{2\sigma b\alpha^2}{1+\alpha^2} & -d_2 n^2 \end{bmatrix} \begin{pmatrix} \phi_1(0) \\ \phi_2(0) \end{pmatrix} + \begin{bmatrix} 0 & \frac{-4\alpha}{1+\alpha^2} \\ 0 & \frac{-\sigma b\alpha}{1+\alpha^2} \end{bmatrix} \begin{pmatrix} \phi_1(-\tau) \\ \phi_2(-\tau) \end{pmatrix},$$

$$F(\phi(\theta)) = \begin{pmatrix} f(\phi(\theta)) \\ g(\phi(\theta)) \end{pmatrix},$$

where $f,\ g : \mathbf{C}^1[-1, 0] \to \mathbb{R}$

$$f(\phi(\theta)) = \frac{4\alpha\left(3-\alpha^2\right)}{\left(1+\alpha^2\right)^2}\phi_1(0)^2 + \frac{4\left(\alpha^2-1\right)}{\left(1+\alpha^2\right)^2}\phi_1(0)\phi_2(-\tau) + \text{h.o.t.},$$

$$g(\phi(\theta)) = \frac{\sigma b}{4} f(\phi(\theta)).$$

Let $U(t) = \begin{pmatrix} u(t) \\ v(t) \end{pmatrix}$ and U_t be two notations such that $U_t = U(t+\theta),\ \theta \in [-\tau, 0]$ so that system (6.40) turns into

$$\frac{\partial U}{\partial t} = L_n U_t + F(U_t).$$

By now taking $t = \tau s$ and $\mu = \tau - \tau_n$, we may write the new scaled system, whose bifurcation value is shifted to 0, as follows:

$$\frac{\partial U}{\partial s} = (\tau_n + \mu) L_n U_s + (\tau_n + \mu) F(U_s), \tag{6.41}$$

where $U_s = U(s+\theta),\ \theta \in [-1, 0]$.

Let

$$L_{n_\mu}(\phi(\theta)) = (\tau_n + \mu) \left(\begin{bmatrix} \frac{3\alpha^2-5}{1+\alpha^2} - d_1 n^2 & 0 \\ \frac{2\sigma b \alpha^2}{1+\alpha^2} & -d_2 n^2 \end{bmatrix} \begin{pmatrix} \phi_1(0) \\ \phi_2(0) \end{pmatrix} + \begin{bmatrix} 0 & \frac{-4\alpha}{1+\alpha^2} \\ 0 & \frac{-\sigma b \alpha}{1+\alpha^2} \end{bmatrix} \begin{pmatrix} \phi_1(-1) \\ \phi_2(-1) \end{pmatrix} \right)$$

and

$$\tilde{F}(\phi(\theta)) = (\tau_n + \mu) F(U_s). \tag{6.42}$$

For convenience, we continue our calculations by taking $s = t$ and $\tilde{F}(\phi(\theta)) = F(\phi(\theta))$ in the rest of the chapter. We rewrite system (6.41) in the following form:

$$\frac{\partial U}{\partial t} = L_{n_\mu} U_t + F(U_t, \mu). \tag{6.43}$$

Notice that system (6.43) has two different unknown functions, namely, $U(x,t)$ and $U_t = U(x, t+\theta)$. Applying the Riesz representation theorem yields the existence of a matrix-valued function $\eta(\cdot, \mu)$, where $\eta(\cdot, \mu) : [-1, 0] \to \mathbb{R}^2$ and $\phi \in \mathbf{C}^1[-1, 0]$, so that

$$L_{n_\mu}(\phi(\theta)) = \int_{-1}^{0} \mathrm{d}\eta(\theta, \mu) \phi(\theta). \tag{6.44}$$

Let us choose $\mathrm{d}\eta(\theta,\mu)$ as follows:

$$\mathrm{d}\eta(\theta,\mu) = (\tau_n+\mu)\left(\begin{array}{c}\begin{bmatrix}\frac{3\alpha^2-5}{1+\alpha^2}-d_1n^2 & 0\\ \frac{2\sigma b\alpha^2}{1+\alpha^2} & -d_2n^2\end{bmatrix}\delta(\theta)\\ +\begin{bmatrix}0 & \frac{-4\alpha}{1+\alpha^2}\\ 0 & \frac{-\sigma b\alpha}{1+\alpha^2}\end{bmatrix}\delta(\theta+1)\end{array}\right)d\theta,$$

where $\delta(\theta)$ is the Dirac delta function. Using them, we define the operators $A(\mu)\phi$ and $R(\mu)\phi$ as follows:

$$A(\mu)\phi = \begin{cases}\frac{\mathrm{d}\phi(\theta)}{\mathrm{d}\theta} & ,\theta\in[-1,0)\\ \int_{-1}^{0}\mathrm{d}\eta(\theta,\mu)\phi(\theta) = L_{n_\mu}(\phi) & ,\theta=0\end{cases} \tag{6.45}$$

and

$$R(\mu)\phi = \begin{cases}0 & ,\theta\in[-1,0)\\ F(\theta) & ,\theta=0\end{cases}. \tag{6.46}$$

Now we can state system (6.43) as follows:

$$\frac{\partial U_t}{\partial t} = A(\mu)U_t + R(\mu)U_t, \tag{6.47}$$

which involves only one unknown function. In order to construct center manifold coordinates, we need to define an inner product. For $\psi,\phi\in\mathbf{C}[-1,0]$, one can define it as follows:

$$<\psi,\phi> = \overline{\psi}(0)\cdot\phi(0) - \int_{\theta=-r}^{0}\int_{\xi=0}^{\theta}\overline{\psi}^T(\xi-\theta)\mathrm{d}\eta(\theta,\mu)\phi(\xi)\mathrm{d}\xi. \tag{6.48}$$

Let $q(\theta)$ be an eigenvector of $A(0)$ corresponding to $\lambda(0)=\mathrm{i}\omega_n$ and $q^*(s)$ be an eigenvector of $A^*(0)$ associated with $\overline{\lambda}(0)=-\mathrm{i}\omega_n$ satisfying

$$<q^*(s),q(\theta)>=1 \text{ and } <q^*(s),\overline{q}(\theta)>=0, \tag{6.49}$$

$$A(0)q(\theta)=\mathrm{i}\omega_n q(\theta) \text{ and } A^*(0)q(s) = -\mathrm{i}\omega_n q^*(s), \tag{6.50}$$

where $A^*(\mu)$ is adjoint operator of $A(\mu)$ defined as

$$A^*(\mu)\phi = \begin{cases}-\frac{\mathrm{d}\phi(s)}{\mathrm{d}s} & ,s\in[-1,0)\\ \int_{-1}^{0}\mathrm{d}\eta^T(s,\mu)\phi(-s) & ,s=0\end{cases}. \tag{6.50a}$$

First, we determine $q(\theta)$ from $A(0)q(\theta) = \mathrm{i}\omega_n q(\theta)$ in (6.50). It will be done in two cases as follows:

Case A1: If $\theta \in [-1, 0)$, then, by (6.45),

$$A(0)q(\theta) = \frac{\mathrm{d}q(\theta)}{\mathrm{d}\theta} = \mathrm{i}\omega_n q(\theta) \tag{6.51}$$

so that we obtain that $q(\theta) = \binom{1}{c} e^{\mathrm{i}\omega_n\theta}$ from (6.51) where c will be determined in case A2.

Case A2: When $\theta = 0$, utilizing (6.45) we have

$$\begin{aligned} A(0)q(\theta) &= \int_{-1}^{0} \mathrm{d}\eta(\theta, \mu) q(\theta) \\ &= \tau_n \begin{bmatrix} m - d_1 n^2 & 0 \\ f & -d_2 n^2 \end{bmatrix} \int_{-1}^{0} \delta(\theta)\, q(\theta)\mathrm{d}\theta \\ &\quad + \tau_n \begin{bmatrix} 0 & g \\ 0 & -k \end{bmatrix} \int_{-1}^{0} \delta(\theta + 1)\, q(\theta)\mathrm{d}\theta \\ &= \tau_n \begin{bmatrix} m - d_1 n^2 & 0 \\ f & -d_2 n^2 \end{bmatrix} q(0) + \tau_n \begin{bmatrix} 0 & g \\ 0 & -k \end{bmatrix} q(-1) \\ &= \begin{bmatrix} \tau_n \left(m - d_1 n^2\right) + \tau_n g \mathrm{e}^{-\mathrm{i}\omega_n} c \\ \tau_n f - \tau_n d_2 n^2 c + \tau_n (-k)\, \mathrm{e}^{-\mathrm{i}\omega_n} c \end{bmatrix} \\ &= \mathrm{i}\omega_n q(0) = \mathrm{i}\omega_n \begin{pmatrix} 1 \\ c \end{pmatrix} \mathrm{e}^{\mathrm{i}\omega_n 0} = \begin{pmatrix} \mathrm{i}\omega_n \\ \mathrm{i}\omega_n c \end{pmatrix}, \end{aligned}$$

where

$$m = \frac{3\alpha^2 - 5}{1 + \alpha^2}, f = \frac{2\sigma b\alpha^2}{1 + \alpha^2}, g = \frac{-4\alpha}{1 + \alpha^2}, k = \frac{\sigma b\alpha}{1 + \alpha^2}. \tag{6.52}$$

From the preceding calculations we obtain c as follows:

$$c = \left(\frac{\mathrm{i}\omega_n - \tau_n \left(m - d_1 n^2\right)}{\tau_n g \mathrm{e}^{-\mathrm{i}\omega_n}} \right). \tag{6.53}$$

Second, we determine $q^*(s)$ from $A(0)q^*(s) = -\mathrm{i}\omega_n q^*(s)$ in (6.50). Once again, it will be done in two cases as follows:

Case B1: If $\theta \in [-1, 0)$, then, by (6.50a), one has

$$A^*(0)q^*(s) = -\frac{\mathrm{d}q^*(s)}{\mathrm{d}s} = -\mathrm{i}\omega_n q^*(s)$$

so that one obtains that $q^*(s) = E\binom{c^*}{1}e^{i\omega_n\theta}$. The constant c^* will be calculated ahead.

Case B2: When $\theta = 0$, we have [see (6.50a)]

$$\begin{aligned}
A^*(0)q^*(s) &= \int_{-1}^{0} d\eta^T(s,\mu)\phi(-s) \\
&= \tau_n \begin{bmatrix} m - d_1 n^2 & f \\ 0 & -d_2 n^2 \end{bmatrix} \int_{-1}^{0} \delta(s)\, q^*(-s)ds \\
&\quad + \tau_n \begin{bmatrix} 0 & 0 \\ g & -k \end{bmatrix} \int_{-1}^{0} \delta(s+1)\, q^*(-s)ds \\
&= \tau_n \begin{bmatrix} m - d_1 n^2 & f \\ 0 & -d_2 n^2 \end{bmatrix} q^*(0) + \tau_n \begin{bmatrix} 0 & 0 \\ g & -k \end{bmatrix} q^*(1) \\
&= E \begin{bmatrix} \tau_n (m - d_1 n^2) c^* + \tau_n f \\ \tau_n e^{i\omega_n} g c^* - \tau_n d_2 n^2 + \tau_n e^{i\omega_n}(-k) \end{bmatrix} \\
&= -i\omega_n q^*(0) = -i\omega_n E \begin{pmatrix} c^* \\ 1 \end{pmatrix} = E \begin{pmatrix} -i\omega_n c^* \\ -i\omega_n . \end{pmatrix}.
\end{aligned}$$

These calculations yield that c^* has the following form:

$$c^* = \left(\frac{-\tau_n f}{\tau_n (m - d_1 n^2) + i\omega_n} \right). \tag{6.54}$$

These two eigenvectors must satisfy the properties given in (6.49). Since $\lambda(\mu)$ is a simple eigenvalue, one can show that $< q^*(s), \overline{q}(\theta) >= 0$ (see [15] and [35]). Let us now choose E such that $< q^*(s), q(\theta) >= 1$. By the definition of inner product [see (6.48)] one has

$$\begin{aligned}
< q^*(s), q(\theta) > &= \overline{q^*}(0) \cdot q(0) - \int_{\theta=-r}^{0} \int_{\xi=0}^{\theta} \overline{q^*}^T(\xi - \theta) d\eta(\theta,\mu) q(\xi) d\xi \\
&= \overline{E}\left(\overline{c^*} + c\right) - \overline{E}\left(\overline{c^*}\ 1\right) \left(\int_{-1}^{0} d\eta(\theta, 0) e^{i\omega_n \theta} \theta \right) \begin{pmatrix} 1 \\ c \end{pmatrix}.
\end{aligned}$$

First, we calculate the integral on the right-hand side of the latter equation as follows:

$$\left(\int_{-1}^{0} d\eta(\theta,0) e^{i\omega_n\theta}\theta \right) = \int_{-1}^{0} \tau_n \left(\begin{array}{c} \begin{bmatrix} m - d_1 n^2 & 0 \\ f & -d_2 n^2 \end{bmatrix} \delta(\theta) \\ + \begin{bmatrix} 0 & g \\ 0 & -k \end{bmatrix} \delta(\theta + 1) \end{array} \right) e^{i\omega_n\theta} \theta d\theta$$

$$= \begin{bmatrix} 0 & -\tau_n g e^{-i\omega_n} \\ 0 & \tau k e^{-i\omega_n} \end{bmatrix}.$$

Second, we substitute the result into the preceding equation to determine $\overline{E}$:

$$< q^*(s), q(\theta) >= \overline{E}\left(\overline{c^*} + c\right) - \overline{E}\left(\overline{c^*}\ 1\right)\begin{bmatrix} 0 & -\tau_n g e^{-i\omega_n} \\ 0 & \tau_n k e^{-i\omega_n} \end{bmatrix}\begin{pmatrix} 1 \\ c \end{pmatrix}$$
$$= \overline{E}\left(\overline{c^*} + c + \tau_n g e^{-i\omega_n} c\overline{c^*} - \tau_n k e^{-i\omega_n} c\right).$$

Finally, since $< q^*(s), q(\theta) >= 1$, we obtain $\overline{E}$ as follows:

$$\overline{E} = \frac{1}{\left(\overline{c^*} + c + \tau_n g e^{-i\omega_n} c\overline{c^*} - \tau_n k e^{-i\omega_n} c\right)}. \tag{6.55}$$

Next, we define center manifold coordinates by using these eigenvectors. Let $\mathbf{X}$ denote the domain of the operator L_{n_μ} [see (6.44)]. We decompose $\mathbf{X} = \mathbf{X}^C + \mathbf{X}^S$ with $\mathbf{X}^C := \{zq + \overline{zq} | \, z \in \mathbb{C}\}$, $\mathbf{X}^S := \{w \in \mathbf{X} | < q^*, w >= 0\}$. For any $U = \begin{pmatrix} u \\ v \end{pmatrix} \in \mathbf{X}$, there exist $z \in \mathbb{C}$ and $w = \begin{pmatrix} w_1 \\ w_2 \end{pmatrix} \in \mathbf{X}^s$ such that

$$U = \begin{pmatrix} u \\ v \end{pmatrix} = zq + \overline{zq} + \begin{pmatrix} w_1 \\ w_2 \end{pmatrix}. \tag{6.56}$$

Thus, at $\mu = 0$, system (6.47) is reduced to the following system in (z, w)-coordinates:

$$\begin{cases} \frac{\partial z}{\partial t} = i\omega_n z + < q^*, F_0 >= \\ i\omega_n z + g(z, \overline{z}), \\ \frac{\partial w}{\partial t} = A(0)w + H(z, \overline{z}, \theta), \end{cases} \tag{6.57}$$

where

$$F_0 := F(zq + \overline{zq} + w, 0), \; < q^*, F_0 >= \overline{q^*}(0) \cdot F_0$$

and

$$H(z, \overline{z}, \theta) = \begin{cases} - < q^*, F_0 > q - < \overline{q^*}, F_0 > \overline{q}, \theta \in [-1, 0) \\ F_0 - < q^*, F_0 > q - < \overline{q^*}, F_0 > \overline{q}, \theta = 0 \end{cases}. \tag{6.58}$$

From (6.42) we have

$$F_0 = F(zq + \overline{zq} + w, 0) = F(U_t, 0)$$
$$= \tau_n \begin{pmatrix} \frac{4\alpha(3-\alpha^2)}{(1+\alpha^2)^2} u_t(0)^2 + \frac{4(\alpha^2-1)}{(1+\alpha^2)^2} u_t(0)v_t(-1) + \text{h.o.t.} \\ \frac{\alpha\sigma b(3-\alpha^2)}{(1+\alpha^2)^2} u_t(0)^2 + \frac{\sigma b(\alpha^2-1)}{(1+\alpha^2)^2} u_t(0)v_t(-1) + \text{h.o.t.} \end{pmatrix}. \tag{6.59}$$

Representing w as $w(z, \overline{z}) = \sum \frac{1}{i!j!} w_{ij}(z)^i (\overline{z})^j$ and using (6.56) we get

$$U_t(\theta) = zq(\theta) + \overline{zq}(\theta) + w_{20}(\theta)\frac{z^2}{2} + w_{11}(\theta)z\overline{z} + w_{02}(\theta)\frac{(\overline{z})^2}{2} + \text{h.o.t.}. \tag{6.60}$$

To get $u_t(0)$, we put $\theta = 0$ in (6.60), which leads to the following equation:

$$u_t(0) = z + \overline{z} + w_{20_1}(0)\frac{z^2}{2} + w_{11_1}(0)z\overline{z} + w_{02_1}(0)\frac{(\overline{z})^2}{2} + \text{h.o.t.}.$$

Similarly, $v_t(-1)$ can be obtained by plugging in $\theta = -1$ into (6.60), so we have the following:

$$v_t(-1) = z + \overline{z} + w_{20_2}(-1)\frac{z^2}{2} + w_{11_2}(-1)z\overline{z} + w_{02_2}(-1)\frac{(\overline{z})^2}{2} + \text{h.o.t.}.$$

Now substituting $u_t(0)$ and $v_t(-1)$ into (6.59), we obtain F_0 as follows:

$$F_0 = F(zq + \overline{zq} + w, 0)$$
$$= \begin{pmatrix} F_{0_1} \\ F_{0_2} \end{pmatrix}$$
$$= \begin{pmatrix} K_{20}z^2 + K_{11}z\overline{z} + K_{02}(\overline{z})^2 + K_{21}z^2\overline{z} \\ \frac{\sigma b}{4}\left(K_{20}z^2 + K_{11}z\overline{z} + K_{02}(\overline{z})^2 + K_{21}z^2\overline{z}\right) \end{pmatrix},$$

where

$$K_{20} = \tau_n\left(p + rce^{-i\omega_n}\right), K_{02} = \tau_n\left(p + r\overline{c}e^{i\omega_n}\right),$$
$$K_{11} = \tau_n\left(2p + r\overline{c}e^{i\omega_n} + rce^{-i\omega_n}\right),$$
$$K_{21} = \tau_n\begin{pmatrix} 2pw_{11_1}(0) + pw_{20_1}(0) + rw_{11_2}(-1) \\ +\frac{1}{2}rw_{20_2}(-1) + \frac{1}{2}r\overline{c}e^{i\omega_n}w_{20_1}(0) + rce^{-i\omega_n}w_{11_1}(0) \end{pmatrix}$$

and

$$p = \frac{4\alpha\left(3-\alpha^2\right)}{(1+\alpha^2)^2}, \quad r = \frac{4\left(\alpha^2-1\right)}{(1+\alpha^2)^2}. \tag{6.61}$$

Since $g(z,\bar{z}) = \overline{q^*}(0) \cdot F_0(z,\bar{z}) = \sum \frac{1}{i!j!} g_{ij}(z)^i(\bar{z})^j$, we have

$$\sum \frac{1}{i!j!} g_{ij}(z)^i(\bar{z})^j = \overline{q^*}(0) \cdot F_0(z,\bar{z})$$
$$= \overline{E}\left[\overline{c^*}\ 1\right]\begin{bmatrix} F_{0_1} \\ \frac{\sigma b}{4} F_{0_1} \end{bmatrix} \tag{6.62}$$
$$= \overline{E c^*} F_{0_1} + \overline{E}\frac{\sigma b}{4} F_{0_1}.$$

In order to determine the stability and direction of the Hopf bifurcation, we need to find the Lyapunov coefficient (see [15]) that is given by the following formula:

$$c_1(\tau_n) = \frac{\mathrm{i}}{2\omega_n}\left(g_{20}g_{11} - 2\left|g_{11}\right|^2 - \frac{1}{3}\left|g_{02}\right|^2\right) + \frac{g_{21}}{2}, \tag{6.63}$$

where from (6.62) we have

$$g_{20} = 2\overline{E}(\overline{c^*} + \frac{\sigma b}{4})K_{20},\ g_{11} = \overline{E}(\overline{c^*} + \frac{\sigma b}{4})K_{11}, \tag{6.64}$$

$$g_{02} = 2\overline{E}(\overline{c^*} + \frac{\sigma b}{4})K_{02},\ g_{21} = 2\overline{E}(\overline{c^*} + \frac{\sigma b}{4})K_{21}. \tag{6.65}$$

To calculate g_{21}, we first need to find w_{20} and w_{11}. We can express $w(z,\bar{z},\theta)$ and $H(z,\bar{z},\theta)$ as $w(z,\bar{z},\theta) = \sum \frac{1}{i!j!} w_{ij}(z)^i(\bar{z})^j$ and $H(z,\bar{z},\theta) = \sum \frac{1}{i!j!} H_{ij}(\theta)(z)^i(\bar{z})^j$, respectively. Substituting these expressions into

$$\frac{\partial w}{\partial t} = Aw + H(z,\bar{z},\theta)$$

yields the following equalities:

$$H_{20} = (2\mathrm{i}\omega_n - A)w_{20},$$
$$H_{11} = -Aw_{11}, \tag{6.66}$$
$$w_{02} = \overline{w}_{20}.$$

First, we find w_{20}. From (6.58) we see that H_{20} equals

$$H_{20}(\theta) = \begin{cases} -g_{20}q(\theta) - \overline{g}_{02}\overline{q}(\theta),\ \theta \in [-1,0) \\ F_0 - g_{20}q(0) - \overline{g}_{02}\overline{q}(0),\ \theta = 0 \end{cases}. \tag{6.67}$$

We analyze the right-hand side of the latter equation with respect to the position of θ as follows:

Case C1: If $\theta \in [-1, 0)$, then using (6.45) we can rewrite (6.66) as follows:

$$H_{20}(\theta) = 2\mathrm{i}\omega_n w_{20}(\theta) - \frac{\mathrm{d}w_{20}(\theta)}{\mathrm{d}\theta}. \tag{6.68}$$

Combining (6.67) and (6.68), one obtains the following differential equation:

$$\frac{\mathrm{d}w_{20}(\theta)}{\mathrm{d}\theta} - 2\mathrm{i}\omega_n w_{20}(\theta) = g_{20}q(\theta) + \overline{g}_{02}\overline{q}(\theta).$$

Its solution is

$$w_{20}(\theta) = -\frac{1}{\mathrm{i}\omega_n}q(0)\mathrm{e}^{\mathrm{i}\omega_n\theta}g_{20} - \frac{1}{3\mathrm{i}\omega_n}\overline{q}(0)\mathrm{e}^{-\mathrm{i}\omega_n\theta}\overline{g}_{02} + S\mathrm{e}^{2\mathrm{i}\omega_n\theta}. \tag{6.69}$$

Case C2: If $\theta = 0$, then from (6.67) we get

$$H_{20}(0) = 2K_{20}\begin{pmatrix} 1 \\ \frac{\sigma b}{4} \end{pmatrix} - g_{20}q(0) - \overline{g}_{02}\overline{q}(0). \tag{6.70}$$

Both (6.66) and (6.70) give us

$$A(0)w_{20}(0) = 2\mathrm{i}\omega_n w_{20}(0) + g_{20}q(0) + \overline{g}_{02}\overline{q}(0) - 2K_{20}\begin{pmatrix} 1 \\ \frac{\sigma b}{4} \end{pmatrix}. \tag{6.71}$$

From case C1 we have a formula for $w_{20}(\theta)$, namely (6.69). By substituting $w_{20}(0)$ into (6.71) we obtain

$$A(0)w_{20}(0) = -g_{20}q(0) + \frac{1}{3}\overline{g}_{02}\overline{q}(0) + 2\mathrm{i}\omega_n S - 2K_{20}\begin{pmatrix} 1 \\ \frac{\sigma b}{4} \end{pmatrix}. \tag{6.72}$$

On the other hand, from the definition of the operator $A(0)$ [see (6.45)] we get

$$A(0)w_{20}(0) = -g_{20}q(0) + \frac{1}{3}\overline{g}_{02}\overline{q}(0) + S\int_{-1}^{0}\mathrm{d}\eta(\theta,0)\mathrm{e}^{2\mathrm{i}\omega_n\theta}, \tag{6.73}$$

so that (6.72) and (6.73) yield the following equality that will give us S:

$$S\left(2\mathrm{i}\omega_n I - \int_{-1}^{0}\mathrm{d}\eta(\theta,0)\mathrm{e}^{2\mathrm{i}\omega_n\theta}\right) = 2K_{20}\begin{pmatrix} 1 \\ \frac{\sigma b}{4} \end{pmatrix}.$$

Evaluating the preceding integral, one obtains the following:

$$\int_{-1}^{0} \mathrm{d}\eta(\theta,0)\mathrm{e}^{2\mathrm{i}\omega_n\theta} = \int_{-1}^{0} \tau_n \left(\begin{array}{c} \begin{bmatrix} m-d_1n^2 & 0 \\ f & -d_2n^2 \end{bmatrix} \delta(\theta) \\ + \begin{bmatrix} 0 & g \\ 0 & -k \end{bmatrix} \delta(\theta+1) \end{array} \right) \mathrm{e}^{2\mathrm{i}\omega_n\theta}\mathrm{d}\theta$$

$$= \tau_n \begin{bmatrix} m-d_1n^2 & g\mathrm{e}^{-2\mathrm{i}\omega_n} \\ f & -d_2n^2-k\mathrm{e}^{-2\mathrm{i}\omega_n} \end{bmatrix}.$$

Hence, S is equal to

$$S = \left(\begin{bmatrix} 2\mathrm{i}\omega_n - \tau_n\left(m-d_1n^2\right) & \tau_n g\mathrm{e}^{-2\mathrm{i}\omega_n} \\ \tau_n f & 2\mathrm{i}\omega_n + \tau_n d_2n^2 + \tau_n k\mathrm{e}^{-2\mathrm{i}\omega_n} \end{bmatrix} \right)^{-1} 2K_{20} \begin{pmatrix} 1 \\ \frac{\sigma b}{4} \end{pmatrix}. \quad (6.74)$$

Similarly, we will find w_{11} so that, from (6.58), H_{11} will be equal to

$$H_{11}(\theta) = \begin{cases} -g_{11}q(\theta) - \overline{g}_{11}\overline{q}(\theta), \theta \in [-1,0) \\ F_0 - g_{11}q(\theta) - \overline{g}_{11}\overline{q}(\theta), \theta = 0 \end{cases}. \quad (6.75)$$

To do this, we consider two cases as follows:

Case D1: If $\theta \in [-1,0)$, then because of the definition of the operator $A(\theta)$ [see(6.45)] the equality (6.66) becomes

$$H_{11}(\theta) = -\frac{\mathrm{d}w_{11}(\theta)}{\mathrm{d}\theta}. \quad (6.76)$$

Both (6.75) and (6.76) give us

$$\frac{\mathrm{d}w_{11}(\theta)}{\mathrm{d}\theta} = g_{11}q(\theta) + \overline{g}_{11}\overline{q}(\theta),$$

so that we have

$$w_{11}(\theta) = \frac{1}{\mathrm{i}\omega_n}q(0)\mathrm{e}^{\mathrm{i}\omega_n\theta}g_{11} - \frac{1}{\mathrm{i}\omega_n}\overline{q}(0)\mathrm{e}^{-\mathrm{i}\omega_n\theta}\overline{g}_{11} + G, \quad (6.77)$$

where G will be determined in case D2.

Case D2: If $\theta = 0$, then from (6.75) we have

$$H_{11}(0) = g_{11}q(0) + \overline{g}_{11}\overline{q}(0) - K_{11}\begin{pmatrix} 1 \\ \frac{\sigma b}{4} \end{pmatrix}. \quad (6.78)$$

Both (6.66) and (6.78) give us

$$A(0)w_{11}(0) = g_{11}q(0) + \overline{g}_{11}\overline{q}(0) - K_{11}\begin{pmatrix} 1 \\ \frac{\sigma b}{4} \end{pmatrix}. \quad (6.79)$$

From the definition of the operator $A(0)$ [see (6.45)] we get

$$A(0)w_{11}(0) = g_{11}q(0) + \overline{g}_{11}\overline{q}(0) + G\int_{-1}^{0} \mathrm{d}\eta(\theta, 0). \tag{6.80}$$

Equating the right-hand sides of Eqs. (6.79) and (6.80), one obtains the following identity, which is used to determine G:

$$G\left(\int_{-1}^{0} d\eta(\theta,0)\right) = K_{11}\begin{pmatrix} 1 \\ \frac{\sigma b}{4} \end{pmatrix}.$$

First, we calculate the preceding integral as follows:

$$\int_{-1}^{0} \mathrm{d}\eta(\theta,0) = \int_{-1}^{0} \tau_n \left(\begin{array}{c} \begin{bmatrix} m - d_1 n^2 & 0 \\ f & -d_2 n^2 \end{bmatrix} \delta(\theta) \\ + \begin{bmatrix} 0 & g \\ 0 & -k \end{bmatrix} \delta(\theta + 1) \end{array} \right) \mathrm{d}\theta$$

$$= \tau_n \begin{bmatrix} m - d_1 n^2 & g \\ f & (-k - d_2 n^2) \end{bmatrix}.$$

So we have

$$G\tau_n \begin{bmatrix} m - d_1 n^2 & g \\ f & (-k - d_2 n^2) \end{bmatrix} = K_{11}\begin{pmatrix} 1 \\ \frac{\sigma b}{4} \end{pmatrix},$$

so that G is equal to

$$G = \left(\tau_n \begin{bmatrix} m - d_1 n^2 & g \\ f & (-k - d_2 n^2) \end{bmatrix}\right)^{-1} K_{11}\begin{pmatrix} 1 \\ \frac{\sigma b}{4} \end{pmatrix}. \tag{6.81}$$

Now we can compute all the unknowns in the equation of $c_1(\tau_n)$, which is given (6.63). In order to determine the direction of the bifurcation, we also need to know sign of the Lyapunov coefficient. It can be determined by using the following formula:

$$\mathrm{Re}(c_1(\tau_n)) = \mathrm{Re}(\frac{g_{21}}{2}) - \frac{1}{2\omega_n}\left(\mathrm{Re}(g_{20})\,\mathrm{Im}(g_{11}) + \mathrm{Im}(g_{20})\,\mathrm{Re}(g_{11})\right). \tag{6.82}$$

From the previous analysis and the general Hopf bifurcation theorem (see [15]), we can deduce the following results.

Theorem 4. *If* $\frac{1}{\alpha'(0)}\mathrm{Re}(c_1(\tau_n)) < 0$ $(\frac{1}{\alpha'(0)}\mathrm{Re}(c_1(\tau_n)) > 0)$*, then the bifurcation is supercritical (subcritical). In addition, if other eigenvalues of* L_n *have negative real parts, then the bifurcating periodic solution is stable (unstable) if* $\mathrm{Re}(c_1(\tau_{n,k})) < 0$ *[* $\mathrm{Re}(c_1(\tau_{n,k})) > 0$*].*

6.3.3 *Analysis of Spatially Homogeneous Lengyel–Epstein System with Delay*

When $n = 0$ in (6.32), the characteristic equation (6.33) becomes

$$\lambda^2 + A\lambda + Be^{-\lambda\tau} + C\lambda e^{-\lambda\tau} = 0, \tag{6.83}$$

where

$$A = -m, \; B = 5k, \; C = k$$

and m and k are given in (6.52). By substituting $\lambda = i\omega_0, \; \omega_0 > 0$, into (6.83) we get the following equation:

$$\omega_0^4 + (A^2 - C^2)\omega_0^2 - B^2 = 0. \tag{6.84}$$

From (6.84) we have

$$\omega_0^2 = \frac{-(A^2 - C^2) \pm \sqrt{(A^2 - C^2)^2 - 4\,(-B^2)}}{2}. \tag{6.85}$$

Let $X_0 = (A^2 - C^2)$ and $Y_0 = -B^2$. Thus, (6.85) can be written as follows:

$$\omega_0^2 = \frac{-X_0 \pm \sqrt{X_0^2 - 4Y_0}}{2}. \tag{6.86}$$

Our aim is to get at least one $\omega_0 \in \mathbb{R}^+$. By analyzing (6.86) we get the following results:

1. If $X_0^2 - 4Y_0 < 0$, then $\omega_0^2 \in \mathbb{C}$. Hence, there is no real root.
2. If $X_0^2 - 4Y_0 = 0$, then $B = 0$ and $A^2 = C^2 \Longrightarrow m = k = b = \sigma = 0$ so that there is no positive real root.
3. If $X_0^2 - 4Y_0 > 0$, then there are three possibilities as follows:

   a. $X_0 < 0 \Longrightarrow$ there is only one positive real root, which is

$$\omega_0 = \sqrt{\frac{-X_0 + \sqrt{X_0^2 - 4Y_0}}{2}}.$$

   b. $X_0 = 0 \Longrightarrow$ there is only one positive real root, namely,

$$\omega_0 = \sqrt{\frac{\sqrt{-4Y_0}}{2}} = \sqrt{B} = \sqrt{5k}.$$

c. $X_0 > 0 \Longrightarrow$ there is only one positive real root, which is

$$\omega_0 = \sqrt{\frac{-X_0 + \sqrt{X_0^2 - 4Y_0}}{2}}.$$

In conclusion, we have only one positive real ω_0 for $3a$, $3b$, and $3c$.

Finally, we need to check the transversality condition. From (6.38) we can obtain $\mathrm{Re}\left(\frac{\mathrm{d}\lambda}{\mathrm{d}\tau}\Big|_{\tau=\tau_0}\right)^{-1}$ as follows:

$$\mathrm{Re}\left(\frac{\mathrm{d}\lambda}{\mathrm{d}\tau}\Big|_{\tau=\tau_0}\right)^{-1} = \frac{2\omega_0^2 + X_0}{B^2 + C^2\omega_0^2}, \tag{6.87}$$

which yields

$$\mathrm{Re}\left(\frac{\mathrm{d}\lambda}{\mathrm{d}\tau}\Big|_{\tau=\tau_0}\right)^{-1} = \frac{\sqrt{X_0^2 - 4Y_0}}{B^2 + C^2\omega_0^2}.$$

Hence, we have $\mathrm{Re}\left(\frac{\mathrm{d}\lambda}{\mathrm{d}\tau}\Big|_{\tau=\tau_0}\right)^{-1} > 0$, which means $\mathrm{Re}\left(\frac{\mathrm{d}\lambda}{d\tau}\Big|_{\tau=\tau_0}\right) > 0$ in all cases $3a$, $3b$, and $3c$. It shows that the transversality condition holds.

By Theorem 3, when $n = 0$, the Hopf bifurcation occurs at $\mu = 0$ for each ω_0 which we found above, and system (6.25) possesses a family of real-valued periodic solutions bifurcating from the equilibrium point $(0, 0)$ at $\mu = 0$.

Next, we find the direction of this bifurcation and analyze the stability of periodic solutions. For these goals we need to know the sign of the Lyapunov coefficient, which is given by (6.82), where

$$\mathrm{Re}(c_1(\tau_n)) = \mathrm{Re}(g_{21}/2) - \frac{1}{2\omega_n}\left(\mathrm{Re}(g_{20})\,\mathrm{Im}(g_{11}) + \mathrm{Im}(g_{20})\,\mathrm{Re}(g_{11})\right).$$

So we need to compute g_{20}, g_{11} and g_{21} for $n = 0$.
From (6.64) we have

$$g_{20} = 2\overline{E}(\overline{c^*} + \frac{\sigma b}{4})K_{20},$$

where

$$\overline{E} = \frac{1}{(\overline{c^*} + c + \tau_0 g e^{-\mathrm{i}\omega_0} c\overline{c^*} - \tau_0 k e^{-\mathrm{i}\omega_0} c)},$$

$$c^* = \left(\frac{-\tau_0 f}{\tau_0 m + \mathrm{i}\omega_0}\right), \quad c = \left(\frac{\mathrm{i}\omega_0 - \tau_0 m}{\tau_0 g e^{-\mathrm{i}\omega_0}}\right), \quad K_{20} = \tau_0\left(p + rce^{-\mathrm{i}\omega_0}\right)$$

in which m, f, g, k, p, and r are given in (6.52) and (6.61). Similarly, the coefficient g_{11} has the following form:

$$g_{11} = \overline{E}(\overline{c^*} + \frac{\sigma b}{4})K_{11},$$

where

$$\overline{E} = \frac{1}{\left(\overline{c^*} + c + \tau_0 g e^{-i\omega_0} c\overline{c^*} - \tau_0 k e^{-i\omega_0} c\right)},$$

$$c^* = \left(\frac{-\tau_0 f}{\tau_0 m + i\omega_0}\right), \quad c = \left(\frac{i\omega_0 - \tau_0 m}{\tau_0 g e^{-i\omega_0}}\right), \quad K_{11} = \tau_0\left(2p + r\overline{c}e^{i\omega_0} + rce^{-i\omega_0}\right).$$

From (6.64) we also have

$$g_{21} = \overline{E}(\overline{c^*} + \frac{\sigma b}{4})K_{21},$$

where

$$\overline{E} = \frac{1}{\left(\overline{c^*} + c + \tau_0 g e^{-i\omega_0} c\overline{c^*} - \tau_0 k e^{-i\omega_0} c\right)},$$

$$c^* = \left(\frac{-\tau_0 f}{\tau_0 m + i\omega_0}\right), \quad c = \left(\frac{i\omega_0 - \tau_0 m}{\tau_0 g e^{-i\omega_0}}\right),$$

$$K_{21} = \tau_0 \begin{pmatrix} 2pw_{11_1}(0) + pw_{20_1}(0) + 2rw_{11_2}(-1) + rw_{20_2}(-1) \\ +r\overline{c}e^{i\omega_0}w_{20_1}(0) + 2rce^{-i\omega_0}w_{11_1}(0) \end{pmatrix}$$

and

$$w_{20}(\theta) = -\frac{1}{i\omega_0}q(0)e^{i\omega_0\theta}g_{20} - \frac{1}{3i\omega_0}\overline{q}(0)e^{-i\omega_0\theta}\overline{g}_{02} + Se^{2i\omega_0\theta},$$

where

$$S = \left(\begin{bmatrix} 2i\omega_0 - \tau_0 m & \tau_0 g e^{-2i\omega_0} \\ \tau_0 f & 2i\omega_0 + \tau_0 k e^{-2i\omega_0} \end{bmatrix}\right)^{-1} 2K_{20}\begin{pmatrix} 1 \\ \frac{\sigma b}{4} \end{pmatrix},$$

and

$$w_{11}(\theta) = \frac{1}{i\omega_0}q(0)e^{i\omega_0\theta}g_{11} - \frac{1}{i\omega_0}\overline{q}(0)e^{-i\omega_0\theta}\overline{g}_{11} + G,$$

where

$$G = \left(\tau_0 \begin{bmatrix} m & g \\ f & -k \end{bmatrix}\right)^{-1} K_{11} \begin{pmatrix} 1 \\ \frac{\sigma b}{4} \end{pmatrix}.$$

Now, repeating a similar calculation as in the former section, we can determine the direction of the Hopf bifurcation with respect to Theorem 4.

6.4 Numerical Simulations

In this section, we present some numerical simulations by using the symbolic mathematical software Matlab to support our theoretical results obtained for the ODE models in Section 6.1.

6.4.1 Numerical Simulations of System (6.4)

The DDE model (6.4) contains three parameters: a, σ, b. First, we choose these parameters as follows:

$$a = 15, \ \sigma = 8 \text{ and } b = 1.2. \tag{6.88}$$

Using them, we write Eq. (6.4) as follows:

$$\begin{cases} \frac{du}{dt} = 15 - u(t) - \frac{4u(t-\tau)v(t)}{1+u^2(t)}, \text{ for } t > 0, \\ \frac{dv}{dt} = 9.6\left(u(t) - \frac{u(t-\tau)v(t)}{1+u^2(t)}\right), \text{ for } t > 0. \end{cases} \tag{6.89}$$

The equilibrium point of Eq. (6.89) is $(u^*, v^*) = (3, 10)$. From Eqs. (6.12) and (6.13) we calculate ω_1 and ω_2 as

$$\omega_1 = 5.0708,$$
$$\omega_2 = 2.8398.$$

In addition, by Eq. (6.14), we calculate $\tau_{n,k}$ (for $n = 1, 2$ and $k = 0, 1, 2, \cdots$) as

$$\tau_{1,k} = 0.1167 + \frac{2k\pi}{\omega_1},$$
$$\tau_{2,k} = 0.2084 + \frac{2k\pi}{\omega_2}.$$

Hence, by Theorem 1, the equilibrium point $(3, 10)$ is stable when $\tau \in [0, 0.1167) \cup (0.2084, 0.1167 + \frac{2\pi}{\omega_1}) \cup \cdots\cdots \cup (0.2084 + \frac{2(s-1)\pi}{\omega_2}, 0.1167 + \frac{2s\pi}{\omega_1})$ and

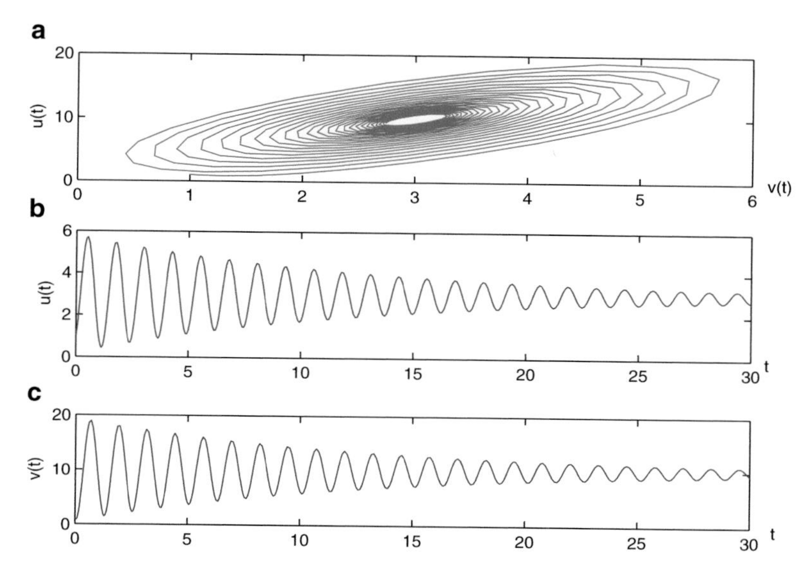

Fig. 6.1 (**a**) Phase portraits of (6.4) with parameters in (6.88), (**b**) trajectory of activator density vs. time, (**c**) trajectory of inhibitor density vs. time; when $\tau = 0.1100 \in [0, \tau_{1,0})$

unstable when $\tau \in (0.1167, 0.2084) \cup (0.1167, 0.2084 + \frac{2\pi}{\omega_2}) \cdots\cdots \cup (0.1167 + \frac{2(s-1)\pi}{\omega_1}, 0.2084 + \frac{2(s-1)\pi}{\omega_2}) \cup (0.1167 + \frac{2s\pi}{\omega_1}, \infty)$ for some positive integer s. In addition, by the Hopf bifurcation theorem in [15], Eq. (6.89) undergoes a Hopf bifurcation at (3, 10) for eigenvalues $\lambda_n = i\omega_n$ $(n = 1, 2)$, where $\omega_1 = 5.0708$ and $\omega_2 = 2.8398$, and possesses a family of real-valued periodic solutions at $\tau_{1,k} = 0.1167 + \frac{2k\pi}{\omega_1}, k = 0, 1, 2, \cdots$, and $\tau_{2,k} = 0.2084 + \frac{2k\pi}{\omega_2}$, $k = 0, 1, 2, \cdots$, respectively.

Now, we illustrate that the Hopf bifurcation occurs at the equilibrium point (3, 10) when $\tau_{1,0} = 0.1167$ for the eigenvalue $\lambda_1 = i\omega_1 = i5.0708$.

In the numerical simulations, the Matlab DDE solver is used to simulate Eq. (6.89). Figure 6.1 shows that the equilibrium point (3, 10) is asymptotically stable when $\tau \in [0, \tau_{1,0})$. In order to demonstrate this, we take $\tau = 0.1100 < \tau_{1,0}$. Matlab simulations in Fig. 6.2 indicate that a stable bifurcating periodic solution occurs at $\tau = \tau_{1,0} = 0.1167$. Finally, Fig. 6.3 presents that the equilibrium point (3, 10) is unstable for $\tau > \tau_{1,0}$ by taking $\tau = 0.1250$. Consequently, these numerical simulations support our analytical results.

6.4.2 *Numerical Simulations of System (6.5)*

The DDE model (6.5) also contains three parameters: a, σ, b. First, we choose the parameters as follows:

$$a = 15,\ \sigma = 8 \text{ and } b = 1.2. \tag{6.90}$$

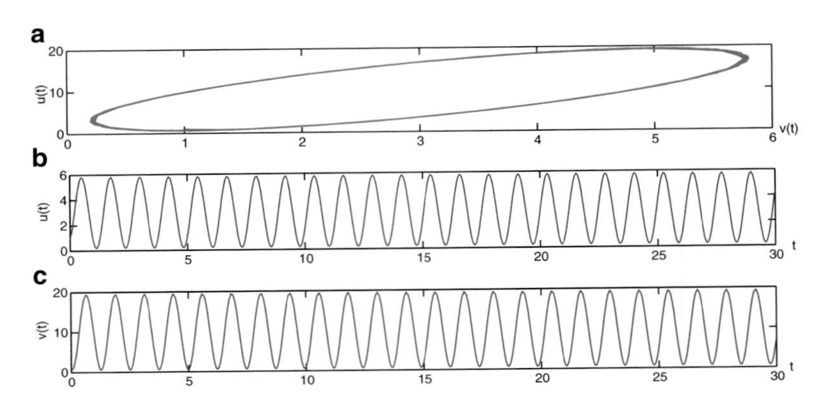

Fig. 6.2 (**a**) Phase portraits of (6.4) with parameters in (6.88), (**b**) trajectory of activator density vs. time, (**c**) trajectory of inhibitor density vs. time; when $\tau = \tau_{1,0} = 0.1167$

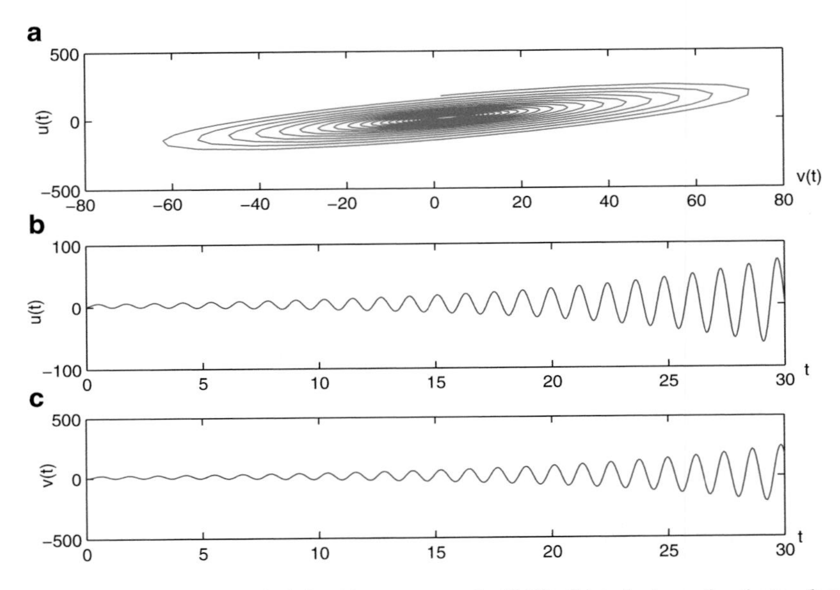

Fig. 6.3 (**a**) Phase portraits of (6.4) with parameters in (6.88), (**b**) trajectory of activator density vs. time, (**c**) trajectory of inhibitor density vs. time; when $\tau = 0.1250 > \tau_{1,0}$

Under the set of parameters in (6.90), Eq. (6.5) turns into

$$\begin{cases} \frac{du}{dt} = 15 - u(t-\tau) - \frac{4u(t-\tau)v(t)}{1+u^2(t-\tau)}, \text{ for } t > 0, \\ \frac{dv}{dt} = 9.6\left(u(t-\tau) - \frac{u(t-\tau)v(t)}{1+u^2(t-\tau)}\right), \text{ for } t > 0. \end{cases} \tag{6.91}$$

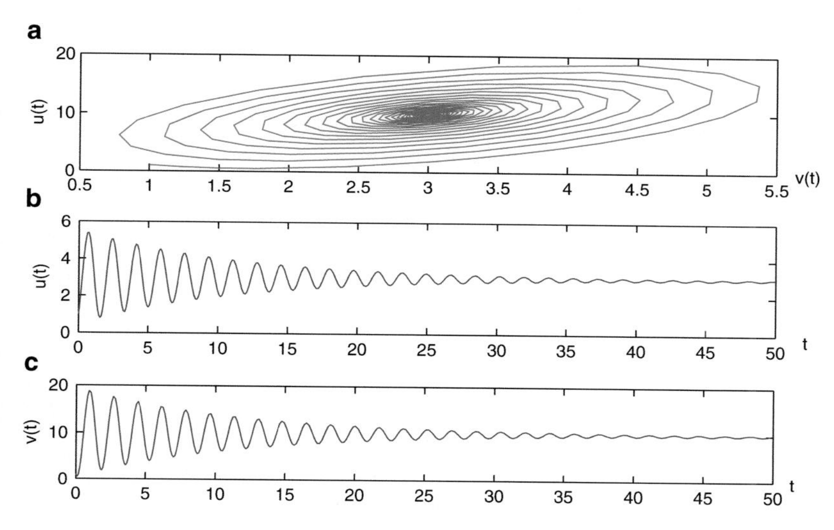

Fig. 6.4 (**a**) Phase portraits of (6.5) with parameters in (6.90), (**b**) trajectory of activator density vs. time, (**c**) trajectory of inhibitor density vs. time; when $\tau = 0.0350 \in [0, \tau_{0,0})$

The equilibrium point of Eq. (6.91) is $(u^*, v^*) = (3, 10)$. Equation (6.21) gives only one ω_0, as follows:

$$\omega_0 = 3.5744.$$

In addition, from Eq. (6.22), we calculate $\tau_{0,k}$ (for $k = 0, 1, 2, \cdots$) as

$$\tau_{0,k} = 0.0499 + \frac{2k\pi}{\omega_0}.$$

Hence, by Theorem 2, the equilibrium point $(3, 10)$ is stable when $\tau \in [0, 0.0499)$ and unstable when $\tau > 0.0499$. In addition, by the Hopf bifurcation theorem in [15], the system (6.91) undergoes a Hopf bifurcation at $(3, 10)$ for the eigenvalue $\lambda_0 = i\omega_0$, where $\omega_0 = 3.5744$, and possesses a family of real-valued periodic solutions at $\tau_{0,k} = 0.0499 + \frac{2k\pi}{\omega_0}$ for $k = 0, 1, 2, \cdots$.

Now, we illustrate the Hopf bifurcation occurring at the equilibrium point $(3, 10)$ when $\tau_{0,0} = 0.0499$ for the eigenvalue $\lambda_0 = i\omega_0 = i3.5744$.

In the numerical simulations, the Matlab DDE solver is again used to simulate Eq. (6.91). Figure 6.4 shows that the equilibrium point $(3, 10)$ is asymptotically stable when $\tau \in [0, \tau_{0,0})$. In order to demonstrate this, we take $\tau = 0.0350 < \tau_{0,0}$. Matlab simulations in Fig. 6.5 indicate that a stable bifurcating periodic solution occurs at $\tau = \tau_{0,0} = 0.0499$. Finally, Fig. 6.6 presents that the equilibrium point $(3, 10)$ is unstable for $\tau > \tau_{0,0}$ by taking $\tau = 0.0600$. Consequently, these numerical simulations support our analytical results.

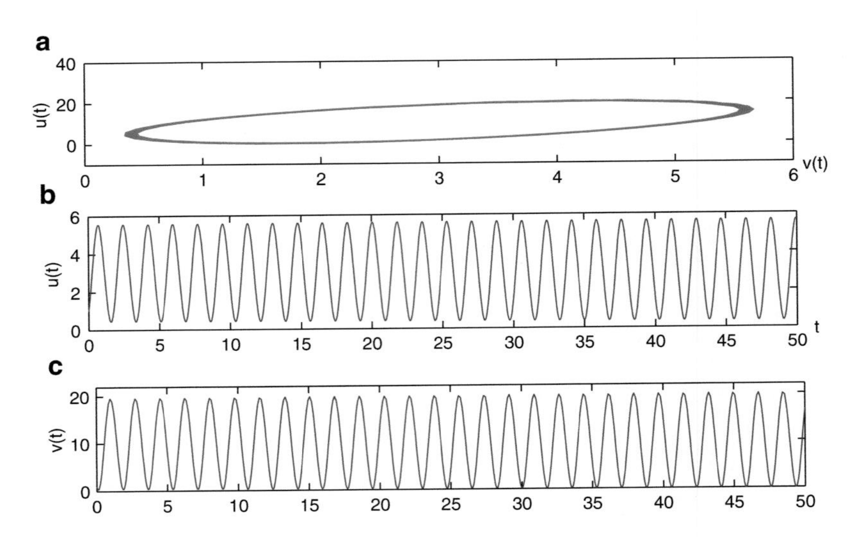

Fig. 6.5 (**a**) Phase portraits of (6.5) with parameters in (6.90), (**b**) trajectory of activator density vs. time, (**c**) trajectory of inhibitor density vs. time; when $\tau = \tau_{0,0} = 0.0499$

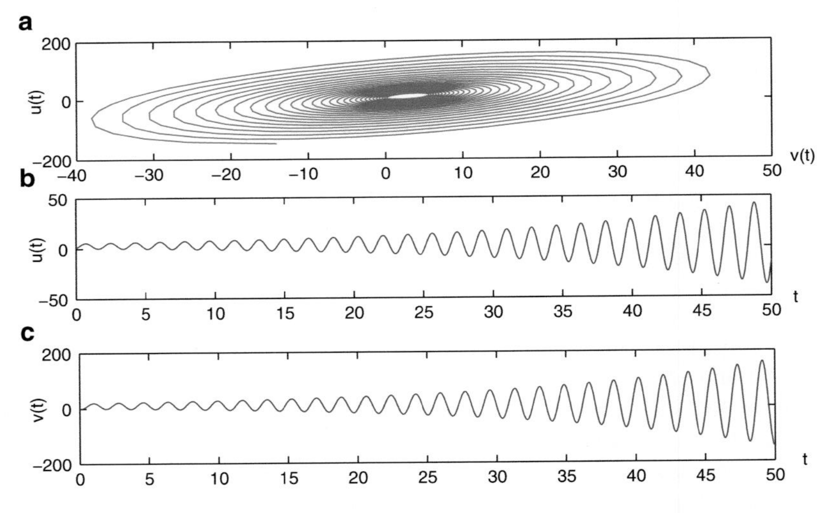

Fig. 6.6 (**a**) Phase portraits of (6.5) with parameters in (6.90), (**b**) trajectory of activator density vs. time, (**c**) trajectory of inhibitor density vs. time; when $\tau = 0.0600 > \tau_{0,0}$

6.5 Conclusion

Former studies have shown that the delay parameter plays an important role in the stability analysis of positive equilibrium points of a dynamical system (see, for example, [1, 2, 4, 9, 15, 26, 27, 30] and references therein). In addition, diffusion-driven instability, which is also known as Turing instability in the literature, has been studied extensively in the last two decades (see, for example, [1, 4, 9–12, 17, 23–27, 29–31, 33, 38, 39] and references therein).

In this chapter we have studied delay effects on the dynamics of the Lengyel–Epstein reaction-diffusion model with Neumann boundary conditions. First, we investigated the necessary conditions at which Hopf bifurcation occurs by choosing the delay parameter τ as a bifurcation parameter for the ODE and PDE models. Using Poincaré's normal form and the center manifold reduction for partial functional differential equations, we obtained the formulas that determine the direction of bifurcation and the stability of periodic solutions for the ODE model. We showed that when the bifurcation parameter τ passes through a critical bifurcation value $\tau_{n,k}$ $(n, k = 0, 1, 2, \ldots)$, stability of the positive equilibrium point of the system changes from stable to unstable or vice versa, and a Hopf bifurcation occurs at these critical values when the associated characteristic equation has only one purely imaginary root. Moreover, if the characteristic equation has two different pure imaginary roots, then not only does Hopf bifurcation occur but the stability switches of equilibrium point also holds.

References

1. Akkocaoğlu, H., Merdan, H., Çelik., C.: Hopf bifurcation analysis of a general non-linear differential equation with delay. J. Comput. Appl. Math. **237**, 565–575 (2013)
2. Allen, L.J.S.: An Introduction to Mathematical Biology. Pearson-Prentice Hall, Upper Saddle River, NJ (2007)
3. Andronov, A.A., Witt, A.: Sur la theórie mathematiques des autooscillations. C. R. Acad. Sci. Paris **190**, 256–258 (1930) [French]
4. Balachandran, B., Kalmar-Nagy, T., Gilsinn, D.E.: Delay Differential Equations: Recent Advances and New Directions. Springer, New York (2009)
5. Bellman, R., Cooke, K.L.: Differential-Difference Equations. Academic Press, New York (1963)
6. Chafee, N.: A bifurcation problem for functional differential equation of finitely retarded type. J. Math. Anal. Appl. **35**, 312–348 (1971)
7. Cooke, K.L., Driessche, P.: On zeroes of some transcendental equations. Funkcialaj Ekvacioj **29**, 77–90 (1986)
8. Cooke, K.L., Grossman, Z.: Discrete delay, distributed delay and stability switches. J. Math. Anal. Appl. **86**, 592–627 (1982)
9. Çelik, C., Merdan, H.: Hopf bifurcation analysis of a system of coupled delayed-differential equations. Appl. Math. Comput. **219**(12), 6605–6617 (2013)
10. De Kepper, P., Castets, V., Dulos, E., Boissonade, J.: Turing-type chemical patterns in the chlorite-iodide-malonic acid reaction. Phys. D **49**, 161–169 (1991)
11. Du, L., Wang, M.: Hopf bifurcation analysis in the 1-D Lengyel–Epstein reaction-diffusion model. J. Math. Anal. Appl. **366**, 473–485 (2010)
12. Epstein, I.R., Pojman, J.A.: An Introduction to Nonlinear Chemical Dynamics. Oxford University Press, Oxford (1998)
13. Hale, J.K.: Theory of Functional Differential Equations. Springer, Berlin (1977)
14. Hale, J.K., Kogak, H.: Dynamics and Bifurcations. Springer, New York (1991)
15. Hassard, B.D., Kazarinoff, N.D., Wan, Y.H.: Theory and Application of Hopf Bifurcation. Cambridge University Press, Cambridge (1981)

16. Hopf, E.: Abzweigung einer periodischen Lösung von einer stationären Lösung eines differential systems. Ber. d. Sachs. Akad. d. Wiss. (Math.-Phys. Kl). Leipzig **94**, 1–22 (1942) [German]
17. Jang, J., Ni, W.M., Tang, M.: Global bifurcation and structure of Turing patterns in 1-D Lengyel–Epstein model. J. Dyn. Diff. Equ. **16**, 297–320 (2004)
18. Jin, J., Shi, J., Wei, J., Yi, F.: Bifurcations of patterned solutions in diffusive Lengyel–Epstein system of CIMA chemical reaction. Rocky Mountain J. Math. **43**(5), 1637–1674 (2013)
19. Karaoglu, E., Merdan, H.: Hopf bifurcation analysis for a ratio-dependent predator-prey system involving two delays. ANZIAM J. **55**, 214–231 (2014)
20. Karaoglu, E., Merdan, H.: Hopf bifurcations of a ratio-dependent predator-prey model involving two discrete maturation time delays. Chaos Soliton Fractals **68**, 159–168 (2014)
21. Kuang, Y.: Delay Differential Equations with Application in Population Dynamics. Academic Press, New York (1993)
22. Kuznetsov, Y.A.: Elements of Applied Bifurcation Theory. Springer, New York (1995)
23. Lengyel, I., Epstein, I.R.: Modeling of Turing structure in the chlorite-iodide-malonic acid-starch reaction system. Science **251**, 650–652 (1991)
24. Lengyel, I., Epstein, I.R.: A chemical approach to designing Turing patterns in reaction-diffusion system. Proc. Natl. Acad. Sci. USA **89**, 3977–3979 (1992)
25. Li, B., Wang, M.: Diffusion-driven instability and Hopf bifurcation in Brusselator system. Appl. Math. Mech. (English Ed.) **29**, 825–832 (2008)
26. Ma, Z.P.: Stability and Hopf bifurcation for a three-component reaction-diffusion population model with delay effect. Appl. Math. Model. **37**(8), 5984–6007 (2013)
27. Mao, X.-C., Hu, H.-Y.: Hopf bifurcation analysis of a four-neuron network with multiple time delays. Nonliear Dyn. **55**(1–2), 95–112 (2009)
28. Marsden, J.E., McCracken, M.: The Hopf Bifurcation and Its Applications. Springer, New York (1976)
29. Merdan, H., Kayan, Ş.: Hopf bifurcations in Lengyel-Epstein reaction-diffusion model with discrete time delay. Nonlinear Dyn. **79**, 1757–1770 (2015)
30. Murray, J.D.: Mathematical Biology. Springer, New York, (2002)
31. Ni, W., Tang, M.: Turing patterns in the Lengyel–Epstein system for the CIMA reaction. Trans. Am. Math. Soc. **357**, 3953–3969 (2005)
32. Rovinsky, A., Menzinger, M.: Interaction of Turing and Hopf bifurcations in chemical systems. Phys. Rev. A **46**(10), 6315–6322 (1998)
33. Ruan, S.: Diffusion-driven instability in the Gierer–Meinhardt model of morphogenesis. Nat. Resour. Model. **11**, 131–132 (1998)
34. Turing, A.M.: The chemical basis of morphogenesis. Philos. Trans. A Ser. B **237**, 37–72 (1952)
35. Wu, J.: Theory and Applications of Partial Differential Equations. Springer, New York (1996)
36. Xu, C., Shao, Y.: Bifurcations in a predator-prey model with discrete and distributed time delay. Nonliear Dyn. **67**(3), 2207–2223 (2012)
37. Yafia, R.: Hopf bifurcation in differential equations with delay for tumor-immune system competition model. SIAM J. Appl. Math. **67**(6), 1693–1703 (2007)
38. Yi, F., Wei, J., Shi, J.: Diffusion-driven instability and bifurcation in the Lengyel–Epstein system. Nonlinear Anal. Real World Appl. **9**(3), 1038–1051 (2008)
39. Yi, F., Wei, J., Shi, J.: Global asymptotical behavior of the Lengyel–Epstein reaction-diffusion system. Appl. Math. Lett. **22**(1), 52–55 (2009)
40. Zang, G., Shen, Y., Chen, B.: Hopf bifurcation of a predator-prey system with predator harvesting and two delays. Nonliear Dyn. **73**(4), 2119–2131 (2013)

Chapter 7
Almost Periodic Solutions of Evolution Differential Equations with Impulsive Action

Viktor Tkachenko

Abstract In an abstract Banach space we study conditions for the existence of piecewise continuous, almost periodic solutions for semilinear impulsive differential equations with fixed and nonfixed moments of impulsive action.

7.1 Introduction

We consider the problem of the existence of piecewise continuous, almost periodic solutions for the linear impulsive differential equation

$$\frac{du}{dt} + (A + A_1(t)u = f(t, u), \quad t \neq \tau_j(u), \tag{7.1}$$

$$u(\tau_j(u) + 0) - u(\tau_j(u)) = B_j u + g_j(u), \quad j \in Z, \tag{7.2}$$

where $u : R \to X,\ X$ is a Banach space, A is a sectorial operator in X, $A_1(t)$ is some operator-valued function, $\{B_j\}$ is a sequence of some closed operators, and $\{\tau_j(u)\}$ is an unbounded and strictly increasing sequence of real numbers for all u from some domain of space X.

We use the concept of piecewise continuous, almost periodic functions proposed in [7]. Points of discontinuities of these functions coincide with points of impulsive actions $\{\tau_j\}$. We mention the remarkable paper [18], where a number of important statements about the almost periodic pulse system were proved. Then these results were included in the well-known monograph [19]. Today there are many articles related to the study of almost periodic impulsive systems (see, for example, [1, 3]). In the papers [8, 23, 27, 28] almost periodic solutions for abstract impulsive differential equations in the Banach space are investigated.

In this chapter we consider the semilinear abstract impulsive differential equation in a Banach space with sectorial operator in the linear part of the equation and

V. Tkachenko (✉)
Institute of Mathematics National Academy of Sciences of Ukraine, Tereshchenkivska str. 3, Kiev, Ukraine
e-mail: vitk@imath.kiev.ua

A.C.J. Luo, H. Merdan (eds.), *Mathematical Modeling and Applications in Nonlinear Dynamics*, Nonlinear Systems and Complexity 14,
DOI 10.1007/978-3-319-26630-5_7

some closed operators in linear parts of impulsive action. Using fractional powers of operator A and corresponding interpolation spaces allows us to consider strong or classical solutions. Note that such equations with periodic right-hand sides were first studied in [17]. In equations with nonfixed moments of impulsive action, points of discontinuity depend on solutions; that is, every solution has its own points of discontinuity. Moreover, a solution can intersect the surface of impulsive action several times or even an infinite number of times. This is the so-called pulsation or beating phenomenon. We will assume that solutions of (7.1) and (7.2) don't have beating at the surfaces $t = \tau_j(u)$; in other words, solutions intersect each surface no more than once. For impulsive systems in the finite-dimensional case, there are several sufficient conditions that allow us to exclude the phenomenon of pulsation (see, [19], [22]). Unfortunately, in a Banach space this conditions cannot easily be verified. In every concrete case one needs a separate investigation.

We assume that the corresponding linear homogeneous equation (if $f \equiv 0$, $g_j \equiv 0$) has an exponential dichotomy. The definition of exponential dichotomy for an impulsive evolution equation corresponds to the definition of exponential dichotomy for continuous evolution equations in an infinite-dimensional Banach space [5, 9, 16]. We require that only solutions of a linear system from an unstable manifold be unambiguously extended to the negative semiaxis.

Robustness is an impotent property of the exponential dichotomy [5, 10, 16]. We mention the papers [4, 14, 25, 26], where the robustness of the exponential dichotomy for impulsive systems by small perturbations of right-hand sides is proved. In this chapter we prove robustness of the exponential dichotomy also by the small perturbation of points of impulsive action. We use a change of time in the system. Then approximation of the impulsive system by difference systems (see [9]) can be used. If a linear homogeneous equation is exponentially stable, we prove stability of the almost periodic solution of nonlinear equations (7.1) and (7.2). Following [17], we use the generalized Gronwall inequality, taking into account singularities in integrals and impulsive influences.

This chapter is organized as follows. In Sect. 7.2 we present some preliminary definitions and results. In Sect. 7.3, we study an exponential dichotomy of impulsive linear equations. Section 7.4 is devoted to studying the existence and stability of almost periodic solutions in linear inhomogeneous equations with impulsive action and semilinear impulsive equations with fixed moments of impulsive action. In Sect. 7.5 we consider impulsive evolution equations with nonfixed moments of impulsive action. In Sect. 7.6 we discuss the case of unbounded operators B_j in linear parts of linear parts of impulsive action.

7.2 Preliminaries

Let $(X, \|.\|)$ be an abstract Banach space and R and Z be the sets of real and integer numbers, respectively.

We consider the space $\mathscr{PC}(J, X)$, $J \subset R$, of all piecewise continuous functions $x : J \to X$ such that

i) the set $\{\tau_j \in J : \tau_{j+1} > \tau_j, j \in Z\}$ of discontinuities of x has no finite limit points;
ii) $x(t)$ is left-continuous $x(\tau_j + 0) = x(\tau_j)$ and there exists $\lim_{t\to\tau_j-0} x(t) = x(\tau_j - 0) < \infty$.

We will use the norm $\|x\|_{PC} = \sup_{t\in J} \|x(t)\|$, in the space $\mathscr{PC}(J, X)$.

Definition 1. The integer p is called an ε-almost period of a sequence $\{x_k\}$ if $\|x_{k+p} - x_k\| < \varepsilon$ for any $k \in Z$. The sequence $\{x_k\}$ is almost periodic if for any $\varepsilon > 0$ there exists a relatively dense set of its ε-almost periods.

Definition 2. The strictly increasing sequence $\{\tau_k\}$ of real numbers has uniformly almost periodic sequences of differences if for any $\varepsilon > 0$ there exists a relatively dense set of ε-almost periods common for all sequences $\{\tau_k^j\}$, where $\tau_k^j = \tau_{k+j} - \tau_k, j \in Z$.

By Samoilenko and Trofimchuk [21], the sequence $\{\tau_k\}$ has uniformly almost periodic sequences of differences if and only if $\tau_k = ak + c_k$, where $\{c_k\}$ is an almost periodic sequence and a is a positive real number.

By Lemma 22 ([19], p. 192), for a sequence $\{\tau_j\}$ with uniformly almost periodic sequences of differences there exists the limit

$$\lim_{T\to\infty} \frac{i(t, t+T)}{T} = p \tag{7.3}$$

uniformly with respect to $t \in R$, where $i(s, t)$ is the number of the points τ_k lying in the interval (s, t). Then for each $q > 0$ there exists a positive integer N such that on each interval of length q there are no more than N elements of the sequence $\{\tau_j\}$; that is, $i(s, t) \leq N(t - s) + N$.

Also, for sequence $\{\tau_j\}$ with uniformly almost periodic sequences of differences there exists $\Theta > 0$ such that $\tau_{j+1} - \tau_j \leq \Theta, j \in Z$.

Definition 3. The function $\varphi \in \mathscr{PC}(R, X)$ is said to be W-almost periodic if

i) the strictly increasing sequence $\{\tau_k\}$ of discontinuities of $\varphi(t)$ has uniformly almost periodic sequences of differences;
ii) for any $\varepsilon > 0$ there exists a positive number $\delta = \delta(\varepsilon)$ such that if the points t' and t'' belong to the same interval of continuity and $|t' - t''| < \delta$, then $\|\varphi(t') - \varphi(t'')\| < \varepsilon$;
iii) for any $\varepsilon > 0$ there exists a relatively dense set Γ of ε-almost periods such that if $\tau \in \Gamma$, then $\|\varphi(t + \tau) - \varphi(t)\| < \varepsilon$ for all $t \in R$ that satisfy the condition $|t - t_k| \geq \varepsilon, k \in Z$.

We consider the impulsive equations (7.1) and (7.2) with the following assumptions:

(H1) A is a sectorial operator acting in X and $\inf\{Re\mu : \mu \in \sigma(A)\} \geq \delta > 0$, where $\sigma(A)$ is the spectrum of A. Consequently, the fractional powers of A are well defined, and one can consider the spaces $X^\alpha = D(A^\alpha)$ for $\alpha \geq 0$ endowed with the norms $\|x\|_\alpha = \|A^\alpha x\|$.

(H2) The function $A_1(t) : R \to L(X^\alpha, X)$ is Bohr almost periodic and Hölder continuous, $\alpha \geq 0$, $L(X^\alpha, X)$ is the space of linear bounded operators $X^\alpha \to X$.

(H3) We shall use the notation $U_\varrho^\alpha = \{x \in X^\alpha : \|x\|_\alpha \leq \varrho\}$. Assume that the sequence $\{\tau_j(u)\}$ of functions $\tau_j : U_\varrho^\alpha \to R$ has uniformly almost periodic sequences of differences uniformly with respect to $u \in U_\varrho^\alpha$ and there exists $\theta > 0$ such that $\inf_u \tau_{j+1}(u) - \sup_u \tau_j(u) \geq \theta > 0$, for all $u \in U_\varrho^\alpha$ and $j \in Z$. Also, there exists $\Theta > 0$ such that $\sup_u \tau_{j+1}(u) - \inf_u \tau_j(u) \leq \Theta$ for all $j \in Z$ and $u \in U_\varrho^\alpha$.

(H4) The sequence $\{B_j\}$ of bounded operators is almost periodic and there exists $b > 0$ such that $\|B_j u\|_\alpha \leq b\|u\|_\alpha$ for $j \in Z, \alpha \geq 0$, and $u \in X^\alpha$.

(H5) The function $f(t, u) : R \times U_\rho^\alpha \to X$ is continuous in u and is Hölder continuous and W-almost periodic in t uniformly with respect to $x \in U_\rho^\alpha$ with some $\rho > 0$.

(H6) The sequence $\{g_j(u)\}$ of continuous functions $U_\rho^\alpha \to X^\alpha$ is almost periodic uniformly with respect to $x \in U_\rho^\alpha$.

Remark 1. We assume that operators B_j are bounded and satisfy assumption **(H4)**. Many of our results are valid if the B_j are unbounded closed operators $X^{\alpha+\gamma} \to X^\alpha$ for $\alpha \geq 0$ and some $\gamma \geq 0$. We discuss this case in the last section.

We use the following generalization of Lemma 7 from [7] (also, see [6] and [19]):

Lemma 1. *Assume that a sequence of real numbers $\{\tau_j\}$ has uniformly almost periodic sequences of differences, the sequence $\{B_j\}$ is almost periodic, and the function $f(t) : R \to X$ is W-almost periodic. Then for any $\varepsilon > 0$ there exist a such $l = l(\varepsilon) > 0$ that for any interval J of length l there are such $r \in J$ and an integer q that the following relations hold:*

$$\|f(t + r) - f(t)\| < \varepsilon, \; t \in R, |t - \tau_j| > \varepsilon, j \in Z,$$

$$\|B_{k+q} - B_k\| < \varepsilon, \; \|\tau_k^q - r\| < \nu, k \in Z.$$

If A is a sectorial operator, then $(-A)$ is an infinitesimal generator of the analytical semigroup e^{-At}. For every $x \in X^\alpha$ we get $e^{-At}A^\alpha x = A^\alpha e^{-At}x$. Further, we shall use the inequalities (see [9])

$$\|A^\alpha e^{-At}\| \leq C_\alpha t^{-\alpha} e^{-\delta t}, \; t > 0, \; \alpha > 0,$$

$$\|(e^{-At} - I)u\| \leq \frac{1}{\alpha} C_{1-\alpha} t^\alpha \|A^\alpha u\|, \; t > 0, \alpha \in (0, 1], u \in X^\alpha,$$

where $C_\alpha \in R$ is nonnegative and bounded as $\alpha \to +0$.

Definition 4. The function $x(t) : [t_0, t_1] \to X^\alpha$ is said to be a solution of the initial-value problem $u(t_0) = u_0 \in X^\alpha$ for Eqs. (7.1) and (7.2) on $[t_0, t_1]$ if

(i) it is continuous in $[t_0, \tau_k], (\tau_k, \tau_{k+1}], \ldots, (t_{k+s}, t_1]$ with the discontinuities of the first kind at the moments $t = \tau_j(u)$ of intersections with impulsive surfaces;
(ii) $x(t)$ is continuously differentiable in each of the intervals (t_0, τ_k), (τ_k, τ_{k+1}), $\ldots, (t_{k+s}, t_1)$ and satisfies Eqs. (7.1) and (7.2) if $t \in (t_0, t_1), t \neq \tau_j$, and $t = \tau_j$, respectively;
(iii) the initial-value condition $u(t_0) = u_0$ is fulfilled.

We assume that solutions $u(t)$ of (7.1) and (7.2) are left-hand-side continuous; hence $u(\tau_j) = u(\tau_j - 0)$ at all points of impulsive action.

Also, we assume that in the domain U_ρ^α solutions of (7.1) and (7.2) don't have beating at the surfaces $t = \tau_j(u)$; in other words, solutions intersect each surface only once.

7.3 Exponential Dichotomy

Together with Eqs. (7.1) and (7.2) we consider the corresponding linear homogeneous equation

$$\frac{du}{dt} + (A + A_1(t))u = 0, \quad t \neq \tau_j, \tag{7.4}$$

$$\Delta u|_{t=\tau_j} = u(\tau_j + 0) - u(\tau_j) = B_j u(\tau_j), \quad j \in Z, \tag{7.5}$$

where $\tau_j = \tau_j(0)$. Denote by $V(t,s)$ the evolution operator of the linear equation without impulses (7.4). It satisfies $V(\tau,\tau) = I,\ V(t,s)V(s,\tau) = V(t,\tau),\ t \geq s \geq \tau.$

By Theorem 7.1.3 [9, p.190], $V(t,\tau)$ is strongly continuous with values in $L(X^\beta)$ for any $0 \leq \beta < 1$ and

$$\|V(t,\tau)x\|_\beta \leq L_Q(t-\tau)^{(\gamma-\beta)_-}\|x\|_\gamma, \tag{7.6}$$

where $(\gamma - \beta)_- = \min(\gamma - \beta, 0),\ t - \tau \leq Q,\ L_Q = L_Q(Q)$. Moreover,

$$\|V(t,\tau)x - x\|_\beta \leq L_{\beta,\nu}(t-\tau)^\nu \|x\|_{\beta+\nu}, \quad \nu > 0, \beta + \nu \leq 1. \tag{7.7}$$

Using the proof of Lemma 7.1.1 from [9], p. 188, one can verify the following generalized Gronwall inequality:

Lemma 2. *$a_1 \geq 0, a_2 \geq 0$, and $y(t)$ is a nonnegative function locally integrable on $0 \leq t < Q$ with*

$$y(t) \leq a_1 + a_2 t^{-\alpha} + b\int_0^t (t-s)^{-\beta} u(s)ds$$

on this interval; then there is a constant $\tilde{C} = \tilde{C}(\beta, b, Q) < \infty$ *such that*

$$y(t) \leq \left(a_1 + \frac{a_2}{(1-\alpha)t^\alpha}\right) \tilde{C}(\beta, b, Q).$$

We will use the following perturbation lemma.

Lemma 3. *Let us consider the perturbation equation*

$$\frac{du}{dt} + (\gamma A + A_2(t))u = 0, \tag{7.8}$$

where $\gamma = Const > 0,\ A_2(t) : R \to L(X^\alpha, X)$.

For $Q > 0$, *there exists* $\varepsilon_0 > 0$ *such that for all* $\varepsilon \leq \varepsilon_0$ *and* $|\gamma - 1| \leq \varepsilon$, $\sup_t \|A_1(t) - A_2(t)\|_{L(X^\alpha, X)} \leq \varepsilon$ *the evolution operators* $V(t,s)$ *of (7.4) and* $V_1(t,s)$ *of (7.8) satisfy*

$$\|V(t,s) - V_1(t,s)\|_\alpha \leq R_1(\varepsilon),\ t - s \leq Q, \tag{7.9}$$

where $R_1(\varepsilon)$ *depends on* Q, α, *and* $R_1(\varepsilon) \to 0$ *as* $\varepsilon \to 0$.

Proof. For definiteness let $\gamma > 1$. Solutions $x(t)$ and $y(t)$ of Eqs. (7.4) and (7.8) satisfy the following integral equations:

$$x(t) = e^{-A(t-t_0)}x_0 + \int_{t_0}^{t} e^{-A(t-s)}A_1(s)x(s)ds$$

and

$$y(t) = e^{-A\gamma(t-t_0)}x_0 + \int_{t_0}^{t} e^{-A\gamma(t-s)}A_2(s)y(s)ds.$$

Then

$$\begin{aligned}
&\|x(t) - y(t)\|_\alpha \leq \|(I - e^{-A(\gamma-1)(t-t_0)})A^\alpha e^{-A(t-t_0)}x_0\| + \\
&+ \int_{t_0}^{t} \|(I - e^{-A(\gamma-1)(t-s)})A^\alpha e^{-A(t-s)}A_1(s)x(s)\|ds + \\
&+ \int_{t_0}^{t} \|A^\alpha e^{-A\gamma(t-s)}(A_1(s) - A_2(s))x(s)\|ds + \\
&+ \int_{t_0}^{t} \|A^\alpha e^{-A\gamma(t-s)}A_2(s)(x(s) - y(s))\|ds \leq \\
&\leq a_1(\varepsilon)\|x_0\|_\alpha + a_2 \int_{t_0}^{t} (t-s)^{-\alpha}\|x(s) - y(s)\|_\alpha ds,
\end{aligned}$$

where $a_2 = C_\alpha \sup_s \|A_1(s)\|_{L(X^\alpha,X)}$ and $a_1(\varepsilon) \to 0$ as $\varepsilon \to 0$. By Lemma 2, there exists a positive constant K_1 depending on α and Q such that

$$\|x(t) - y(t)\|_\alpha \leq K_1 a_1(\varepsilon)\|x_0\|_\alpha = R_2(\varepsilon)\|x_0\|_\alpha.$$

Lemma 4. *Let us consider Eq. (7.4) and*

$$\frac{dv}{dt} + (A + A_2(t))v = 0, \tag{7.10}$$

such that $A_2 : R \to L(X^\alpha, X)$ is a bounded and Hölder continuous function.
Then for $Q > 0$, there exists $\varepsilon_0 > 0$ such that for all $\varepsilon \leq \varepsilon_0$ and

$$\sup_t \|A_1(t) - A_2(t)\|_{L(X^\alpha,X)} \leq \varepsilon$$

the evolution operators $V(t,s)$ of (7.4) and $V_1(t,s)$ of (7.10) satisfy

$$\|(V(t,s) - V_1(t,s))u\|_\alpha \leq R_3(\varepsilon)|t - t_0|^{1-2\alpha+\delta}\|u\|_\delta,\ t - s \leq Q, \tag{7.11}$$

where $R_3(\varepsilon) = R_3(\varepsilon, Q, \alpha)$ and $R_3(\varepsilon) \to 0$ as $\varepsilon \to 0$.

Proof. Denote by $u(t)$ and $v(t)$ solutions of (7.4) and (7.10) with initial value $u(t_0) = u(t_0) = u_0$. They satisfy the inequalities

$$\|u(t) - v(t)\|_\alpha \leq \int_{t_0}^t \|A^\alpha e^{-A(t-s)}(A_1(s) - A_2(s))u(s)\|ds +$$

$$+ \int_{t_0}^t \|A^\alpha e^{-A(t-s)}A_2(s)(u(s) - v(s))\|ds \leq$$

$$\leq C_\alpha L_Q \varepsilon \|u_0\|_\delta \int_{t_0}^t \frac{ds}{(t-s)^\alpha (s-t_0)^{\alpha-\delta}} + C_\alpha \|A_1\|_L \int_{t_0}^t \frac{\|u(s) - v(s)\|_\alpha ds}{(t-s)^\alpha} \leq$$

$$\leq \varepsilon \|u_0\|_\delta R_4 + C_\alpha \|A_1\|_L \int_{t_0}^t \frac{\|u(s) - v(s)\|_\alpha ds}{(t-s)^\alpha}. \tag{7.12}$$

Applying Lemma 2 to (7.12), we obtain (7.11).

We define the evolution operator for Eqs. (7.4) and (7.5) as

$$U(t,s) = V(t,s) \text{ if } \tau_k < s \leq t \leq \tau_{k+1}$$

and

$$U(t,s) = V(t,\tau_k)(I + B_k)V(\tau_k,\tau_{k-1})\ldots(I + B_m)V(\tau_m,s) \tag{7.13}$$

if $\tau_{m-1} < s < \tau_m < \tau_{m+1} < \ldots < \tau_k \leq t \leq \tau_{k+1}$.

It it easy to verify that for fixed $t > s$ the operator $U(t,s)$ is bounded in the space X^α.

Definition 5. We say that the system (7.4)–(7.5) has an exponential dichotomy on R with exponent $\beta > 0$ and bound $M \geq 1$ (with respect to X^α) if there exist projections $P(t), t \in R$, such that

(i) $U(t,s)P(s) = P(t)U(t,s),\ t \geq s$;
(ii) $U(t,s)|_{Im(P(s))}$ for $t \geq s$ is an isomorphism on $Im(P(s))$, and then $U(s,t)$ is defined as an inverse map from $Im(P(t))$ to $Im(P(s))$;
(iii) $\|U(t,s)(1-P(s))u\|_\alpha \leq Me^{-\beta(t-s)}\|u\|_\alpha,\ t \geq s,\ u \in X^\alpha$;
(iv) $\|U(t,s)P(s)\|_\alpha \leq Me^{\beta(t-s)}\|u\|_\alpha,\ t \leq s,\ u \in X^\alpha$.

If the system (7.4)–(7.5) has an exponential dichotomy on R, then the nonhomogeneous equation

$$\frac{du}{dt} + (A + A_1(t))u = f(t), \quad t \neq \tau_j, \tag{7.14}$$

$$\Delta u|_{t=\tau_j} = u(\tau_j + 0) - u(\tau_j) = B_j u(\tau_j) + g_j, \quad j \in Z, \tag{7.15}$$

has a unique solution bounded on R

$$u_0(t) = \int_{-\infty}^{\infty} G(t,s)f(s)(x)ds + \sum_{j\in Z} G(t,\tau_j)g_j, \tag{7.16}$$

where

$$G(t,s) = \begin{cases} U(t,s)(I - P(s)), & t \geq s, \\ -U(t,s)P(s), & t < s, \end{cases}$$

is the Green function such that

$$\|G(t,s)u\|_\alpha \leq Me^{-\beta|t-s|}\|u\|_\alpha,\ t,s \in R. \tag{7.17}$$

Analogous to [9], p. 250, it can be proven that a function $u(t)$ is a bounded solution on the semiaxis $[t_0, +\infty)$ if and only if

$$u(t) = U(t,t_0)(I - P(t_0))u(t_0) + \int_{t_0}^{+\infty} G(t,s)f(s)ds + \sum_{t_0 \leq \tau_j} G(t,\tau_j)g_j,\ t \geq t_0. \tag{7.18}$$

A function $u(t)$ is a bounded solution on the semiaxis $(-\infty, t_0]$ if and only if

$$u(t) = U(t,t_0)P(t_0)u(t_0) + \int_{-\infty}^{t_0} G(t,s)f(s)ds + \sum_{t_0 > \tau_j} G(t,\tau_j)g_j,\ t \leq t_0. \tag{7.19}$$

Now we estimate $\|G(t,s)u\|_\alpha$ for $u \in X$. Let $t > s$ and $\tau_{m-1} < s < \tau_m,\ \tau_k < t < \tau_{k+1}$. Then

$$\begin{aligned}
&\|G(t,s)u\|_\alpha = \|U(t,s)(I - P(s))u\|_\alpha \le \\
&\le \|U(t,\tau_m)(I - P(\tau_m))\|_\alpha \|U(\tau_m, s)u\|_\alpha \le \\
&\le Me^{-\beta(t-\tau_m)} L_\Theta(\tau_m - s)^{-\alpha}\|u\| \le \tilde{M}e^{-\beta(t-s)}|\tau_m - s|^{-\alpha}\|u\|
\end{aligned} \tag{7.20}$$

and

$$\begin{aligned}
&\|G(s,t)u\|_\alpha = \|U(s,t)P(t)u\|_\alpha \le \\
&\le \|U(s,t+1)P(t+1)\|_\alpha \|A^\alpha U(t+1,t)u\| \le \tilde{M}e^{-\beta(t-s)}\|u\|.
\end{aligned} \tag{7.21}$$

If t_1 and t_2 belong to the same interval of continuity, then

$$\|P(t_1)u - P(t_2)u\|_\gamma \le \tilde{M}_1\|t_1 - t_2|^\nu \|u\|_{\gamma+\nu} \tag{7.22}$$

since as in [9], p. 247,

$$\begin{aligned}
&\|P(t+h)u - P(t)u\|_\gamma \le \|P(t)u - V(t+h,t)P(t)u\|_\gamma + \\
&+\|V(t+h,t)P(t)u - P(t+h)u\|_\gamma \le \\
&\le \|(I - V(t+h,t))P(t)u\|_\gamma + \|P(t+h)(V(t+h,t)u - u)\|_\gamma.
\end{aligned}$$

Lemma 5. *Let the impulsive system (7.4) and (7.5) be exponentially dichotomous with positive constants β and M. Then there exists $\varepsilon > 0$ such that the perturbed systems*

$$\frac{du}{dt} + (A + \tilde{A}(t))u = 0, \quad t \ne \tilde{\tau}_j, \tag{7.23}$$

$$\Delta u|_{t=\tilde{\tau}_j} = u(\tilde{\tau}_j + 0) - u(\tilde{\tau}_j) = \tilde{B}_j u(\tilde{\tau}_j), \quad j \in Z, \tag{7.24}$$

with $\sup_j |\tau_j - \tilde{\tau}_j| \le \varepsilon$, $\sup_j \|B_j - \tilde{B}_j\| \le \varepsilon$, $\sup_t \|A_1(t) - \tilde{A}(t)\|_{L((X^\alpha,X)} \le \varepsilon$, are also exponentially dichotomous with some constants $\beta_1 \le \beta$ and $M_1 \ge M$.

Proof. In system (7.4) and (7.5), we introduce the change of time $t = \vartheta(t')$ such that $\tau_j = \vartheta(\tilde{\tau}_j), j \in Z$, and the function ϑ is continuously differentiable and monotonic on each interval $(\tilde{\tau}_j, \tilde{\tau}_{j+1})$.

The function ϑ can be chosen in piecewise linear form:

$$t = a_j t' + b_j,\ a_j = \frac{\tau_{j+1} - \tau_j}{\tilde{\tau}_{j+1} - \tilde{\tau}_j},\ b_j = \frac{\tau_j \tilde{\tau}_{j+1} - \tau_{j+1}\tilde{\tau}_j}{\tilde{\tau}_{j+1} - \tilde{\tau}_j} \text{ if } t' \in (\tilde{\tau}_j, \tilde{\tau}_{j+1}). \quad (7.25)$$

The function $\vartheta(t')$ satisfies the conditions

$$|\vartheta(t') - t'| \le \varepsilon,\ |\frac{d\vartheta(t')}{dt'} - 1| \le 2\varepsilon/\theta.$$

The system (7.4) and (7.5) in the new coordinates $v(t') = u(\vartheta(t'))$ has the form

$$\frac{dv}{dt'} + \frac{d\vartheta(t')}{dt'}\left(A + A_1(\vartheta(t'))\right) v = 0, \quad t \ne \tilde{\tau}_j, \quad (7.26)$$

$$\Delta v|_{t'=\tilde{\tau}_j} = v(\tilde{\tau}_j + 0) - v(\tilde{\tau}_j) = B_j v(\tilde{\tau}_j), \quad j \in Z. \quad (7.27)$$

The system (7.26) and (7.27) has the evolution operator $U_1(t', s') = U(\vartheta(t'), \vartheta(s'))$. If the system (7.4) and (7.5) has an exponential dichotomy with projector $P(t)$ at point t, then the system (7.26) and (7.27) has an exponential dichotomy with projector $P_1(t') = P(\vartheta(t'))$ at point t'. Really,

$$\|U_1(t', s')(1 - P_1(s'))\|_\alpha = \|U(\vartheta(t'), \vartheta(s'))(1 - P(\vartheta(s'))\|_\alpha \le$$
$$\le Me^{-\beta(\vartheta(t') - \vartheta(s'))} \le Me^{2\varepsilon}e^{-\beta(t'-s')},\ t \ge s.$$

The inequality for an unstable manifold is proved analogously.

The linear systems (7.26), (7.27) and (7.23), (7.24) have the same points of impulsive actions $\tilde{\tau}_j, j \in Z$, and

$$\|\frac{d\vartheta(t')}{dt'}A_1(\vartheta(t')) - \tilde{A}(t')\| \le \|\frac{d\vartheta(t')}{dt'}A_1(\vartheta(t')) - A_1(\vartheta(t'))\| +$$
$$+\|A_1(\vartheta(t')) - A_1(t')\| + \|A_1(t') - \tilde{A}(t')\| \le K_2(\varepsilon),$$

where $K_2(\varepsilon) \to 0$ as $\varepsilon \to 0$.

Let $\tilde{U}(t', s')$ be the evolution operator for the system (7.23) and (7.24). To show that for sufficiently small δ_0 the system (7.23) and (7.24) is exponentially dichotomous, we use the following variant of Theorem 7.6.10 [9]:

Assume that the evolution operator $U_1(t', s')$ has an exponential dichotomy on R and satisfies

$$\sup_{0 \le t' - s' \le d} \|U_1(t', s')\|_\alpha < \infty \quad (7.28)$$

for some positive d. Then there exists $\eta > 0$ such that

$$\|\tilde{U}(t',s') - U_1(t',s')\|_\alpha < \eta, \text{ whenever } t - s \le d;$$

the evolution operator $\tilde{U}(t',s')$ also has an exponential dichotomy on R with some constants $\beta_1 \le \beta, M_1 \ge M$.

To prove this statement, we set for $n \in Z$

$$t_n = s' + dn, \quad T_n = U_1(s' + d(n+1), s' + dn),\ \tilde{T}_n = \tilde{U}(s' + d(n+1), s' + dn).$$

If the evolution operator $U_1(t,s)$ has an exponential dichotomy, then $\{T_n\}$ has a discrete dichotomy in the sense of [9, Definition 7.6.4].

According to Henry [9], Theorem 7.6.7, there exists $\eta > 0$ such that $\{\tilde{T}_n\}$ with $\sup_n \|T_n - \tilde{T}_n\|_\alpha \le \eta$ has a discrete dichotomy.

Now we are in the conditions of [9], Exercise 10, pp. 229–230 (see also a more general statement [5, Theorem 4.1]), which finishes the proof.

Let us estimate the difference $\|\tilde{T}_k - T_k\|_\alpha$. There exists a positive integer N such that each interval of length d contains no more than N elements of sequence $\{\tau_j\}$. Let the interval $[\xi_n, \xi_{n+1}]$ contain points of impulses $\tilde{\tau}_m, \ldots, \tilde{\tau}_k$ where $k - m \le N$. Denote by $V_1(t,s)$ and $\tilde{V}(t,s)$ the evolution operators of equations without impulses (7.26) and (7.23), respectively. Then

$$\begin{aligned}
&\|T_n - \tilde{T}_n\|_\alpha = \|U_1(\xi_{n+1}, \xi_n) - \tilde{U}(\xi_{n+1}, \xi_n)\|_\alpha \\
&\le \|(V_1(\xi_{n+1}, \tilde{\tau}_k) - \tilde{V}(\xi_{n+1}, \tilde{\tau}_k))(I + B_k)V_1(\tilde{\tau}_k, \tilde{\tau}_{k-1})\ldots(I + B_m)V_1(\tilde{\tau}_m, \xi_n)\|_\alpha + \\
&+\|\tilde{V}(\xi_{n+1}, \tilde{\tau}_k)(B_k - \tilde{B}_k)V_1(\tilde{\tau}_k, \tilde{\tau}_{k-1})\ldots(I + B_m)V_1(\tilde{\tau}_m, \tilde{\xi}_n)\|_\alpha + \ \ldots\ + \\
&+\|\tilde{V}(\xi_{n+1}, \tilde{\tau}_k)(I + \tilde{B}_k)\tilde{V}(\tilde{\tau}_k, \tilde{\tau}_{k-1})\ldots(I + \tilde{B}_m)(V_1(\tilde{\tau}_m, \xi_n) - \tilde{V}(\tilde{\tau}_m, \xi_n))\|_\alpha. \quad (7.29)
\end{aligned}$$

Using (7.9), we get that

$$\sup_n \|T_n - \tilde{T}_n\|_\alpha \le K_3(\varepsilon)$$

with some $K_3(\varepsilon) \to 0$ as $\varepsilon \to 0$.

The exponentially dichotomous system (7.23) and (7.24) has Green's function

$$\tilde{G}(t,s) = \begin{cases} \tilde{U}(t,s)(I - \tilde{P}(s)),\ t \ge s, \\ -\tilde{U}(t,s)\tilde{P}(s),\ t < s, \end{cases}$$

such that

$$\|\tilde{G}(t,s)u\|_\alpha \le M_1 e^{-\beta_1|t-s|}\|u\|_\alpha,\ t,s \in R,\ u \in X^\alpha.$$

The sequence of bounded operators $T_n : X^\alpha \to X^\alpha$ defines the difference equation

$$u_{n+1} = T_n u_n,\ n \in Z, \quad (7.30)$$

with evolution operator $T_{n,m} = T_{n-1} \ldots T_m,\ n \geq m,\ T_{m,m} = I$. It is exponentially dichotomous with Green's function

$$G_{n,m} = \begin{cases} T_{n,m}(I - P_m),\ n \geq m, \\ -T_{n,m}P_m,\ n < m, \end{cases}$$

where $P_m = P(\xi_m)$.

The second difference equation

$$u_{n+1} = \tilde{T}_n u_n,\ n \in Z, \tag{7.31}$$

has the evolution operator $\tilde{T}_{n,m} = \tilde{T}_{n-1} \ldots \tilde{T}_m,\ n \geq m,\ \tilde{T}_{m,m} = I$.

By sufficiently small $\sup_n \|T_n - \tilde{T}_n\|_\alpha$, Eq. (7.31) is exponentially dichotomous with Green's function

$$\tilde{G}_{n,m} = \begin{cases} \tilde{T}_{n,m}(I - \tilde{P}_m),\ n \geq m, \\ -\tilde{T}_{n,m}\tilde{P}_m,\ n < m. \end{cases}$$

According to Henry [9], p. 233, the difference between two Green's functions satisfies equality:

$$\tilde{G}_{n,m} - G_{n,m} = \sum_{k \in Z} G_{n,k+1}(\tilde{T}_k - T_k)\tilde{G}_{k,m} \tag{7.32}$$

and estimation

$$\|\tilde{G}_{n,m} - G_{n,m}\|_\alpha = M_2 e^{-\beta_2 d|n-m|} \sup_k \|\tilde{T}_k - T_k\|_\alpha,\ n, m \in Z \tag{7.33}$$

with some constants $\beta_2 \leq \beta_1, M_2 \geq M_1$.

Now we can consider the difference of two Green's functions $\tilde{G}(t,s) - G_1(t,s)$. Let $t = s + nd + t_1, t_1 \in [0, d)$. Then

$$\begin{aligned} &\|\tilde{G}(t,s) - G_1(t,s)\|_\alpha = \\ &= \|\tilde{U}(s + nd + t_1, s + nd)\tilde{G}(s + nd, s) - U(s + nd + t_1, s + nd)G(s + nd, s)\|_\alpha \leq \\ &\leq \|(\tilde{U}(s + nd + t_1, s + nd) - U(s + nd + t_1, s + nd))\tilde{G}(s + nd, s)\|_\alpha + \\ &+ \|U(s + nd + t_1, s + nd)(\tilde{G}(s + nd, s) - G(s + nd, s))\|_\alpha. \end{aligned}$$

Using (7.33) and an estimation of the difference $\tilde{U} - U_1$ at a bounded interval as is done in (7.29), we get

$$\|\tilde{G}(t,\tau) - G_1(t,\tau)\|_\alpha \leq \tilde{M}_2(\varepsilon)e^{-\beta_2|t-\tau|}, \quad t, \tau \in R, \tag{7.34}$$

with $\tilde{M}_2(\varepsilon) \to 0$ as $\varepsilon \to 0$.

By the definition of Green's function, we have

$$\|\tilde{P}(\tau) - P_1(\tau)\|_\alpha \le \tilde{M}_2(\varepsilon) \;\text{ for all }\; \tau \in R. \tag{7.35}$$

Corollary 1. *Let the conditions of Lemma 5 be satisfied. Then for $t \in R, |t - \tau_j| \ge \varepsilon, j \in Z$, we have*

$$\|(P(t) - \tilde{P}(t))u\|_\alpha \le \tilde{M}_3(\varepsilon)\|u\|_{\alpha+\nu}, \tag{7.36}$$

where $\nu > 0, \alpha + \nu < 1$, and $\tilde{M}_3(\varepsilon) \to 0$ as $\varepsilon \to 0$.

Proof. Using (7.22) and (7.35), we get

$$\|(P(t) - \tilde{P}(t))u\|_\alpha \le \|(P(t) - P(\vartheta(t)))u\|_\alpha +$$
$$+\|(P(\vartheta(t)) - \tilde{P}(\vartheta(t)))u\|_\alpha + \|(\tilde{P}(\vartheta(t)) - \tilde{P}(t))u\|_\alpha \le \tilde{M}_3(\varepsilon)\|u\|_{\alpha+\nu}.$$

7.4 Almost Periodic Solutions of Equations with Fixed Moments of Impulsive Action

Consider the linear inhomogeneous equation

$$\frac{du}{dt} + (A + A_1(t))u = f(t), \quad t \ne \tau_j, \tag{7.37}$$

$$\Delta u|_{t=\tau_j} = u(\tau_j + 0) - u(\tau_j) = B_j u(\tau_j) + g_j, \quad j \in Z. \tag{7.38}$$

We assume that

(H7) the function $f(t) : R \to X$ is W-almost periodic and locally Hölder continuous with points of discontinuity at moments $t = \tau_j, j \in Z$, at which it is continuous from the left;

(H8) the sequence $\{g_j\}$ of $g_j \in X^{\alpha_1}, \alpha_1 > \alpha > 0$, is almost periodic.

Theorem 1. *Assume that Eqs. (7.37) and (7.38) satisfy conditions* **(H1)–(H3)**, **(H7)**, *and* **(H8)** *and that the corresponding homogeneous equation is exponentially dichotomous.*

Then the equation has a unique W-almost periodic solution $u_0(t) \in \mathscr{PC}(R, X^\alpha)$.

Proof. We show that an almost periodic solution is given by the formula (7.16). For $t \in (\tau_i, \tau_{i+1}]$, it satisfies

$$\|u_0(t)\|_\alpha \le \int_{-\infty}^{t} \|A^\alpha U(t,s)(I - P(s))f(s)\|ds +$$
$$+ \int_t^\infty \|A^\alpha U(t,s)P(s)f(s)\|ds + \sum_{j\in Z} \|G(t,\tau_j)g_j\|_\alpha \le$$

$$\leq \sum_{j\in Z} \|G(t,\tau_j)g_j\|_\alpha + \int_{\tau_i}^{t} \|A^\alpha V(t,s)(I-P(s))f(s)\|ds +$$

$$+ \sum_{k=0}^{\infty} \int_{\tau_{i-k-1}}^{\tau_{i-k}} \|U(t,\tau_{i-k})(I-P(\tau_{i-k}))\|_\alpha \|A^\alpha U(\tau_{i-k},s)f(s)\|ds +$$

$$+ \sum_{k=1}^{\infty} \int_{\tau_{i+k}}^{\tau_{i+k+1}} \|U(t,\tau_{i+k+1})P(\tau_{i+k+1})\|_\alpha \|A^\alpha U(\tau_{i+k+1},s)f(s)\|ds +$$

$$+ \int_{t}^{\tau_{i+1}} \|A^\alpha V(t,s)P(s)f(s)\|ds \leq \frac{2M}{1-e^{-\theta\beta}} \frac{C_\alpha \Theta^{1-\alpha}}{1-\alpha} \|f\|_{PC} +$$

$$+ \frac{2M}{1-e^{-\theta\beta}} \sup_j \|g_j\|_\alpha \leq \tilde{M}_0 \max\{\|f(t)\|_{PC}, \|g_j\|_\alpha\} \tag{7.39}$$

with some constant $\tilde{M}_0 > 0$.

Take an ε-almost period h for the right-hand side of the equation, which satisfies the conditions of Lemma 1; that is, there exists a positive integer q such that $\tau_{j+q} \in (s+h, t+h)$ if $\tau_j \in (s,t)$ and $|\tau_j + h - \tau_{j+q}| < \varepsilon$, $\|B_{j+q} - B_j\| < \varepsilon$.

Let $t \in (\tau_i + \varepsilon, \tau_{i+1} - \varepsilon)$. We define points $\eta_k = (\tau_k + \tau_{k-1})/2, k \in Z$. Then

$$\|u_0(t+h) - u_0(t)\|_\alpha \leq \sum_{j\in Z} \|G(t+h,\tau_{j+q})g_{j+q} - G(t,\tau_j)g_j\|_\alpha +$$

$$+ \int_{-\infty}^{\infty} \|G(t+h,s+h)f(s+h) - G(t,s)f(s)\|_\alpha ds \leq$$

$$\leq \int_{-\infty}^{\infty} \|(G(t+h,s+h) - G(t,s))f(s+h)\|_\alpha ds +$$

$$+ \int_{-\infty}^{\infty} \|G(t,s))(f(s+h) - f(s))\|_\alpha ds + \sum_{j\in Z} \|G(t,\tau_j))(g_{j+q} - g_j)\|_\alpha +$$

$$+ \sum_{j\in Z} \|(G(t+h,\tau_{j+q}) - G(t,\tau_j))g_{j+q}\|_\alpha. \tag{7.40}$$

Denote $U_2(t,s) = U(t+h,s+h)$. If $u(t) = U(t,s)u_0$, $u(s) = u_0$, is a solution of the impulsive equations (7.4) and (7.5), then $u_2(t) = U(t+h,s+h)u_0$, $u_2(s) = u_0$, is a solution of the equation

$$\frac{du}{dt} + (A + A_1(t+h))u = 0, \quad t \neq \tau_{j+q} - h, \tag{7.41}$$

$$\Delta u|_{t+h=\tau_{j+q}} = u(\tau_{j+q}+0) - u(\tau_{j+q}) = B_{j+q}u(\tau_{j+q}), \quad j \in Z. \tag{7.42}$$

We will use the notation $V_2(t,s) = V(t+h, s+h)$ for the evolution operator of an equation without impulses (7.41). Denote also $\tilde{\tau}_n = \tau_{n+q} - h, \tilde{B}_n = B_{n+q}$. Since Eqs. (7.4) and (7.5) are exponentially dichotomous, Eqs. (7.41) and (7.42) are exponentially dichotomous also with projector $P_2(s) = P(s+h)$.

The first integral in (7.40) is the sum of two integrals:

$$\int_{-\infty}^{\infty} \|(G(t+r, s+r) - G(t,s))f(s+r)\|_\alpha ds =$$

$$= \int_{-\infty}^{t} \|(U_2(t,s)(I - P_2(s)) - U(t,s)(I - P(s)))f(s+r)\|_\alpha ds +$$

$$+ \int_{t}^{\infty} \|(U_2(t,s)P_2(s) - U(t,s)P(s))f(s+r)\|_\alpha ds. \tag{7.43}$$

We estimate the first integral in (7.43); the second integral is considered analogously.

$$\int_{-\infty}^{t} \|(U_2(t,s)(I - P_2(s)) - U(t,s)(I - P(s)))f(s+r)\|_\alpha ds \leq$$

$$\leq \int_{\tau_i+\varepsilon}^{t} \|A^\alpha(V_2(t,s)(I - P_2(s)) - V(t,s)(I - P(s)))f(s+r)\| ds +$$

$$+ \int_{\tau_i-\varepsilon}^{\tau_i+\varepsilon} \|A^\alpha(U_2(t,s)(I - P_2(s)) - U(t,s)(I - P(s)))f(s+r)\| ds +$$

$$+ \int_{\eta_i}^{\tau_i-\varepsilon} \|A^\alpha(U_2(t,s)(I - P_2(s)) - U(t,s)(I - P(s)))f(s+r)\| ds +$$

$$+ \sum_{k=1}^{\infty} \int_{\eta_{i-k}}^{\eta_{i-k+1}} \|A^\alpha(U_2(t,s)(I - P_2(s)) - U(t,s)(I - P(s)))f(s+r)\| ds. \tag{7.44}$$

Let us consider all integrals in (7.44) separately. By (7.36) and (7.11) we have

$$I_{11} = \int_{\tau_i+\varepsilon}^{t} \|A^\alpha(V_2(t,s)(I - P_2(s)) - V(t,s)(I - P(s)))f(s+r)\| ds =$$

$$= \int_{\tau_i+\varepsilon}^{t} \|A^\alpha((I - P_2(t))V_2(t,s) - (I - P(t))V(t,s))f(s+r)\| ds \leq$$

$$\leq \int_{\tau_i+\varepsilon}^{t} \|A^\alpha(P_2(t) - P(t))V_2(t,s)f(s+r)\| ds +$$

$$+ \int_{\tau_i+\varepsilon}^{t} \|A^\alpha(I - P(t))(V_2(t,s) - V(t,s))f(s+r)\| ds \leq$$

$$\leq \left(\int_{\tau_i+\varepsilon}^{t} \frac{\tilde{M}_3(\varepsilon) L_Q ds}{(t-s)^{\alpha}} + \int_{\tau_i+\varepsilon}^{t} \frac{R_3(\varepsilon) ds}{(t-s)^{2\alpha-1}} \right) \|f\|_{PC} \leq \Gamma_1(\varepsilon) \|f\|_{PC}.$$

$$I_{12} = \int_{\tau_i-\varepsilon}^{\tau_i+\varepsilon} \|A^{\alpha} U(t,s)(I-P(s))f(s+h)\| ds \leq$$

$$\leq \int_{\tau_i}^{\tau_i+\varepsilon} \|A^{\alpha}(I-P(t))V(t,s)f(s+h)\| ds +$$

$$+ \int_{\tau_i-\varepsilon}^{\tau_i} \| \|A^{\alpha}(I-P(t))V(t,\tau_i)(I+B_i)U(\tau_i,s)f(s+h)\| ds \leq$$

$$\leq \Big(\int_{\tau_i}^{\tau_i+\varepsilon} \frac{C_{\alpha} ds}{(t-s)^{\alpha}} + M\|I+B_i\| \int_{\tau_i-\varepsilon}^{\tau_i} \frac{C_{\alpha} ds}{(s-\tau_i)^{\alpha}} \Big) \|f\|_{PC} \leq$$

$$\leq \Gamma_2(\varepsilon) \|f\|_{PC}.$$

Analogously,

$$I_{13} = \int_{\tau_i-\varepsilon}^{\tau_i+\varepsilon} \|A^{\alpha} U_2(t,s)(I-P_2(s))f(s+h)\| ds \leq \Gamma_3(\varepsilon) \|f\|_{PC},$$

where $\Gamma_j(\varepsilon) \to 0$ as $\varepsilon \to 0$, $j = 1, 2, 3$.
Using (7.11) and (7.36), we get

$$I_{14} = \int_{\eta_i}^{\tau_i-\varepsilon} \|A^{\alpha}(U_2(t,s)(I-P_2(s)) - U(t,s)(I-P(s)))f(s+r)\| ds =$$

$$= \int_{\eta_i}^{\tau_i-\varepsilon} \Big\| \Big((I-P_2(t))V_2(t,\tilde{\tau}_i)(I+\tilde{B}_i)V_1(\tilde{\tau}_i,s) -$$

$$-(I-P(t))V(t,\tau_i)(I+B_i)V(\tau_i,s) \Big) f(s+h) \Big\|_{\alpha} ds \leq$$

$$\leq \int_{\eta_i}^{\tau_i-\varepsilon} \|(P_2(t)-P(t))V_2(t,\tilde{\tau}_i)(I+B_i)V_2(\tilde{\tau}_i,s)f(s+h)\|_{\alpha} ds +$$

$$+ \int_{\eta_i}^{\tau_i-\varepsilon} \|(I-P(t))(V_2(t,\tilde{\tau}_i)-V(t,\tau_i))(I+B_i)V_2(\tilde{\tau}_i,s)f(s+h)\|_{\alpha} ds +$$

$$+ \int_{\eta_i}^{\tau_i-\varepsilon} \|(I-P(t))V(t,\tau_i)(\tilde{B}_i-B_i)V_2(\tilde{\tau}_i,s)f(s+h)\|_{\alpha} ds +$$

$$+ \int_{\eta_i}^{\tau_i-\varepsilon} \|(I-P(t))V(t,\tau_i)(I-B_i)(V_2(\tilde{\tau}_i,s)-V(\tau_i,s))f(s+h)\|_{\alpha} ds \leq$$

$$\leq \Gamma_4(\varepsilon) \|f\|_{PC},$$

where $\Gamma_4(\varepsilon) \to 0$ as $\varepsilon \to 0$.

The last sum in (7.44) is transformed as follows:

$$I_{15} = \sum_{k=1}^{\infty} \int_{\eta_{i-k}}^{\eta_{i-k+1}} \|A^{\alpha}(U_2(t,s)(I - P_2(s)) - U(t,s)(I - P(s)))f(s+r)\| ds =$$

$$= \sum_{k=1}^{\infty} \int_{\eta_{i-k}}^{\eta_{i-k+1}} \|(U(t,\eta_i)(I - P(\eta_i))U(\eta_i, \eta_{i-k+1})U(\eta_{i-k+1}, s) -$$

$$-U_2(t,\eta_i)(I - P_2(\eta_i))U_2(\eta_i, \eta_{i-k+1})U_2(\xi_{i-k+1}, s))f(s+h)\|_{\alpha} ds \leq$$

$$\leq \sum_{k=1}^{\infty} \int_{\eta_{i-k}}^{\eta_{i-k+1}} \Big\| \Big(U(t,\eta_i) - U_2(t,\eta_i))(I - P(\eta_i))U(\eta_i, \eta_{i-k+1})U(\eta_{i-k+1}, s) +$$

$$+U_2(t,\eta_i)((I - P(\eta_i))U(\eta_i, \eta_{i-k+1}) - (I - P_2(\eta_i))U_2(\eta_i, \eta_{i-k+1}))U(\eta_{i-k+1}, s) +$$

$$+U_2(t,\eta_i)(I - P_2(\eta_i))U_2(\eta_i, \eta_{i-k+1})(U(\eta_{i-k+1}, s) - U_2(\eta_{i-k+1}, s))\Big) f(s+h)\Big\|_{\alpha} ds.$$

As in the proof of Lemma 5, we construct in space X^{α} two sequences of bounded operators

$$S_n = U(\eta_{n+1}, \eta_n), \quad \tilde{S}_n = U_2(\eta_{n+1}, \eta_n), \quad n \in Z,$$

and corresponding difference equations

$$u_{n+1} = S_n u_n, \quad v_{n+1} = \tilde{S}_n v_n, \quad n \in Z.$$

Per our assumption, these difference equations are exponentially dichotomous with corresponding evolution operators

$$S_{n,m} = S_{n-1} \ldots S_m, \quad \tilde{S}_{n,m} = \tilde{S}_{n-1} \ldots \tilde{S}_m, \quad n \geq m,$$

and Green's functions

$$G_{n,m} = \begin{cases} S_{n,m}(I - P_m), \ n \geq m, \\ -S_{n,m}P_m, \ n < m, \end{cases} \qquad \tilde{G}_{n,m} = \begin{cases} \tilde{S}_{n,m}(I - \tilde{P}_m), \ n \geq m, \\ -\tilde{S}_{n,m}\tilde{P}_m, \ n < m, \end{cases}$$

where $P_m = P(\eta_m), \tilde{P}_m = P_2(\eta_m)$.

Analogous to (7.32) and (7.33), we obtain

$$\tilde{G}_{n,m} - G_{n,m} = \sum_{k \in Z} G_{n,k+1}(\tilde{S}_k - S_k)\tilde{G}_{k,m}$$

and

$$\|\tilde{G}_{n,m} - G_{n,m}\|_\alpha = M_1 e^{-\beta_1 \theta |n-m|} \sup_k \|\tilde{S}_k - S_k\|_\alpha, \ n, m \in Z \tag{7.45}$$

with some constants $\beta_1 \leq \beta, M_1 \geq M$.

$$\begin{aligned}
&\|S_n - \tilde{S}_n\|_\alpha = \|U(\eta_{n+1}, \eta_n) - U_2(\eta_{n+1}, \eta_n)\|_\alpha = \\
&= \|V(\eta_{n+1}, \tau_n)(I + B_n)V(\tau_n, \eta_n) - V_2(\eta_{n+1}, \tilde{\tau}_n)(I + \tilde{B}_n)V_2(\tilde{\tau}_n, \eta_n)\|_\alpha \leq \\
&\leq \|(V(\eta_{n+1}, \tau_n) - V_2(\eta_{n+1}, \tilde{\tau}_n))(I + B_n)V(\tau_n, \eta_n)\|_\alpha + \\
&+ \|V_2(\eta_{n+1}, \tilde{\tau}_n))(B_n - \tilde{B}_n)V(\tau_n, \eta_n)\|_\alpha + \\
&+ \|V_2(\eta_{n+1}, \tilde{\tau}_n))(I + \tilde{B}_n)(V(\tau_n, \eta_n) - V_2(\tilde{\tau}_n, \eta_n))\|_\alpha
\end{aligned}$$

Here we assume for definiteness that $\tilde{\tau}_n \geq \tau_n$. We have

$$\begin{aligned}
&\|(V(\eta_{n+1}, \tau_n) - V_2(\eta_{n+1}, \tilde{\tau}_n))y\|_\alpha \leq \|V(\eta_{n+1}, \tilde{\tau}_n)(V(\tilde{\tau}_n, \tau_n) - I)y\|_\alpha + \\
&+ \|(V(\eta_{n+1}, \tilde{\tau}_n) - V_2(\eta_{n+1}, \tilde{\tau}_n))y\|_\alpha \leq \\
&\leq \Gamma_5(\varepsilon)\|y\|_\alpha
\end{aligned}$$

and

$$\begin{aligned}
&\|(V_2(\tilde{\tau}_n, \eta_n) - V(\tau_n, \eta_n))y\|_\alpha \leq \|(V_2(\tilde{\tau}_n, \tau_n) - I)V_2(\tau_n, \eta_n)y\|_\alpha + \\
&+ \|V_2(\tau_n, \eta_n) - V(\tau_n, \eta_n)y\|_\alpha \leq \Gamma_6(\varepsilon)\|y\|_\alpha,
\end{aligned}$$

where $\Gamma_5(\varepsilon) \to 0$ and $\Gamma_6(\varepsilon) \to 0$ as $\varepsilon \to 0$.

Now we get

$$\begin{aligned}
&\|S_n - \tilde{S}_n\|_\alpha \leq \Gamma_5(\varepsilon)\|I + B_n\| \, \|U(\tau_n, \eta_n)\|_\alpha + \\
&+ \varepsilon\|U_2(\eta_n, \tau_n)\|_\alpha \|U(\tau_n, \eta_n)\|_\alpha + \Gamma_6(\varepsilon)\|U_2(\eta_{n+1}, \tilde{\tau}_n)\|_\alpha \|I + \tilde{B}_n\| \leq \Gamma_7(\varepsilon)
\end{aligned}$$

and by (7.45)

$$\|U(\eta_i, \eta_{i-k}) - U_2(\eta_i, \eta_{i-k})\|_\alpha \leq M_1 e^{-\beta_1 \theta k}\Gamma_7(\varepsilon), \tag{7.46}$$

where $\Gamma_7(\varepsilon) \to 0$ as $\varepsilon \to 0$.

Continuing to evaluate I_{15}, we can obtain the inequalities

$$\begin{aligned}
&\|U_2(t, \eta_i)g\|_\alpha \leq M_2\|g\|_\alpha, \\
&\|(U(t, \eta_i) - U_2(t, \eta_i))g\|_\alpha \leq \Gamma_8(\varepsilon)\|g\|_\alpha,
\end{aligned}$$

$$\int_{\xi_{i-k}}^{\eta_{i-k+1}} \|(U(\eta_{i-k+1}, s) - U_2(\eta_{i-k+1}, s))f(s+h)\|_\alpha ds \le \Gamma_9(\varepsilon)\|f\|_{PC},$$

where $\Gamma_8(\varepsilon) \to 0$ and $\Gamma_9(\varepsilon) \to 0$ as $\varepsilon \to 0$, M_2 is some positive constant. Note that as earlier, $t \in (\tau_i + \varepsilon, \tau_{i+1} - \varepsilon)$.

Taking into account the last inequalities, we conclude that series I_{15} is convergent and there exists $\Gamma_{10}(\varepsilon)$ such that $I_{15} \le \Gamma_{10}(\varepsilon)\|f\|_{PC}$ and $\Gamma_{10}(\varepsilon) \to 0$ as $\varepsilon \to 0$.

Using estimations for $I_{11}, \ldots, I_{15}$, we get that there exists $\Gamma_{11}(\varepsilon)$ such that

$$\int_{-\infty}^{\infty} \|(G(t+r, s+r) - G(t,s))f(s+r)\|_\alpha ds \le \Gamma_{11}(\varepsilon)\|f\|_{PC} \tag{7.47}$$

and $\Gamma_{11}(\varepsilon) \to 0$ as $\varepsilon \to 0$.

By Lemma 1, $|\tau_{j+q} - \tau_j - h| < \varepsilon$; therefore, $\tau_j + h + \varepsilon > \tau_{j+q}$ (we assume that $h > 0$ for definiteness). The difference $G(t, \tau_j) - G(t+h, \tau_{j+q})$ is estimated as follows. Let $t - \tau_j \ge \varepsilon$. Then

$$\begin{aligned}
&\|(G(t, \tau_j) - G(t+h, \tau_{j+q}))g_{j+q}\|_\alpha = \\
&= \|(U(t, \tau_j)(I - P(\tau_j)) - U(t+h, \tau_{j+q})(I - P(\tau_{j+q})))g_{j+q}\|_\alpha \le \\
&\le \|(U(t, \tau_j)(I - P(\tau_j)) - U(t, \tau_j + \varepsilon)(I - P(\tau_j + \varepsilon)))g_{j+q}\|_\alpha + \\
&+\|(U(t, \tau_j + \varepsilon)(I - P(\tau_j + \varepsilon)) - U(t+h, \tau_j + \varepsilon + h) \times \\
&\times(I - P(\tau_j + \varepsilon + h)))g_{j+q}\|_\alpha + \|U(t+h, \tau_{j+q})(I - P(\tau_{j+q})))g_{j+q} - \\
&-(U(t+h, \tau_j + \varepsilon + h)(I - P(\tau_j + \varepsilon + h))g_{j+q}\|_\alpha.
\end{aligned} \tag{7.48}$$

The first and third differences are small due to the continuity of function $U(t,s)$ at intervals between impulse points:

$$\begin{aligned}
&\|(U(t, \tau_j)(I - P(\tau_j)) - U(t, \tau_j + \varepsilon)(I - P(\tau_j + \varepsilon)))g_{j+q}\|_\alpha \le \\
&\le \|U(t, \tau_j + \varepsilon)(I - P(\tau_j + \varepsilon))(U(\tau_j + \varepsilon, \tau_j) - I)g_{j+q}\|_\alpha \le \\
&\le \|(I - P(t))U(t, \tau_j + \varepsilon)\|_\alpha \|(U(\tau_j + \varepsilon, \tau_j) - I)g_{j+q}\|_\alpha \le \\
&\le Me^{-\beta(t-\tau_j-\varepsilon)}C_{1-\alpha_1+\alpha}\varepsilon^{\alpha_1-\alpha}\|g_{j+q}\|_{\alpha_1},
\end{aligned}$$

$$\begin{aligned}
&\|(U(t+h, \tau_j + \varepsilon + h)(I - P(\tau_j + \varepsilon + h)) - U(t+h, \tau_{j+q})(I - P(\tau_{j+q})))g_{j+q}\|_\alpha = \\
&= \|\|U(t+h, \tau_j + \varepsilon + h)(I - P(\tau_j + \varepsilon + h))(U(\tau_j + \varepsilon + h, \tau_{j+q}) - I)g_{j+q}\|_\alpha \le \\
&\le Me^{-\beta(t-\tau_j-\varepsilon)}C_{1-\alpha_1+\alpha}\varepsilon^{\alpha_1-\alpha}\|g_{j+q}\|_{\alpha_1}.
\end{aligned}$$

The second difference in (7.48) is estimated using inequality (7.46) and the following transformation:

$$\|U(t, \tau_j + \varepsilon)(I - P(\tau_j + \varepsilon)) - U(t+h, \tau_j + \varepsilon + h)(I - P(\tau_j + \varepsilon + h))\|_\alpha =$$

$$= \|U(t,\tau_j+\varepsilon)(I-P(\tau_j+\varepsilon)) - U_2(t,\tau_j+\varepsilon)(I-P_2(\tau_j+\varepsilon))\|_\alpha =$$
$$= \|U(t,\eta_i)(I-P(\eta_i))U(\eta_i,\eta_{j+1})U(\eta_{j+1},\tau_j+\varepsilon) -$$
$$-U_2(t,\eta_i)(I-P(\eta_i)U_2(\eta_i,\eta_{j+1})U_2(\eta_{j+1},\tau_j+\varepsilon)\|_\alpha \le$$
$$\le \|(U(t,\eta_i)-U_2(t,\eta_i))(I-P(\eta_i))U(\eta_i,\eta_{j+1})U(\eta_{j+1},\tau_j+\varepsilon)\|_\alpha +$$
$$+\|U_1(t,\eta_i)(P(\eta_i)U(\eta_i,\eta_{j+1}) - P_2(\eta_i)U_2(\eta_i,\eta_{j+1})U(\eta_{j+1},\tau_j+\varepsilon)\|_\alpha +$$
$$+\|U_2(t,\eta_i)P_2(\eta_i)U_2(\eta_i,\eta_{j+1})(U(\eta_{j+1},\tau_j+\varepsilon) - U_2(\eta_{j+1},\tau_j+\varepsilon))\|_\alpha.$$

Therefore,

$$\sum_{j\in Z} \|(G(t+h,\tau_{j+q}) - G(t,\tau_j))g_{j+q}\|_\alpha \le \Gamma_{12}(\varepsilon)\sup_j \|g_j\|_{\alpha_1}, \tag{7.49}$$

where $\Gamma_{12}(\varepsilon) \to 0$ as $\varepsilon \to 0$.

The second integral and first sum in (7.40) are estimated as in (7.39):

$$\int_{-\infty}^{\infty} \|G(t,s))(f(s+h)-f(s))\|_\alpha ds + \sum_{j\in Z} \|U(t,\tau_j)(g_{j+q}-g_j)\|_\alpha \le M_3\varepsilon$$

since h is ε-almost periodic of the right-hand side of the equation.

As a result of these evaluations, we get

$$\|u_0(t+h) - u_0(t)\|_\alpha \le \Gamma(\varepsilon) \;\text{ for }\; t \in R,\; |t-\tau_j| > \varepsilon,\; j \in Z,$$

with $\Gamma(\varepsilon) \to 0$ as $\varepsilon \to 0$. The last inequality implies that the function $u_0(t)$ is W-almost periodic as function $R \to X^\alpha$.

Corollary 2. *Assume that Eqs. (7.37) and (7.38) satisfy the following:*

i) conditions **(H1)**–**(H3)**, **(H7)**;
ii) the sequence $\{g_j\}$ *of* $g_j \in X^\alpha$ *is almost periodic;*
iii) the corresponding homogeneous equation is exponentially dichotomous.

Then the equation has a unique W-almost periodic solution $u_0(t) \in \mathscr{PC}(R, X^\gamma)$ *with* $\gamma < \alpha$.

Now we consider a nonlinear equation with fixed moments of impulsive action:

$$\frac{du}{dt} + (A + A_1(t))u = f(t,u), \quad t \ne \tau_j, \tag{7.50}$$

$$\Delta u|_{t=\tau_j} = u(\tau_j+0) - u(\tau_j) = B_j u(\tau_j) + g_j(u(\tau_j)), \quad j \in Z. \tag{7.51}$$

Theorem 2. *Let us consider Eqs. (7.50) and (7.51) in some domain* $U_\rho^\alpha = \{x \in X^\alpha : \|x\|_\alpha \le \rho\}$ *of space* X^α. *Assume that*

1) the equation satisfies assumptions **(H1)–(H4)**, $\tau_j = \tau_j(0)$;
2) the corresponding linear equation is exponentially dichotomous;
3) the function $f(t, u) : R \times U^\alpha_\rho \to X$ *is continuous in* u, *W-almost periodic, and Hölder continuous in* t *uniformly with respect to* $u \in U^\alpha_\rho$ *with some* $\rho > 0$, *and there exist constants* $N_1 > 0$ *and* $\nu > 0$ *such that*

$$\|f(t_1, u_1) - f(t_2, u_2)\| \le N_1(|t_1 - t_2|^\nu + \|u_1 - u_2\|_\alpha);$$

4) the sequence $\{g_j(u)\}$ *of continuous functions* $U^\alpha_\rho \to X^{\alpha_1}$ *is almost periodic uniformly with respect to* $u \in U^\alpha_\rho$ *and*

$$\|g_j(u_1) - g_j(u_2)\|_\alpha \le N_1\|u_1 - u_2\|_\alpha, \ j \in Z,$$

for $t_1, t_2 \in R$, $u_1, u_2 \in U^\alpha_\rho$ *and some* $\alpha_1 > \alpha$;
5) the functions $f(t, 0)$ *and* $g_j(0)$ *are uniformly bounded for* $t \in R, j \in Z$.

Then in domain U^α_ρ *for sufficiently small* $N_1 > 0$ *there exists a unique W-almost periodic solution* $u_0(t)$ *of Eqs. (7.50) and (7.51).*

Proof. Denote by $\mathscr{M}_\varrho$ the set of all W-almost periodic functions $\varphi : R \to X^\alpha$ with discontinuity points $\tau_j, j \in Z$, satisfying the inequality $\|\varphi\|_{PC} \le \varrho$. In $\mathscr{M}_\varrho$, we define the operator

$$(\mathscr{F}\varphi)(t) = \int_{-\infty}^{\infty} G(t, s)f(s, \varphi(s))ds + \sum_{j\in Z} G(t, \tau_j)g_j(\varphi(\tau_j)).$$

Proceeding in the same way as in the proof of Theorem 1, we prove that $(\mathscr{F}\varphi)(t)$ is a W-almost periodic function and $\mathscr{F} : \mathscr{M}_\varrho \to \mathscr{M}_\varrho$ for some $\varrho > 0$.

Next, $\mathscr{F}$ is a contracting operator in $\mathscr{M}_\varrho$ by sufficiently small $N_1 > 0$.

Hence, there exists $\varphi_0 \in \mathscr{M}_\varrho$ such that

$$\varphi_0(t) = \int_{-\infty}^{\infty} G(t, s)f(s, \varphi_0(s))ds + \sum_{j\in Z} G(t, \tau_j)g_j(\varphi_0(\tau_j)).$$

The function $\varphi_0(t)$ is locally Hölder continuous on every interval $(\tau_j, \tau_{j+1}), j \in Z$. Actually,

$$\varphi_0(t + \delta) - \varphi_0(t) = \int_{-\infty}^{\infty} G(t + \delta, s)f(s, \varphi_0(s))ds - \int_{-\infty}^{\infty} G(t, s)f(s, \varphi_0(s))ds +$$

$$+ \sum_{j\in Z} G(t + \delta, \tau_j)g_j(\varphi_0(\tau_j)) - \sum_{j\in Z} G(t, \tau_j)g_j(\varphi_0(\tau_j)) =$$

$$= \int_{-\infty}^{t} (V(t + \delta, t) - I)U(t, s)(I - P(s))f(s, \varphi_0(s))ds -$$

$$- \int_{t+\delta}^{\infty} (V(t+\delta,t) - I)U(t,s)P(s)f(s,\varphi_0(s))ds +$$

$$+ \int_{t}^{t+\delta} V(t+\delta,s)(I-P(s))f(s,\varphi_0(s))ds + \int_{t}^{t+\delta} V(t,s)P(s)f(s,\varphi_0(s))ds$$

$$+ \sum_{\tau_j<t} (V(t+\delta,t) - I)U(t,\tau_j)(I-P(\tau_j))g_j(\varphi_0(\tau_j)) +$$

$$+ \sum_{\tau_j>t} (V(t+\delta,t) - I)U(t,\tau_j)P(\tau_j)g_j(\varphi_0(\tau_j)).$$

Applying (7.7), (7.20), (7.21), and (7.39), we conclude that for every interval $t \in (t', t'')$ not containing impulse points τ_j, there exists a positive constant C such that $\|\varphi_0(t+\delta) - \varphi_0(t)\|_\alpha \leq C\delta^{\alpha_1-\alpha}$.

The local Hölder continuity of $f(t,\varphi_0(t))$ follows from

$$\|f(t,\varphi_0(t)) - f(s,\varphi_0(s))\| \leq N_1 \left(|t-s|^\nu + \|\varphi_0(t) - \varphi_0(s)\|_\alpha\right) \leq$$
$$\leq C_1 \left(|t-s|^\nu + |t-s|^{\alpha_1-\alpha}\right).$$

By Lemma 37, [19], p. 214, if $\varphi_0(t)$ is W-almost periodic and $\inf_k(\tau_{k+1} - \tau_k) > 0$, then $\{\varphi_0(\tau_k)\}$ is an almost periodic sequence.

The linear inhomogeneous equation

$$\frac{du}{dt} + (A + A_1(t))u = f(t,\varphi_0(t)), \quad t \neq \tau_j, \tag{7.52}$$

$$\Delta u|_{t=\tau_j} = u(\tau_j + 0) - u(\tau_j) = B_j u(\tau_j) + g_j(\varphi_0(\tau_j)), \quad j \in Z, \tag{7.53}$$

has a unique W-almost periodic solution in the sense of Definition 4. Due to the uniqueness, it coincides with $\varphi_0(t)$.

Hence, the W-almost periodic function $\varphi_0(t) : R \to X^\alpha$ satisfies Eq. (7.50) for $t \in (\tau_j, \tau_{j+1})$ and the difference equation (7.51) for $t = \tau_j$.

Now we study the stability of the almost periodic solution assuming exponential stability of the linear equation. First, using ideas in [17], we prove the following generalized Gronwall inequality for impulsive systems.

Lemma 6. *Assume that* $\{t_j\}$ *is an increasing sequence of real numbers such that* $Q \geq t_{j+1} - t_j \geq \theta > 0$ *for all* j, M_1, M_2, *and* M_3 *are positive constants, and* $\alpha \in (0,1)$. *Then there exists a positive constant* $\tilde{C}$ *such that the positive piecewise continuous function* $u : [t_0, t] \to R$ *satisfying*

$$z(t) \leq M_1 z_0 + M_2 \sum_{j=1}^{m} \int_{t_{j-1}}^{t_j} (t_j - s)^{-\alpha} z(s)ds + M_2 \int_{t_m}^{t} (t-s)^{-\alpha} z(s)ds +$$

$$+M_3 \sum_{j=1}^{m} z(t_j) \quad \text{for} \quad t \in (t_m, t_{m+1}] \tag{7.54}$$

also satisfies

$$z(t) \le M_1 z_0 \tilde{C} \left(1 + M_2 \tilde{C} \frac{Q^{1-\alpha}}{1-\alpha} + M_3 \tilde{C} \right)^m . \tag{7.55}$$

Proof. We apply the method of mathematical induction. At the interval $t \in [t_0, t_1]$ the inequality (7.54) has the form

$$z(t) \le M_1 z_0 + M_2 \int_{t_0}^{\tau_1} (\tau_1 - s)^{-\alpha} z(s) ds.$$

By Lemma 2 there exists $\tilde{C}$ such that

$$0 \le z(t) \le M_1 z_0 \tilde{C}, \;\; t \in [t_0, t], \tilde{C} = \tilde{C}(M_1, M_2, Q). \tag{7.56}$$

Hence, (7.55) is true for $t \in [t_0, t_1]$. Assume (7.55) is true for $t \in [t_0, t_n]$ and prove it for $t \in (t_n, t_{n+1}]$. Hence, for $t \in (t_n, t_{n+1}]$ we have

$$z(t) \le M_1 z_0 + M_2 \int_{t_0}^{t_1} (t_1 - s)^{-\alpha} z(s) ds + M_3 z(t_1) +$$

$$+ M_2 \sum_{j=2}^{n} \int_{t_{j-1}}^{t_j} (t_j - s)^{-\alpha} z(s) ds + M_3 \sum_{j=1}^{n} z(t_j) + M_2 \int_{t_n}^{t} (t - s)^{-\alpha} z(s) ds \le$$

$$\le M_1 z_0 + M_2 \frac{Q^{1-\alpha}}{1-\alpha} M_1 z_0 \tilde{C} + M_3 M_1 z_0 \tilde{C} + M_2 \int_{t_n}^{t} (t - s)^{-\alpha} z(s) ds +$$

$$+ \sum_{j=2}^{n} \left(1 + M_2 \tilde{C} \frac{Q^{1-\alpha}}{1-\alpha} + M_3 \tilde{C} \right)^j \left(M_2 \tilde{C} \frac{Q^{1-\alpha}}{1-\alpha} + M_3 \tilde{C} \right) M_1 z_0 =$$

$$= M_1 z_0 + M_2 \frac{Q^{1-\alpha}}{1-\alpha} M_1 z_0 \tilde{C} + M_3 M_1 z_0 \tilde{C} + M_2 \int_{t_n}^{t} (t - s)^{-\alpha} z(s) ds +$$

$$+ \sum_{j=2}^{n} \left(1 + M_2 \tilde{C} \frac{Q^{1-\alpha}}{1-\alpha} + M_3 \tilde{C} \right)^{j-1} \left[\left(1 + M_2 \tilde{C} \frac{Q^{1-\alpha}}{1-\alpha} + M_3 \tilde{C} \right) - 1 \right] M_1 z_0 =$$

$$\le M_1 z_0 \left(1 + M_2 \frac{Q^{1-\alpha}}{1-\alpha} \tilde{C} + M_3 \tilde{C} \right)^n + M_2 \int_{t_n}^{t} (t - s)^{-\alpha} z(s) ds.$$

Hence, for $t \in [t_n, t_{n+1})$, the function $z(t)$ satisfies the inequality

$$z(t) \leq C_1 + M_2 \int_{t_n}^{t} (t-s)^{-\alpha} z(s) ds,$$

where $C_1 = M_1 z_0 \left(1 + M_2 \frac{Q^{1-\alpha}}{1-\alpha} \tilde{C} + M_3 \tilde{C}\right)^n$. Applying (7.56) at the interval $(t_n, t_{n+1}]$, we obtain (7.55). The lemma is proved.

Theorem 3. *Let Eqs. (7.50) and (7.51) satisfy assumptions of Theorem 2 and let the corresponding linear equation be exponentially stable.*

Then for sufficiently small $N_1 > 0$, the equation has a unique W-almost periodic solution $u_0(t)$, and this solution is exponentially stable.

Proof. The existence and uniqueness of the W-almost periodic solution $u_0(t)$ follows from Theorem 2. We prove its asymptotic stability. Let $u(t)$ be an arbitrary solution of the equation satisfying $\|u(t_0) - u_0(t_0)\|_\alpha \leq \delta$, where δ is a small positive number.

Then by $t \geq t_0$ the difference of these solutions satisfies

$$u(t) - u_0(t) = U(t, t_0)(u(t_0) - u_0(t_0)) + \int_{t_0}^{t} U(t,s)\Big(f(s, u(s)) -$$
$$-f(s, u_0(s))\Big)ds + \sum_{t_0 \leq \tau_k < t} U(t, \tau_k)\left(g_k(u(\tau_k)) - g_k(u_0(\tau_k))\right).$$

Then for $t_0 \in (\tau_0, \tau_1)$ and $t \in (\tau_j, \tau_{j+1}]$ we have

$$\|u(t) - u_0(t)\|_\alpha \leq \|U(t, t_0)\|_\alpha \|u(t_0) - u_0(t_0)\|_\alpha +$$
$$+ \int_{t_0}^{\tau_1} \|U(t, \tau_1)\|_\alpha \|V(\tau_1, s)(f(s, u(s)) - f(s, u_0(s)))\|_\alpha ds + \cdots +$$
$$+ \int_{\tau_{j-1}}^{\tau_j} \|U(t, \tau_j)\|_\alpha \|V(\tau_j, s)(f(s, u(s)) - f(s, u_0(s)))\|_\alpha ds +$$
$$+ \int_{\tau_j}^{t} \|V(t, s)(f(s, u(s)) - f(s, u_0(s)))\|_\alpha ds +$$
$$+ \sum_{t_0 \leq \tau_k < t} \|U(t, \tau_k)\left(g_k(u(\tau_k)) - g_k(u_0(\tau_k))\right)\|_\alpha \leq$$
$$\leq Me^{-\beta(t-t_0)}\|u(t_0) - u_0(t_0)\|_\alpha + Me^{-\beta(t-\tau_1)} \int_{t_0}^{\tau_1} \frac{L_Q N_1}{(\tau_1 - s)^\alpha} \|u(s) - u_0(s)\|_\alpha ds +$$
$$+ \cdots + Me^{-\beta(t-\tau_j)} \int_{\tau_{j-1}}^{\tau_j} \frac{L_Q N_1}{(\tau_j - s)^\alpha} \|u(s) - u_0(s)\|_\alpha ds +$$
$$+ \int_{\tau_j}^{t} \frac{L_Q N_1}{(t-s)^\alpha} \|u(s) - u_0(s)\|_\alpha ds + \sum_{t_0 \leq \tau_k < t} Me^{-\beta(t-\tau_k)} N_1 \|u(\tau_k) - u_0(\tau_k)\|_\alpha.$$

Denote $v(t) = e^{\beta t}\|u(t) - u_0(t)\|_\alpha, M_2 = e^{\beta Q}ML_QN_1, M_3 = MN_1$. Then

$$v(t) \le Mv(t_0) + M_2\int_{t_0}^{\tau_1}\frac{v(s)ds}{(\tau_1 - s)^\alpha} + \cdots + M_2\int_{t_j}^{t}\frac{v(s)ds}{(\tau_j - s)^\alpha} + M_3\sum_{k=1}^{j}v(\tau_k).$$

Then by Lemma 6 we get

$$\|u(t) - u_0(t)\|_\alpha \le M\tilde{C}e^{-\beta(t-t_0)}\left(1 + M_2\tilde{C}\frac{Q^{1-\alpha}}{1-\alpha} + M_3\tilde{C}\right)^{i(t,t_0)}\|u(t_0) - u_0(t_0)\|_\alpha.$$

Therefore, if

$$\beta > p\ln\left(1 + M_2\tilde{C}\frac{Q^{1-\alpha}}{1-\alpha} + M_3\tilde{C}\right),$$

where p is defined by (7.3), then the W-almost periodic solution $u_0(t)$ of Eqs. (7.50) and (7.51) is asymptotically stable. This can be achieved by sufficiently small N_1.

7.5 Almost Periodic Solutions of Equations with Nonfixed Moments of Impulsive Action

We consider the following equation with points of impulsive action depending on solutions

$$\frac{du}{dt} + Au = f(t,u), \quad t \ne \tau_j(u), \tag{7.57}$$

$$u(\tau_j(u) + 0) - u(\tau_j(u)) = B_ju + g_j(u), \quad j \in Z. \tag{7.58}$$

Definition 6 ([11]). A solution $u_0(t)$ of Eqs. (7.57) and (7.58) defined for all $t \ge t_0$, is called Lyapunov stable in space X^α if, for an arbitrary $\varepsilon > 0$ and $\eta > 0$, there exists such a number $\delta = \delta(\varepsilon,\eta)$ that, for any other solution $u(t)$ of the system, $\|u_0(t_0) - u(t_0)\|_\alpha < \delta$ implies that $\|u_0(t) - u(t)\|_\alpha < \varepsilon$ for all $t \ge t_0$ such that $|t - \tau_j^0| > \eta$, where τ_j^0 are the times during which the solution $u_0(t)$ intersects the surfaces $t = \tau_j(u), j \in Z$.

A solution $u_0(t)$ is said to be attractive if for each $\varepsilon > 0, \eta > 0$, and $t_0 \in R$, there exist $\delta_0 = \delta_0(t_0)$ and $T = T(\delta_0,\varepsilon,\eta) > 0$ such that for any other solution $u(t)$ of the system, $\|u_0(t_0) - u(t_0)\| < \delta_0$ implies $\|u_0(t) - u(t)\|_\alpha < \varepsilon$ for $t \ge t_0 + T$ and $|t - \tau_k^0| > \eta$.

A solution $u_0(t)$ is called asymptotically stable if it is stable and attractive.

Theorem 4. *Assume that in some domain $U_\rho^\alpha = \{u \in X^\alpha, \|u\|_\alpha \le \rho\}$, Eqs. (7.57) and (7.58) satisfy conditions (H1), (H3)–(H6), and*

1) all solutions in domain U^α_ρ intersect each surface $t = \tau_j(u)$ no more than once;
2) $\|f(t_1,u) - f(t_2,u)\| \le H_1|t_1 - t_2|^\nu$, $\nu > 0$, $H_1 > 0$;
3) $\|f(t,u_1) - f(t,u_2)\| + \|g_j(u_1) - g_j(u_2)\|_\alpha + |\tau_j(u_1) - \tau_j(u_2)| \le N_1\|u_1 - u_2\|_\alpha$, *uniformly to $t \in R, j \in Z$,*
4) $AB_j = B_jA$, $\|f(t,0)\| \le M_0$, $\|g_j(0)\|_1 \le M_0, j \in Z$
5) the linear homogeneous equation

$$M_* = \frac{M_1}{1 - e^{-\beta_1\theta}}\left(1 + \frac{C_\alpha Q^{1-\alpha}}{1-\alpha}\right).$$

$$\frac{du}{dt} + Au = 0, \quad t \ne \tau_j, \tag{7.59}$$

$$\Delta u|_{t=\tau_j} = u(\tau_j + 0) - u(\tau_j) = B_j u(\tau_j), \quad j \in Z, \tag{7.60}$$

is exponentially stable in space X^α

$$\|U(t,s)u\|_\alpha \le Me^{-\beta(t-s)}\|u\|_\alpha,\ t \ge s, u \in X^\alpha \tag{7.61}$$

where $\tau_j = \tau_j(0)$, $\beta > 0$ and $M \ge 1$.
6) $N_1M_* < 1$ and $\rho \ge \rho_0 = M_0M_*/(1 - N_1M_*)$, where

Then for sufficiently small values of the Lipschitz constant N_1, Eqs. (7.57) and (7.58) have in U^α_ρ a unique W-almost periodic solution and this solution is exponentially stable.

Proof. 1. First, using the method proposed in [6], we prove the existence of the W-almost periodic solution. Let $y = \{y_j\}$ be an almost periodic sequence of elements $y_j \in X^\alpha$, $\|y_j\|_\alpha \le \varrho$. We consider the equation with fixed moments of impulsive action

$$\frac{du}{dt} + Au = f(t,u), \quad t \ne \tau_j(y), \tag{7.62}$$

$$u(\tau_j(y_j) + 0) - u(\tau_j(y_j)) = B_j u(\tau_j(y_j)) + g_j(y_j), \quad j \in Z. \tag{7.63}$$

By Lemma 5, if a constant N_1 sufficiently small, then corresponding to (7.62) and (7.63) the linear impulsive equation [if $f \equiv 0, g_j(y_j) \equiv 0, j \in Z$,] is exponentially stable. Its evolution operator $U(t,\tau,y)$ satisfies estimate

$$\|U(t,\tau,y)u\|_\alpha \le M_1e^{-\beta_1(t-\tau)}\|u\|_\alpha,\ t \ge \tau, \tag{7.64}$$

with some positive constants $M_1 \ge M, \beta_1 \le \beta$.

Equations (7.62) and (7.63) have a unique solution bounded on the axis which satisfies the integral equation

$$\tilde{u}(t,y) = \int_{-\infty}^{t} U(t,\tau,y)f(\tau,\tilde{u}(\tau,y))d\tau + \sum_{\tau_j(y_j)<t} U(t,\tau_j(y_j),y)g_j(y_j). \quad (7.65)$$

We choose $u_0(t,y) \equiv 0$ and construct the sequence of W-almost periodic functions

$$u_{n+1}(t,y) = \int_{-\infty}^{t} U(t,\tau,y)f(\tau,u_n(\tau,y))d\tau + \sum_{\tau_j(y_j)<t} U(t,\tau_j(y_j),y)g_j(y_j),\ n=0,1,\ldots.$$

The proof of the W-almost periodicity of $u_{n+1}(t,y)$ in space X^α is similar to the proof of Theorem 1.

One can verify that for sufficiently small $N_1 > 0$ the sequence $\{u_n(t,y)\}$ converges to the W-almost periodic solution $u^*(t,y) : R \to X^\alpha$ of Eq. (7.65). As in the proof to Theorem 2, we prove that $u^*(t,y)$ is the W-almost periodic solution of impulsive equations (7.62) and (7.63).

Let $t \in (\tilde{\tau}_i, \tilde{\tau}_{i+1})$, where $\tilde{\tau}_i = \tau_i(y_i)$. As in (7.39), we obtain

$$\|u^*(t,y)\|_\alpha \le \int_{-\infty}^{t} \|A^\alpha U(t,s,y)(f(s,0) + f(s,u^*(s,y)) - f(s,0))\|ds +$$
$$+ \sum_{\tau_j(y_j)<t} \|U(t,\tilde{\tau}_j,y)(g_j(0) + g_j(y_j) - g_j(0))\|_\alpha \le$$
$$\le \frac{M_1}{1-e^{-\beta_1\theta}}\left(\frac{C_\alpha\Theta^{1-\alpha}}{1-\alpha}\left(M_0 + N_1\sup_t\|u^*(t,y)\|_\alpha\right) + M_0 + N_1\sup_j\|y_j\|_\alpha\right).$$

Hence, by sufficiently small $N_1 > 0$

$$\sup_t \|u^*(t,y)\| \le \rho_0. \quad (7.66)$$

If we choose the almost periodic sequence $y^* = \{y_j^*\}, y_j^* \in X^\alpha$, such that

$$u^*(\tau_j(y_j^*), y^*) = y_j^*$$

for all $j \in Z$, then the function $u^*(t,y^*)$ will be exactly the W-almost periodic solution of Eqs. (7.57) and (7.58).

We consider the space $\mathscr{N}$ of sequences $y = \{y_j\},\ y_j \in X^\alpha$, with norm $\|y\|_S = \sup_j \|y_j\|_\alpha$ and map $S : \mathscr{N} \to \mathscr{N}$,

$$S(y) = \{u^*(\tau_j(y_j), y)\}_{j\in Z}.$$

By (7.66), S maps the domain $U_\varrho^\alpha \subset \mathscr{N}$ onto itself for $\rho = \rho_0$.
Now we prove that S is a contraction:

$$\|S(y)_j - S(z)_j\|_\alpha = \|u^*(\tau_j(y_j),y) - u^*(\tau_j(z_j),z)\|_\alpha \le$$

$$\leq \|u^*(\tilde{\tau}_j^1, y) - u^*(\tilde{\tau}_j^1, z)\|_\alpha + \|u^*(\tilde{\tau}_j^1, z) - u^*(\tilde{\tau}_j^2, z)\|_\alpha, \tag{7.67}$$

where $\tilde{\tau}_j^1 = \tau_j(y_j), \ \tilde{\tau}_j^2 = \tau_j(z_j)$.

Denote $\mathscr{J} = \cup \mathscr{J}_j$,

$$\mathscr{J}_j = (\max\{\tilde{\tau}_{j-1}^1, \tilde{\tau}_{j-1}^2\}, \min\{\tilde{\tau}_j^1, \tilde{\tau}_j^2\}] = (\tau_{j-1}'', \tau_j'].$$

Denote also $\xi_i = (\tau_i' - \tau_{j-1}'')/2, \ i \in Z$.

To estimate the difference $\|u^*(\tilde{\tau}_j^1, y) - u^*(\tilde{\tau}_j^1, z)\|_\alpha$, we apply iteration on n. Put $u_0(t, y) = u_0(t, z) = 0$. Then for $t \in (\tilde{\tau}_i'', \tilde{\tau}_{i+1}']$ we get

$$\begin{aligned}
&\|u_1(t, y) - u_1(t, z)\|_\alpha = \\
&= \| \sum_{k \leq i} A^\alpha U(t, \tilde{\tau}_k^1, y) g_k(y_k) - \sum_{k \leq i} A^\alpha U(t, \tilde{\tau}_k^2, z) g_k(z_k)\| \leq \\
&\leq \sum_{k \leq i} \|A^\alpha U(t, \tilde{\tau}_k^1, y) \left(g_k(y_k) - g_k(z_k)\right) \| + \|A^\alpha \left(U(t, \tilde{\tau}_i^1, y) - U(t, \tilde{\tau}_i^2, z)\right) g_i(z_i)\| + \\
&+ \sum_{k<i} \|A^\alpha \left(U(t, \tilde{\tau}_k^1, y) - U(t, \tilde{\tau}_k^2, z)\right) g_k(z_k)\| \leq \\
&\leq \sum_{k<i} M_1 e^{-\beta_1 |t - \tilde{\tau}_k^1|} N_1 \|y_k - z_k\|_\alpha + \|A^\alpha e^{-A(t - \tilde{\tau}_i'')} (e^{-A(\tilde{\tau}_i'' - \tilde{\tau}_i')} - I) g_i(z_i)\| + \\
&+ \sum_{k<i} \|(U(t, \xi_i, y)(U(\xi_i, \xi_{k+1}, y) U(\xi_{k+1}, \tilde{\tau}_k^1, y) - \\
&- U(t, \xi_i, z)(U(\xi_i, \xi_{k+1}, z) U(\xi_{k+1}, \tilde{\tau}_k^2, z)) g_k(z_k)\|_\alpha \leq \\
&\leq \frac{M_1 N_1}{1 - e^{-\beta_1 \theta}} \|y - z\|_S + C_\alpha C_0 (t - \tilde{\tau}_i'')^{-\alpha} |\tilde{\tau}_k'' - \tilde{\tau}_k'| \|g_i(z_i)\|_1 + \\
&+ \sum_{k<i} \|A^\alpha (U(t, \xi_i, y) - U(t, \xi_i, z)) U(\xi_i, \xi_{k+1}, y) U(\xi_{k+1}, \tilde{\tau}_k^1, y) g_k(z_k)\| + \\
&+ \sum_{k<i} \|A^\alpha U(t, \xi_i, z)(U(\xi_i, \xi_{k+1}, y) - U(\xi_i, \xi_{k+1}, z)) U(\xi_k, \tilde{\tau}_k^1, y) g_k(z_k)\| + \\
&+ \sum_{k<i} \|A^\alpha U(t, \xi_i, z) U(\xi_i, \xi_{k+1}, z)(U(\xi_{k+1}, \tilde{\tau}_k^1, y) - U(\xi_{k+1}, \tilde{\tau}_k^2, z)) g_k(z_k)\|.
\end{aligned} \tag{7.68}$$

To evaluate the difference $U(\xi_i, \xi_{k+1}, y) - U(\xi_i, \xi_{k+1}, z)$ we construct two sequences of bounded operators $X^\alpha \to X^\alpha$ defined by

$$T_n = U(\xi_{n+1}, \xi_n, y), \ \tilde{T}_n = U(\xi_{n+1}, \xi_n, z), \ n \in Z.$$

The corresponding difference equations $u_{n+1} = T_n u_n$ and $u_{n+1} = \tilde{T}_n u_n$ are exponentially stable. Their evolution operators

$$T_{n,m} = T_{n-1} \ldots T_m,\ n \geq m,\ T_{m,m} = I,$$

and

$$\tilde{T}_{n,m} = \tilde{T}_{n-1} \ldots \tilde{T}_m,\ n \geq m,\ \tilde{T}_{m,m} = I,$$

satisfy equality

$$\tilde{T}_{n,m} - T_{n,m} = \sum_{k<n} T_{n,k+1}(\tilde{T}_k - T_k)\tilde{T}_{k,m},\ n \geq m. \tag{7.69}$$

Analogous to (7.32) and (7.33), we obtain

$$\|\tilde{T}_{n,m} - T_{n,m}\|_\alpha \leq M_2 e^{-\beta_2\theta(n-m)} \sup_k \|\tilde{T}_k - T_k\|_\alpha,\ n \geq m, \tag{7.70}$$

with some $\beta_2 \leq \beta_1, M_2 \geq M_1$.

Now we estimate the difference $\|\tilde{T}_n - T_n\|_\alpha$:

$$\begin{aligned}
&\|T_n - \tilde{T}_n\|_\alpha = \|U(\xi_{n+1}, \xi_n, y) - U(\xi_{n+1}, \xi_n, z)\|_\alpha = \\
&= \|e^{-A(\xi_{n+1}-\tilde{\tau}_n^1)}(I + B_n)e^{-A(\tilde{\tau}_n^1-\xi_n)} - e^{-A(\xi_{n+1}-\tilde{\tau}_n^2)}(I + B_n)e^{-A(\tilde{\tau}_n^2-\xi_n)}\|_\alpha \leq \\
&\leq \|(e^{-A(\xi_{n+1}-\tilde{\tau}_n^1)} - e^{-A(\xi_{n+1}-\tilde{\tau}_n^2)})(I + B_n)e^{-A(\tilde{\tau}_n^1-\xi_n)}\|_\alpha + \\
&+\|e^{-A(\xi_{n+1}-\tilde{\tau}_n^2)}(I + B_n)(e^{-A(\tilde{\tau}_n^1-\xi_n)} - e^{-A(\tilde{\tau}_n^2-\xi_n)})\|_\alpha \leq \\
&\leq 2C_\alpha C_1(\theta/2)^{-1-\alpha}|\tilde{\tau}_n^1 - \tilde{\tau}_n^2|.
\end{aligned} \tag{7.71}$$

Therefore,

$$\begin{aligned}
&\|(\tilde{T}_{n,m} - T_{n,m})u\|_\alpha = \|(U(\xi_n, \xi_m, y) - U(\xi_n, \xi_m, z))u\|_\alpha \leq \\
&\leq M_2 e^{-\beta_2\theta(n-m)} 2C_\alpha C_1(\theta/2)^{-1-\alpha} \sup_j |\tilde{\tau}_i^1 - \tilde{\tau}_i^2|\|u\|_\alpha,\ n \geq m.
\end{aligned} \tag{7.72}$$

To finish the estimation of (7.68), we consider the following two differences:

$$\begin{aligned}
&\|(U(t, \xi_i, y) - U(t, \xi_i, z)))u\|_\alpha \leq \|A^\alpha(e^{-A(t-\tau_i')}(I + B_i)e^{-A(\tau_i'-\xi_i)} - \\
&-e^{-A(t-\tau_i'')}(I + B_i)e^{-A(\tau_i''-\xi_i)})u\| \leq \frac{4C_0C_{1-\alpha}}{\theta(t-\tau_i'')^\alpha}|\tau_i'' - \tau_i'|\|u\|_\alpha.
\end{aligned} \tag{7.73}$$

$$\begin{aligned}
&\|(U(\xi_k, \tilde{\tau}_k^1, y) - U(\xi_k, \tilde{\tau}_k^2, z))u\|_\alpha = \|A^\alpha(I - e^{-A(\tau_k''-\tau_k')})e^{-A(\xi_{k+1}-\tau_k'')}u\| \leq \\
&\leq C_0C_1(\theta/2)^{-\alpha}|\tau_i'' - \tau_i'|\|u\|_\alpha.
\end{aligned} \tag{7.74}$$

Taking into account (7.70), (7.73), and (7.74), by (7.68) we obtain for $t \in (\tau_i'', \tau_{i+1}']$

$$\|u_1(t,y) - u_1(t,z)\|_\alpha \le N_1 \|y - z\|_S \left(K_1' + K_2''(t - \tau_i'')^{-\alpha}\right), \tag{7.75}$$

where the positive constants K_1' and K_2'' don't depend on i.

Now we consider the $(n+1)$st iteration

$$\begin{aligned}
&\|u_{n+1}(t,y) - u_{n+1}(t,z)\|_\alpha = \\
&= \| \int_{-\infty}^{t} A^\alpha U(t,\tau,y) f(\tau, u_n(\tau,y)) d\tau + \sum_{k \le i} A^\alpha U(t, \tilde{\tau}_k^1, y) g_k(y_k) - \\
&- \int_{-\infty}^{t} A^\alpha U(t,\tau,z) f(\tau, u_n(\tau,z)) d\tau - \sum_{k \le i} A^\alpha U(t, \tilde{\tau}_k^2, z) g_k(z_k) \| \le \\
&\le \int_{-\infty}^{t} \|A^\alpha U(t,\tau,y) \left(f(\tau, u_n(\tau,y)) - f(\tau, u_n(\tau,z))\right)\| d\tau + \\
&+ \int_{-\infty}^{t} \|A^\alpha (U(t,\tau,y) - U(t,\tau,z)) f(\tau, u_n(\tau,z))\| d\tau + \\
&+ \sum_{k \le i} \|A^\alpha U(t, \tilde{\tau}_k^1, y) \left(g_k(y_k) - g_k(z_k)\right)\| + \\
&+ \sum_{k \le i} \|A^\alpha \left(U(t, \tilde{\tau}_k^1, y) - U(t, \tilde{\tau}_k^2, z)\right) g_k(z_k)\|.
\end{aligned} \tag{7.76}$$

Similar to (7.39), we get

$$\begin{aligned}
&\int_{\tau_i''}^{t} \|A^\alpha e^{-A(t-s)} \left(f(\tau, u_n(\tau,y)) - f(\tau, u_n(\tau,z))\right)\| d\tau + \\
&\sum_{k<i} \int_{\tau_k''}^{\tau_{k+1}'} \|A^\alpha U(t,\tau,y) \left(f(\tau, u_n(\tau,y)) - f(\tau, u_n(\tau,z))\right)\| d\tau + \\
&\le \frac{M_1}{1 - e^{-\theta\beta_1}} \frac{C_\alpha \Theta^{1-\alpha}}{1-\alpha} N_1 \sup_{\tau \in \mathscr{J}} \|u_n(\tau,y) - u_n(\tau,z)\|,
\end{aligned}$$

$$\sum_{k \le i} \|A^\alpha U(t, \tilde{\tau}_k^1, y) \left(g_k(y_k) - g_k(z_k)\right)\| \le \frac{M_1}{1 - e^{-\theta\beta_1}} N_1 \|y - z\|_\alpha. \tag{7.77}$$

If $\|u_n(\tau,y)\|_\alpha \le \rho$ and $\|u_n(\tau,z)\|_\alpha \le \rho$, then for $t \in (\tau_i'', \tau_{i+1}']$

$$\sum_{k \le i} \int_{\tau_k'}^{\tau_k''} \|A^\alpha U(t,s,y) \left(f(s, u_n(s,y)) - f(s, u_n(s,z))\right)\| ds \le$$

$$\leq \sum_{k\leq i}\int_{\tau_k'}^{\tau_k''} \|U(t,s,y)f(s,u_n(s,y))\|_\alpha ds + \sum_{k\leq i}\int_{\tau_k'}^{\tau_k''} \|U(t,s,y)f(s,u_n(s,z))\|_\alpha ds \leq$$

$$\leq 2\sum_{k<i} M_1 e^{-\beta_1|t-\tau_k''|}(M_0+N_1\rho) + 2\int_{\tau_i'}^{\tau_i''} \|A^\alpha U(t,s,y)\|(M_0+N_1\rho)ds \leq$$

$$\leq \left(\frac{2M_1}{1-e^{-\beta_1\theta}} + \frac{2M_1}{1-\alpha}(t-\tau_i'')^{-\alpha}\right)(M_0+N_1\rho)N_1\|y-z\|_S, \tag{7.78}$$

since for $t > \tau_2 > \tau_1$

$$\int_{\tau_1}^{\tau_2} \frac{ds}{(t-s)^\alpha} \leq \frac{\tau_2-\tau_1}{(1-\alpha)((t-\tau_2)^\alpha}.$$

The second integral in (7.76) satisfies the following inequality:

$$I_2 = \int_{-\infty}^{t} \|A^\alpha(U(t,s,y)-U(t,s,z))f(s,u_n(s,z))\|ds \leq$$

$$\leq \int_{\tau_i''}^{t} \|A^\alpha(e^{-A(t-s)} - e^{-A(t-s)})f(s,u_n(s,z))\|ds +$$

$$+\int_{\tau_i'}^{\tau_i''} \|A^\alpha(U(t,s,y)-U(t,s,z))f(s,u_n(s,z))\|ds +$$

$$+\int_{\xi_i}^{\tau_i'} \|A^\alpha(U(t,s,y)-U(t,s,z))f(s,u_n(s,z))\|ds +$$

$$+\sum_{k<i}\int_{\xi_k}^{\xi_{k+1}} \|A^\alpha(U(t,s,y)-U(t,s,z))f(s,u_n(s,z))\|ds. \tag{7.79}$$

We consider all integrals in (7.79) separately.

$$I_{21} = \int_{\tau_i'}^{\tau_i''} \|A^\alpha U(t,s,y)f(s,u_n(s,z))\|ds \leq \frac{C_\alpha\|I+B_i\|(M_0+N_1\rho)}{(1-\alpha)(t-\tau_i'')^\alpha}|\tau_i''-\tau_i'|,$$

$$I_{22} = \int_{\tau_i'}^{\tau_i''} \|A^\alpha U(t,s,z)f(s,u_n(s,z))\|ds \leq \frac{C_\alpha\|I+B_i\|(M_0+N_1\rho)}{(1-\alpha)(t-\tau_i'')^\alpha}|\tau_i''-\tau_i'|,$$

$$I_{23} = \int_{\xi_i}^{\tau_i'} \|A^\alpha(U(t,s,y)-U(t,s,z))f(s,u_n(s,z))\|ds =$$

$$= \int_{\xi_i}^{\tau_i'} \|A^\alpha(U(t,\tilde\tau_i^1,y)U(\tilde\tau_i^1,s,y) - U(t,\tilde\tau_i^2,z)U(\tilde\tau_i^2,s,z))f(s,u_n(s,z))\|ds \leq$$

$$\leq \int_{\xi_i}^{\tau_i'} \|A^\alpha\Big((e^{-A(t-\tilde{\tau}_i^1)} - e^{-A(t-\tilde{\tau}_i^2)})(I+B_i)e^{-A(\tilde{\tau}_i^1-s)} -$$

$$-A^\alpha e^{-A(t-\tilde{\tau}_i^2)})(I+B_i)(e^{-A(\tilde{\tau}_i^1-s)} - e^{-A(\tilde{\tau}_i^2-s)})\Big)f(s,u_n(s,z))\|ds \leq$$

$$\leq \frac{2C_0C_{\alpha_1}C_{1+\alpha-\alpha_1}}{(t-\tau_i'')^{\alpha_1}}\|I+B_i\|\frac{(\tau_i'-\xi_i)^{\alpha_1-\alpha}}{\alpha_1-\alpha}|\tau_i''-\tau_i'|.$$

The last sum in (7.79) is transformed as follows:

$$I_{24} = \sum_{k<i}\int_{\xi_k}^{\xi_{k+1}} \|A^\alpha(U(t,s,y)-U(t,s,z))f(s,u_n(s,z))\|ds =$$

$$= \sum_{k<i}\int_{\xi_k}^{\xi_{k+1}} \|(U(t,\xi_i,y)U(\xi_i,\xi_{k+1},y)U(\xi_{k+1},s,y) -$$

$$-U(t,\xi_i,z)U(\xi_i,\xi_{k+1},z)U(\xi_{k+1},s,z))f(s,u_n(s,z))\|_\alpha ds \leq$$

$$\leq \sum_{k<i}\int_{\xi_k}^{\xi_{k+1}}\Big(\|(U(t,\xi_i,y)-U(t,\xi_i,z))U(\xi_i,\xi_{k+1},y)U(\xi_{k+1},s,y)f(s,u_n(s,z))\|_\alpha +$$

$$+\|U(t,\xi_i,z)(U(\xi_i,\xi_{k+1},y)-U(\xi_i,\xi_{k+1},z))U(\xi_{k+1},s,y)f(s,u_n(s,z))\|_\alpha +$$

$$+\|U(t,\xi_i,z)U(\xi_i,\xi_{k+1},z)(U(\xi_{k+1},s,y)-U(\xi_{k+1},s,z))f(s,u_n(s,z))\|_\alpha\Big)ds.$$

To finish the estimation of integral I_{24} we use (7.72), (7.73), and (7.74):

$$\int_{\xi_k}^{\xi_{k+1}} \|A^\alpha(U(\xi_{k+1},s,y)-U(\xi_{k+1},s,z))f\|_\alpha ds \leq$$

$$\leq \int_{\tau_k''}^{\xi_{k+1}} \|A^\alpha(e^{-A(\xi_{k+1}-s)} - e^{-A(\xi_{k+1}-s)})f\|ds +$$

$$+ \int_{\tau_k'}^{\tau_k''} \|A^\alpha(e^{-A(\xi_{k+1}-\tau_k'')}(I+B_k)e^{-A(\tau_k''-s)} - e^{-A(\xi_{k+1}-s)})f\|ds +$$

$$+ \int_{\xi_k}^{\tau_k'} \|(e^{-A(\xi_{k+1}-\tau_k')}(I+B_k)e^{-A(\tau_k'-s)} - e^{-A(\xi_{k+1}-\tau_k'')}(I+B_k)e^{-A(\tau_k''-s)})f\|_\alpha ds \leq$$

$$\leq \tilde{K}C_{\alpha_1}(\xi_{k+1}-\tau_k'')^{-\alpha_1}\|I+B_k\||\tau_k''-\tau_k'|\|f\|$$

with some positive constant $\tilde{K}$. Therefore,

$$I_2 \leq \left(K_2'N_1 + \frac{K_2''N_1}{(t-\tau_i'')^{\alpha_1}}\right)\|y-z\|_S \tag{7.80}$$

with $\alpha_1 > \alpha$ and positive constants K_1' and K_2'' independent of i, k.

By (7.75), (7.79), and (7.80) we obtain for $t \in (\tau_i'', \tau_{i+1}']$

$$\|u_{n+1}(t,y) - u_{n+1}(t,z)\|_\alpha \leq$$

$$\leq \sum_{k<i} \int_{\tau_k''}^{\tau_{k+1}'} \|A^\alpha U(t,\tau,y)(f(\tau,u_n(\tau,y)) - f(\tau,u_n(\tau,z)))\| d\tau +$$

$$+ \int_{\tau_i''}^{t} \|A^\alpha U(t,\tau,y)(f(\tau,u_n(\tau,y)) - f(\tau,u_n(\tau,z)))\| d\tau +$$

$$+ \left(K_3' + \frac{K_3''}{(t-\tau_i'')^{\alpha_1}}\right) N_1 \|y-z\|_S, \tag{7.81}$$

where the constants K_3' and K_3'' don't depend on n.

Let the nth iteration satisfy the inequality

$$\|u_n(t,y) - u_n(t,z)\|_\alpha \leq \left(L_n' + \frac{L_n''}{(t-\tau_i'')^{\alpha_1}}\right) N_1 \|y-z\|_S,\ t \in (\tau_i'', \tau_{i+1}'],$$

with positive constants L_n' and L_n''. We estimate the $(n+1)$st iteration.

$$\|u_{n+1}(t,y) - u_{n+1}(t,z)\|_\alpha \leq \left(K_3' + \frac{K_3''}{(t-\tilde{\tau}_i'')^{\alpha_1}}\right) N_1 \|y-z\|_S +$$

$$+ N_1^2 \|y-z\|_S \sum_{k<i} \int_{\tau_k''}^{\tau_{k+1}'} \|A^\alpha U(t,s)\| \left(L_n' + \frac{L_n''}{(s-\tau_k'')^{\alpha_1}}\right) ds +$$

$$+ N_1^2 \|y-z\|_S \int_{\tau_i''}^{t} \|A^\alpha U(t,s)\| \left(L_n' + \frac{L_n''}{(s-\tau_i'')^{\alpha_1}}\right) ds \leq$$

$$\leq N_1^2 \|y-z\|_S \left(\sum_{k\leq i} \int_{\tau_k''}^{\tau_{k+1}'} M_1 e^{-\beta_1|t-s|} \left(L_n' + \frac{L_n''}{(s-\tau_k'')^{\alpha_1}}\right) ds + \right.$$

$$\left. + \int_{\tau_i''}^{t} M_1 (t-s)^{-\alpha_1} \left(L_n' + \frac{L_n''}{(s-\tau_i'')^{\alpha_1}}\right) ds \right) +$$

$$+ \left(K_3' + \frac{K_3''}{(t-\tau_i'')^{\alpha_1}}\right) N_1 \|y-z\|_S \leq$$

$$\leq \left(\frac{M_1}{1-e^{-\beta_1\theta}} \left(L_n' Q + \frac{L_n'' Q^{1-\alpha_1}}{1-\alpha_1}\right) + \frac{L_n'' M_1 2^{2\alpha}}{1-\alpha_1} (t-\tau_i'')^{1-2\alpha_1} + \right.$$

$$\left. + \frac{L_n' M_1}{1-\alpha_1} (t-\tau_i'')^{1-\alpha_1} \right) N_1^2 \|y-z\|_S + \left(K_3' + \frac{K_3''}{(t-\tau_i'')^{\alpha_1}}\right) N_1 \|y-z\|_S \leq$$

$$= \left(L'_{n+1} + \frac{L''_{n+1}}{(t - \tau''_i)^{\alpha_1}} \right) N_1 \|y - z\|_S. \tag{7.82}$$

One can verify that for sufficiently small N_1 the sequences L'_n and L''_n are uniformly bounded by some constants L'_* and L''_*.

Since the sequences $u_n(t, y)$ and $u_n(t, z)$ tend to limit the functions $u_*(t, y)$ and $u_*(t, z)$, respectively, we conclude by (7.82) for $t \in (\tau''_i, \tau'_{i+1}]$ that

$$\|u_*(t, y) - u_*(t, z)\|_\alpha \le \left(L'_* + \frac{L''_*}{(t - \tau''_{i+1})^{\alpha_1}} \right) N_{i+1} \|y - z\|_S$$

and

$$\|u^*(\tau'_{i+1}, y) - u^*(\tau'_{i+1}, z)\|_\alpha \le \left(L'_* + \frac{L''_*}{\theta^{\alpha_1}} \right) N_1 \|y - z\|_S. \tag{7.83}$$

Now we estimate the second summand in (7.67). Note that by our assumption $\tilde{\tau}^1_j < \tilde{\tau}^2_j$.

$$\|u^*(\tilde{\tau}^1_j, z) - u^*(\tilde{\tau}^2_j, z)\|_\alpha = \left\| \int_{\tilde{\tau}^1_j}^{\tilde{\tau}^2_j} \frac{d}{ds} u^*(s, z) ds \right\|_\alpha.$$

By Theorem 3.5.2, [9], at the interval $(\tilde{\tau}^2_{j-1}, \tilde{\tau}^2_j)$ the derivative satisfies

$$\left\| \frac{d}{ds} u^*(s, z) \right\|_\gamma \le \tilde{K}_1 (s - \tilde{\tau}^2_{j-1})^{\alpha - \gamma - 1}$$

with some positive constant $\tilde{K}_1$ independent of j and initial value from U^α_ρ.

Then for $t \in (\tilde{\tau}^1_j, \tilde{\tau}^2_j)$

$$\left\| \frac{d}{ds} u^*(s, z) \right\|_\gamma \le \tilde{K}_1 \left(\frac{\theta}{2} \right)^{\alpha - \gamma - 1} = \tilde{K}_2$$

and

$$\|u^*(\tilde{\tau}^1_j, z) - u^*(\tilde{\tau}^2_j, z)\|_\alpha \le \tilde{K}_2 |\tilde{\tau}^1_j - \tilde{\tau}^2_j| \le \tilde{K}_2 N_1 \|y - z\|_S. \tag{7.84}$$

By (7.83) and (7.84) we have

$$\|u^*(\tilde{\tau}^1_j, z) - u^*(\tilde{\tau}^2_j, z)\|_\alpha = \Gamma_9 \|y - z\|_S, \tag{7.85}$$

where $\Gamma_9 < 1$ uniformly for j and $y, z \in \mathcal{N}_{\varrho_0}$.

By (7.67), (7.83), and (7.85) we conclude that the map $S : \mathscr{N}_{\varrho_0} \to \mathscr{N}_{\varrho_0}$ is a contraction. Therefore, there exists a unique almost periodic sequence $y^* = \{y_j^*\}$ such that $u^*(\tau_j(y_j^*), y^*) = y_j^*$ for all $j \in Z$. The function $u^*(t, y^*)$ is the W-almost periodic solution of Eqs. (7.57) and (7.58).

2. Now we prove the stability of the almost periodic solution. Fix arbitrary $\varepsilon > 0$ and $\eta > 0$. Let $t_0 \in [\tau_0(0) + \eta, \tau_1(0) - \eta]$.

 The W-almost periodic solution $u_0(t)$ satisfies the integral equation

$$u_0(t) = U_0(t, t_0)u_0 + \int_{t_0}^{t} U_0(t, s)f(s, u_0(s))ds + \sum_{t_0<\tau_j^0<t} U_0(t, \tau_j^0)g_j(\tau_j^0), \quad (7.86)$$

where $\tau_j^0 = \tau_j(u_0(\tau_j^0))$ and $U_0(t, s)$ is the evolution operator of the linear equation

$$\frac{du}{dt} + Au = 0, \ u(\tau_j^0 + 0) - u(\tau_j^0) = B_j u(\tau_j^0), \ j = 1, 2, \ldots.$$

Let $u_1 \in X^\alpha$ such that $\|u_0 - u_1\|_\alpha < \delta$. The solution $u_1(t)$ with initial value $u_1(t_0) = u_1$ satisfies equation

$$u_1(t) = U_1(t, t_0)u_1 + \int_{t_0}^{t} U_1(t, s)f(s, u_1(s))ds + \sum_{t_0<\tau_j^1<t} U_1(t, \tau_j^1)g_j(\tau_j^1), \quad (7.87)$$

where $\tau_j^1 = \tau_j(u_1(\tau_j^1))$ and $U_1(t, s)$ is the evolution operator of the linear equation

$$\frac{du}{dt} + Au = 0, \ u(\tau_j^1 + 0) - u(\tau_j^1) = B_j u(\tau_j^1), \ j = 1, 2, \ldots.$$

By Lemma 5, for a sufficiently small Lipschitz constant N_1 the evolution operator $U_0(t, s)$ satisfies the inequality

$$\|U_0(t, s)u\|_\alpha \le M_1 e^{-\beta_1(t-s)}\|u\|_\alpha, \ t \ge s, \quad (7.88)$$

with some positive constants $\beta_1 \le \beta, M_1 \ge M$. Moreover, one can verify that for some domain $U_{\tilde{\rho}}^\alpha, \tilde{\rho} \le \rho$, and $N_1 \le N_0$ the evolution operator satisfies

$$\|U_1(t, s)u\|_\alpha \le M_1 e^{-\beta_1(t-s)}\|u\|_\alpha, \ t \ge s, \ t, s \in [t_0, t_0 + T], \quad (7.89)$$

if the values $u_1(t)$ belong to $U_{\tilde{\rho}}^\alpha$ for $\tau_j^1 \in [t_0, t_0 + T]$.

At the interval without impulses, the difference between solutions $u_0(t) - u_1(t)$ satisfies the inequality

$$\|u_1(t) - u_0(t)\|_\alpha \le \|e^{-A(t-t_1)}(u_0(t_1) - u_1(t_1))\|_\alpha +$$

$$+\int_{t_1}^{t}\|A^{\alpha}e^{-A(t-t_1)}(f(s,u_1(s))-f(s,u_0(s)))\|ds\leq$$

$$\leq M_1e^{-\beta_1(t-t_1)}\|u_0(t_1)-u_1(t_1)\|_{\alpha}+\int_{t_1}^{t}\frac{M_1N_1e^{-\beta_1(t-s)}}{(t-s)^{\alpha}}\|u_1(s)-u_0(s)\|_{\alpha}ds.$$

Then by Lemma 2,

$$\|u_1(t)-u_0(t)\|_{\alpha}\leq M_1\tilde{C}e^{-\beta_1(t-t_1)}\|u_1(t_1)-u_0(t_1)\|_{\alpha},\ t-t_1\leq Q. \tag{7.90}$$

Hence, if initial values belong to the bounded domain from X^{α}, then the corresponding solutions are uniformly bounded for t from the bounded interval.

Assume for definiteness that $\tau_j^0\geq\tau_j^1$ and estimate $|\tau_j^1-\tau_j^0|$ by $(u_1(\tau_j^1)-u_0(\tau_j^1))$.

$$\|(u_1(\tau_j^1)-u_0(\tau_j^0)\|_{\alpha}\leq\|(u_0(\tau_j^1)-u_0(\tau_j^0)\|_{\alpha}+\|u_0(\tau_j^1)-u_1(\tau_j^1)\|_{\alpha}\leq$$

$$\leq\left\|\int_{\tau_j^1}^{\tau_j^0}\frac{d}{d\xi}u_0(\xi)d\xi\right\|_{\alpha}+\|u_0(\tau_j^1)-u_1(\tau_j^1)\|_{\alpha}\leq$$

$$\leq\tilde{K}_2|\tau_j^1-\tau_j^0|+\|u_0(\tau_j^1)-u_1(\tau_j^1)\|_{\alpha}.$$

Hence,

$$|\tau_j^1-\tau_j^1|\leq\|u_0(\tau_j^0)-u_1(\tau_j^1)\|_{\alpha}\leq\frac{N_1}{1-\tilde{K}_2N_1}\|u_0(\tau_j^1)-u_1(\tau_j^1)\|_{\alpha}. \tag{7.91}$$

We assume that $t\in(\tau_i'',\tau_{i+1}']$ and estimate the difference

$$\|u_0(t)-u_1(t)\|_{\alpha}=\|U_0(t,t_0)(u_0-u_1)\|_{\alpha}+\|(U_0(t,t_0)-U_1(t,t_0))u_1\|_{\alpha}$$

$$+\int_{t_0}^{t}\|U_0(t,s)f(s,u_0(s))-U_1(t,s)f(s,u_1(s))\|_{\alpha}ds+$$

$$+\|\sum_{t_0<\tau_j^1<t}U(t,\tau_j^1)g_j(\tau_j^1)-\sum_{t_0<\tau_j^0<t}U(t,\tau_j^0)g_j(\tau_j^0)\|_{\alpha}\leq$$

$$\leq\|U_0(t,t_0)(u_0-u_1)\|_{\alpha}+\|(U_0(t,t_0)-U_1(t,t_0))u_1\|_{\alpha}+$$

$$+\int_{t_0}^{\tau_1'}\|U_0(t,s)f(s,u_0(s))-U_1(t,s)f(s,u_1(s))\|_{\alpha}ds+$$

$$+\sum_{j=1}^{i-1}\int_{\tau_j''}^{\tau_{j+1}'}\|U_0(t,s)(f(s,u_0(s))-f(s,u_1(s)))\|_{\alpha}ds+$$

$$+\sum_{j=1}^{i-1}\int_{\tau_j''}^{\tau_{j+1}'}\|(U_0(t,s)-U_1(t,s))f(s,u_1(s))\|_{\alpha}ds+$$

$$+\sum_{j=1}^{i}\int_{\tau_j'}^{\tau_j''}\|U_0(t,s)f(s,u_0(s))-U_1(t,s)f(s,u_1(s))\|_\alpha ds+$$

$$+\int_{\tau_i''}^{t}\|U_0(t,s)f(s,u_0(s))-U_1(t,s)f(s,u_1(s))\|_\alpha ds+$$

$$+\sum_{j=1}^{i}\|U_0(t,\tau_j^0)g_j(\tau_j^0)-U_1(t,\tau_j^1)g_j(\tau_j^1)\|_\alpha. \tag{7.92}$$

Denote $v(t)=\|u_0(t)-u_1(t)\|_\alpha$. Assume that for $t\in[t_0,\tau_i']$ the values $u(t)$ belong to $U^\alpha_{\tilde{\rho}}$; hence, the evolution operators $U_0(t,\tau)$ and $U_1(t,\tau)$ satisfy (7.88) and (7.89) at this interval. By (7.92), analogous to the proof of (7.75), (7.79), and (7.80), we conclude that there exist positive constants M_2 and P_1 independent of i such that for $t\in\mathcal{J}_{i+1}$

$$v(t)\le M_1e^{-\beta_1(t-t_0)}v(t_0)+\int_{t_0}^{\tau_1'}\frac{M_2N_1}{(\tau_1'-s)^\alpha}e^{-\beta_1(t-\tau_1'')}v(s)ds+$$

$$+\sum_{j=2}^{i-1}\int_{\tau_{j-1}''}^{\tau_j'}M_2N_1e^{-\beta_1(t-\tau_j'')}v(s)ds+\sum_{j=1}^{i-1}P_1N_1e^{-\beta_1(t-\tau_j'')}v(\tau_j')+$$

$$+\frac{1}{(t-\tau_i'')^{\alpha_1}}\left(\int_{\tau_{i-1}''}^{\tau_i'}M_2N_1e^{-\beta_1(t-\tau_i'')}v(s)ds+P_1N_1e^{-\beta_1(t-\tau_i'')}v(\tau_i')\right)+$$

$$+\int_{\tau_i''}^{t}M_2N_1e^{-\beta_1(t-s)}(t-s)^{-\alpha_1}v(s)ds \tag{7.93}$$

with $\alpha_1>\alpha$. By (7.90), at the interval $[t_0,\tau_1']$ $v(t)$ satisfies

$$v(t)\le M_1\tilde{C}e^{-\beta(t-t_0)}v(t_0),\ t\in[t_0,\tau_1']. \tag{7.94}$$

By (7.93) and (7.94), for $t\in(\tau_1'',\tau_2']$ we get

$$v(t)\le M_1e^{-\beta_1(t-t_0)}v(t_0)+\frac{1}{(t-\tau_1'')^{\alpha_1}}\int_{t_0}^{\tau_1'}M_2N_1e^{-\beta_1(t-\tau_1'')}v(s)ds+$$

$$+P_1N_1e^{-\beta_1(t-\tau_1'')}(t-\tau_1'')^{-\alpha_1}v(\tau_1')+\int_{\tau_1''}^{t}M_2N_1e^{-\beta_1(t-s)}(t-s)^{-\alpha_1}v(s)ds.$$

Hence, for $M_3=M_2e^{\beta_1 Q},\tilde{C}_1=\tilde{C}/(1-\alpha),\ v_1(t)=e^{\beta_1 t}v(t)$ and $P_2=P_1e^{\beta_1\sup_j|\tau_j''-\tau_j'|}$

$$v_1(t) \le M_1 v_1(t_0)\left(1 + \frac{N_1\tilde{C}(M_3\tilde{Q}+P_2)}{(t-\tau_1'')^{\alpha_1}}\right) + \int_{\tau_1''}^t M_2N_1(t-s)^{-\alpha_1}v_1(s)ds.$$

By Lemma 2

$$v(t) \le M_1\tilde{C}_1 v(t_0)e^{-\beta_1(t-t_0)}\left(1 + \frac{N_1\tilde{C}(M_3\tilde{Q}+P_2)}{(t-\tau_1'')^{\alpha_1}}\right), \ t \in (\tau_1'', \tau_2']. \quad (7.95)$$

Denote $\tilde{Q} = \max_j\{1, (\tau_{j+1}' - \tau_j'')\}$ and $\tilde{\theta} = \min_j\{1, (\tau_{j+1}' - \tau_j'')\}$. Let us prove that

$$v(t) \le M_1\tilde{C}_1 v(t_0)e^{-\beta_1(t-t_0)}\left(1+\frac{N_1\tilde{C}_1(M_3\tilde{Q}+P_2)}{(t-\tau_j'')^{\alpha_1}}\right)\left(1+\frac{N_1\tilde{C}_1(M_3\tilde{Q}+P_2)}{(1-\alpha_1)\tilde{\theta}^{\alpha_1}}\right)^{i-1} \quad (7.96)$$

for $t \in (\tau_i'', \tau_{i+1}']i \ge 2$. We apply the method of mathematical induction. Assume that (7.96) is true for $t \in [\tau_{i-1}'', \tau_i']$ and prove it for $t \in [\tau_i'', \tau_{i+1}']$. Really, by (7.93) for $t \in [\tau_i'', \tau_{i+1}']$ we have

$$\begin{aligned}
v(t) \le M_1 e^{-\beta_1(t-t_0)}v(t_0)\Bigg(&\left(1 + (M_3\tilde{Q}+P_2)N_1\tilde{C}\right) + \\
&+\sum_{j=2}^{i-1}\mathscr{A}^j M_3N_1\tilde{Q}\tilde{C}_1 + \sum_{j=2}^{i-1}\mathscr{A}^{j-1}\left(1+\frac{N_1\tilde{C}_1(M_3\tilde{Q}+P_2)}{\tilde{\theta}^{\alpha_1}}\right)N_1P_2\tilde{C}_1 + \\
&+\mathscr{A}^{i-2}\Bigg(N_1M_3\tilde{C}_1\left((\tau_i' - \tau_{i-1}'') + \frac{N_1\tilde{C}_1(M_3\tilde{Q}+P_2)(\tau_i'-\tau_{i-1}'')}{(1-\alpha_1)(\tau_i'-\tau_{i-1}'')^{\alpha_1}}\right) + \\
&+N_1P_2\tilde{C}_1\left(1+\frac{N_1\tilde{C}_1(M_3\tilde{Q}+P_2)}{(\tau_i'-\tau_{i-1}'')^{\alpha_1}}\right)\Bigg)\Bigg) + \mathscr{B}(t) \le \\
&\le \mathscr{A} + \sum_{j=2}^{i-1}\mathscr{A}^{j-1}(1 + N_1\tilde{C}_1(M_3\tilde{Q}+P_2) - 1) + \frac{\mathscr{A}^{i-1}N_1\tilde{C}_1(M_3\tilde{Q}+P_2)}{(t-\tau_i'')^{\alpha_1}} + \\
&+\mathscr{B}(t) \le \mathscr{A}^{i-1}\left(1+\frac{N_1\tilde{C}_1(M_3\tilde{Q}+P_2)}{(t-\tau_i'')^{\alpha_1}}\right) + \mathscr{B}(t).
\end{aligned}$$

where

$$\mathscr{A} = \left(1 + \frac{N_1\tilde{C}_1(M_3\tilde{Q}+P_2)}{(1-\alpha)\tilde{\theta}^{\alpha_1}}\right), \ \mathscr{B}(t) = \int_{\tau_i''}^t \frac{M_2N_1}{(t-s)^{\alpha_1}}e^{-\beta_1(t-s)}v(s)ds.$$

Hence, for $t \in (\tau_i'', \tau_{i+1}']$, the function $v_1(t) = e^{\beta_1 t} v(t)$ satisfies the inequality

$$v_1(t) \leq \mathscr{A}^{i-1} \left(1 + \frac{N_1 \tilde{C}_1 (M_3 \tilde{Q} + P_2)}{(t - \tau_i'')^{\alpha_1}} \right) + M_2 N_1 \int_{\tau_i''}^{t} (t-s)^{-\alpha_1} v_1(s) ds.$$

Applying Lemma 2, we obtain (7.96).

Let $N_1 > 0$ be such that $\mathscr{A}^{i(t_0,t)} e^{-\beta_1 (t-t_0)} < e^{-\delta_1 (t-t_0)}$ for some positive δ_1. For the given $\varepsilon > 0$ and $\eta > 0$ we choose $v(t_0) = v_0$ such that

$$M_1 \tilde{C}_1 v_0 \left(1 + \frac{N_1 \tilde{C}_1 (M_3 \tilde{Q} + P_2)}{\eta^{\alpha_1}} \right) < \varepsilon.$$

This proves the asymptotic stability of solution u_0.

Example 1. Let us consider the parabolic equation with impulses in variable moments of time:

$$u_t = u_{xx} + a(t) u_x + b(t,x), \tag{7.97}$$

$$\Delta u \Big|_{t=\tau_j(u)} = u(\tau_j(u)+0, x) - u(\tau_j(u), x) = -a_j u(\tau_j(u), x), \tag{7.98}$$

with boundary conditions

$$u(t,0) = u(t,\pi) = 0, \tag{7.99}$$

where the sequence of hypersurfaces τ_j is defined by

$$\tau_j(u) = \theta_j + b_j \int_0^{\pi} u^2(\xi) d\xi, \tag{7.100}$$

where the sequence of real numbers $\{\theta_j\}$ has uniformly almost periodic sequences of differences and $\theta_{j+1} - \theta_j \geq \theta \geq 1/2$,

$\{a_j\}$ and $\{b_j\}$ are almost periodic sequences of positive numbers,

$a(t)$ is a Bohr almost periodic function,

$b(t,x)$ is a Bohr almost periodic function in t uniformly with respect to $x \in [0,\pi]$.

Denote

$$X = L_2(0,\pi),\ A = -\frac{\partial^2}{\partial x^2},\ X^1 = D(A) = H^2(0,\pi) \cap H_0^1(0,\pi).$$

The operator A is sectorial with simple eigenvalues $\lambda_k = k^2$ and corresponding eigenfunctions

$$\varphi_k(x) = \left(\frac{2}{\pi}\right)^{1/2} \sin kx, \ k = 1, 2, \ldots.$$

Operator $-A$ generates an analytic semigroup e^{-At}.

Let $u = \sum_{k=1}^{\infty} a_k \sin kx, \ a_k = \frac{1}{\pi}\int_0^{\pi} u(x) \sin kx dx$. Then

$$Au = \sum_{k=1}^{\infty} k^2 a_k \sin kx, \ A^{\alpha} u = \sum_{k=1}^{\infty} k^{2\alpha} a_k \sin kx, \ e^{-At} = \sum_{k=1}^{\infty} e^{-k^2 t} a_k \sin kx.$$

Hence,

$$X^{1/2} = D(A^{1/2}) = H_0^1(0, \pi).$$

Let us consider Eqs. (7.97)–(7.99) in space $X^{1/2} = D(A^{1/2}) = H_0^1(0, \pi)$:

$$\frac{du}{dt} + Au = f(t, u), \ u(\tau_j(u) + 0) = (1 - a_j)u(\tau_j(u)), \ j = 0, \ldots,$$

where $f(t, u) : R \times X^{1/2} \to X, \ f(t, u)(x) = a(t)u_x + b(t, x)$.

We verify that in some domain $\mathscr{D} = \{u \geq 0, \|u\| \leq \rho\}$ solutions of (7.97)–(7.99) don't have beating at the surfaces $t = \tau_j(u)$. Assume to the contrary that solution $u(t)$ intersects the surface $t = \tau_j(u)$ at two points t_j^1 and t_j^2, $t_j^1 < t_j^2$.

Denote $u(t_j^1) = u_1, u(t_j^2) = u_2, \tilde{u} = e^{-A(t_j^2 - t_j^1)} u(t_j^1 + 0)$. Then $u(t_j^1 + 0) = (1 - a_j)u_1, \ \tau_j(u_1)) = t_j^1, \ \tau_j(u_2)) = t_j^2$, and

$$u_2 = e^{-A(t_j^2 - t_j^1)} u(t_j^1 + 0) + \int_{t_j^1}^{t_j^2} e^{-A(t_j^2 - s)} f(s, u(s)) ds.$$

We have

$$|\tau_j(u_2) - \tau_j(\tilde{u})| \leq b_j | \int_0^t (u_2(t, x) - \tilde{u}(t, x))(u_2(t, x) + \tilde{u}(t, x)) dx \leq$$

$$\leq b_j \|u_2(t, x) - \tilde{u}(t, x)\|_{L_2} \|u_2(t, x) + \tilde{u}(t, x)\|_{L_2} \leq$$

$$\leq b_j \| \int_{t_j^1}^{t_j^2} e^{-A(t_j^2 - s)} f(s, u(s)) ds \|_{L_2} \|u_2(t, x) + \tilde{u}(t, x)\|_{L_2}.$$

The function $f(t, u)$ satisfies $\|f(t, u)\|_X \leq K(1 + \|u\|_{X^{1/2}})$; hence, solutions of the equation without impulses exist for all $t \geq t_0$ and there exist positive constants M_1 and M_2 such that $M_2 \geq \sup_{u \in \mathscr{D}} \|f(t, u)\|_{L_2}, \ M_3 \geq \sup_{u \in \mathscr{D}} \|u_2(t, x) + \tilde{u}(t, x)\|_{L_2}$. Therefore, $\tau_j(u_2) - \tau_j(\tilde{u}) \leq b_j |t_j^2 - t_j^1| M_2 M_3$. By sufficiently small $b = \sup_j b_j$ we have $b M_2 M_3 < 1$ and

$$0 < t_j^2 - t_j^1 = \tau_j(u_2) - \tau_j(u_1) \leq \tau_j(u_2) - \tau_j(\tilde{u}) + \tau_j(\tilde{u}) - \tau_j(u_1),$$

$$t_j^2 - t_j^1 \leq \frac{1}{1 - bM_2M_3}(\tau_j(\tilde{u}) - \tau_j(u_1)) \leq \frac{b_j((1-a_j)^2 - 1)}{1 - bM_2M_3}\|u_1\|_{L_2}^2 < 0.$$

This contradicts our assumption.

Corresponding to (7.97)–(7.99), the linear impulsive equation is exponentially stable in space $X^{1/2}$. By Theorem 4, for sufficiently small $b = \sup_j b_j$ and $a = \sup_t |a(t)|$ the equation has an asymptotically stable W-almost periodic solution.

7.6 Equations with Unbounded Operators B_j

Many results in our chapter remain true if operators B_j in linear parts of impulsive action are unbounded. We refer to [27], where the following semilinear impulsive differential equation

$$\frac{du}{dt} = Au + f(t,u), \quad t \neq \tau_j, \tag{7.101}$$

$$\Delta u|_{t=\tau_j} = u(\tau_j) - u(\tau_j - 0) = B_j u(\tau_j - 0) + g_j(u(t_j - 0)), \quad j \in Z, \tag{7.102}$$

was studied. Here $u : R \to X$, X is a Banach space, A is a sectorial operator in X, $\{B_j\}$ is a sequence of some closed operators, and $\{\tau_j\}$ is an unbounded and strictly increasing sequence of real numbers. Assume that the equation satisfies conditions $(H1)$, $(H3)$, $(H5)$, $(H6)$, and

(H4u) the sequence $\{B_j\}$ of closed linear operators $B_j \in L(X^{\alpha+\gamma}, X^\alpha)$ is almost periodic in the space $L(X^{\alpha+\gamma}, X^\alpha)$, for $\alpha \geq 0$ and some $\gamma \geq 0$.

As in [17], we assume that solutions $u(t)$ of (7.1), (7.2) are right-hand-side continuous; hence, $u(\tau_j) = u(\tau_j + 0)$ at all points of impulsive action. Due to such a selection we avoid considering operators $e^{-A(t-\tau_j)}(I + B_j)$ with unbounded operator B_j and can work with the family of bounded operators $e^{-A(t-\tau_j)}$.

Since the operator A is sectorial and operators B_j are subordinate to A, an evolution operator of a corresponding linear impulsive equation is constructed correctly. Now analogs of the theorems 2 and 3 can be proven.

Example 2 ([27]). We consider the following parabolic equation with impulsive action:

$$u_t = u_{xx} + f(t,x), \tag{7.103}$$

$$\Delta u\Big|_{t=\tau_j} = u(\tau_j, x) - u(\tau_j - 0, x) = b_k(\sin x)u_x + c_k x(\pi - x), \tag{7.104}$$

with boundary conditions

$$u(t, 0) = u(t, \pi) = 0, \tag{7.105}$$

where $\{\tau_j\}$ is a sequence of real numbers with uniformly almost periodic sequences of differences, $\tau_{j+1} - \tau_j \geq \theta \geq 1/2$,

$\{b_j\}$ and $\{c_j\}$ are almost periodic sequences of real numbers,

$f(t, x)$ is almost periodic and locally Hölder continuous with respect to t and for every fixed t belongs to $L_2(0, \pi)$.

As in Example 1, denote

$$X = L_2(0, \pi),\ A = -\frac{\partial^2}{\partial x^2},\ X^1 = D(A) = H^2(0, \pi) \cap H_0^1(0, \pi).$$

The operator A is sectorial with simple eigenvalues $\lambda_k = k^2$ and corresponding eigenfunctions $\varphi_k(x) = \sin kx, k = 1, 2, \ldots$.

Operators B_j have form $B_j = b_j \sin x \frac{\partial}{\partial x}$.

If $u = \sum_{k=1}^{\infty} a_k \sin kx,\ a_k = \frac{1}{\pi} \int_0^{\pi} u(x) \sin kx dx$, then

$$B_j u = b_j \sin x u_x = b_j \sin x \sum_{k=1}^{\infty} a_k k \cos kx = \frac{b_j}{2}(R - L)A^{1/2}u = b_j T A^{1/2} u,$$

where $Ru = \sum_{k=1}^{\infty} a_k \sin(k-1)x$ and $Lu = \sum_{k=1}^{\infty} a_k \sin(k+1)x$ are bounded shift operators in X. Hence, operators $B_j : X^{\alpha+1/2} \to X^{\alpha}$ are linear continuous, $\alpha \geq 0$.

By (7.13), the evolution operator for homogeneous equations (7.103) and (7.104) is

$$U(t, s) = e^{-A(t-s)},\ \text{if } \tau_k \leq s \leq t < \tau_{k+1},$$

and

$$U(t, s) = e^{-A(t-\tau_k)}(I + B_k)e^{-A(\tau_k - \tau_{k-1})} \ldots (I + B_m)e^{-A(\tau_m - s)}$$

if $\tau_{m-1} \leq s < \tau_m < \tau_{m+1} \ldots \tau_k \leq t < \tau_{k+1},\ m < k,\ k, m \in Z$.

Theorem 5. *Let $p \ln(1+b) < 1$, where p is defined by (7.3) and $b = \sup_j |b_j|$. Then Eqs. (7.103) and (7.104) with boundary conditions (7.105) have a unique W-almost periodic solution which is asymptotically stable.*

Proof. We show that the unique almost periodic solution of (7.103) and (7.104) is given as function $R \to L_2(0, \pi)$ by formula

$$u_0(t) = \int_{-\infty}^{t} U(t, s)\tilde{f}(s) ds + \sum_{\tau_j \leq t} U(t, \tau_j)\tilde{g}_j, \tag{7.106}$$

where $\tilde{f}(t) \equiv f(t, .) : R \to L_2(0, \pi)$, $g_j(x) = c_j x(\pi - x)$, $\tilde{g}_j = g_j(.) : Z \to L_2(0, \pi)$.

First, $u_0(t)$ is bounded in space X^α :

$$\int_{-\infty}^{t} \|U(t,s)\tilde{f}(s)\|_\alpha ds \leq \int_{\tau_i}^{t} \|A^\alpha e^{-A(t-s)}\tilde{f}(s)\|ds +$$

$$+ \int_{\tau_{i-1}}^{\tau_i} \|A^\alpha e^{-A(t-\tau_i)}(I+B_i)e^{-A(\tau_i-s)}\tilde{f}(s)\|ds +$$

$$+ \sum_{k=2}^{\infty} \int_{\tau_{i-k}}^{\tau_{i-k+1}} \|A^\alpha e^{-A(t-\tau_i)}(I+B_i)e^{-A(\tau_i-\tau_{i-1})}\| \times$$

$$\times \prod_{j=i-1}^{i-k+2} \|(I+B_j)e^{-A(\tau_j-\tau_{j-1})}\| \|(I+B_{i-k+1})e^{-A(\tau_{i-k+1}-s)}\tilde{f}(s)\|ds, \quad (7.107)$$

where $t \in [\tau_i, \tau_{i+1})$. The first integral in (7.107) has upper bound

$$\int_{\tau_i}^{t} \|A^\alpha e^{-A(t-s)}\tilde{f}(s)\|ds \leq \frac{C_\alpha}{1-\alpha}\|\tilde{f}\|_{PC}.$$

Next, we need the following inequality (see [17], p. 35):

$$\|A^\alpha T A^\beta e^{-At}\| = \frac{1}{2}\|A^\alpha(R-L)A^\beta e^{-At}\| \leq \frac{4^\alpha+1}{2}\|A^{\alpha+\beta}e^{-At}\|. \quad (7.108)$$

Then by (7.108),

$$\|A^\alpha e^{-A(t-\tau_i)}(I+B_i)e^{-A(\tau_i-s)}\| \leq \|A^\alpha e^{-A(t-s)}\| + \frac{5}{2}\|A^{\alpha+1/2}e^{-A(t-s)}\| \leq$$

$$\leq \left(C_\alpha(t-s)^{-\alpha} + \frac{5}{2}C_{\alpha+1/2}(t-s)^{-(\alpha+1/2)}\right)e^{-\delta(t-s)} \quad (7.109)$$

From Henry [9], p. 25, we have

$$\|A^\alpha e^{-At}\psi\| < b_\alpha(t)\|\psi\|,$$

where $b_\alpha(t) = (te/\alpha)^{-\alpha}$ if $0 < t \leq \alpha/\lambda_1$, and $b_\alpha(t) = \lambda_1^\alpha e^{-\lambda_1 t}$ if $t \geq \alpha/\lambda_1$. Since $\|T\| = 1$ and $\lambda_1 = 1$, we have

$$\|(I+B_j)e^{-A(\tau_j-\tau_{j-1})}\| \leq \|e^{-A(\tau_j-\tau_{j-1})}\| + |b_j|\|A^{1/2}e^{-A(\tau_j-\tau_{j-1})}\| \leq$$

$$\leq (1+|b_j|)e^{-(\tau_j-\tau_{j-1})} \quad (7.110)$$

if $\theta \geq 1/2$.

Let $0 < \varepsilon_1 < 1 - p\ln(1+b)$. Then there exists a positive integer k_1 such that for $k \geq k_1$

$$\frac{i(\tau_{i-k}, \tau_i)}{\tau_i - \tau_{i-k}} \ln(1+b) - 1 < -\varepsilon_1.$$

Denote

$$N_1 = \max_{1 \leq k \leq k_1} \exp\left(i(\tau_{i-k}, \tau_i)\ln(1+b) - (\tau_i - \tau_{i-k})\right).$$

Then

$$\prod_{j=i-k+1}^{i} \|(I+B_j)e^{-A(\tau_j - \tau_{j-1})}\| \leq (1+b)^{i(\tau_{i-k},\tau_i)} e^{-(\tau_i - \tau_{i-k})} \leq$$

$$\leq N_1 e^{-\varepsilon_1(\tau_i - \tau_{i-k})} \leq N_1 e^{-\varepsilon_1 \theta k}. \tag{7.111}$$

For $t \in (\tau_i, \tau_{i+1})$, by (7.109) and (7.111) we get

$$\|U(t, \tau_{i-k})\|_\alpha \leq \|A^\alpha e^{-A(t-\tau_i)}(I+B_i)e^{-A(\tau_i - \tau_{i-1})}\| \times$$

$$\times \prod_{j=i-k+1}^{i-1} \|(I+B_j)e^{-A(\tau_j - \tau_{j-1})}\| \leq K_1 e^{-\varepsilon_1(t-\tau_{i-k})}$$

with constant K_1 independent of t and τ_{i-k}.

Using the last inequality, we obtain the boundedness of $\|u_0(t)\|_\alpha$. We can now proceed analogously to the proof of Theorem 1 and show the almost periodicity of $u_0(t)$.

References

1. Akhmet, M.: Principles of Discontinuous Dynamical Systems. Springer, New York (2010)
2. Akhmetov, M.U., Perestyuk, N.A.: Periodic and almost periodic solutions of strongly nonlinear impulse systems. J. Appl. Math. Mech. **56**, 829–837 (1992)
3. Akhmetov, M.U., Perestyuk, N.A.: Almost periodic solutions of nonlinear systems. Ukr. Math. J. **41**, 291–296 (1989)
4. Barreira, L., Valls, C.: Robustness for impulsive equations. Nonlinear Anal. **72**, 2542–2563 (2010)
5. Chow, S.-N., Leiva, H.: Existence and roughness of the exponential dichotomy for skew-product semiflow in Banach spaces. J. Differ. Equ. **120**, 429–477 (1995)
6. Hakl, R., Pinto, M., Tkachenko, V., Trofimchuk S.: Almost periodic evolution systems with impulse action at state-dependent moments. ArXiv preprint math/1511.07833 (2015)
7. Halanay, A., Wexler, D.: The Qualitative Theory of Systems with Impulse. Nauka, Moscow (1971)

8. Henriquez, H.R., De Andrade, B., Rabelo, M.: Existence of almost periodic solutions for a class of abstract impulsive differential equations. ISRN Math. Anal. Article ID 632687, 21 pp. (2011)
9. Henry, D.: Geometric Theory of Semilinear Parabolic Equations. Lecture Notes in Mathematics, vol. 840. Springer, Berlin (1981)
10. Kmit, I., Recke, L., Tkachenko, V.I.: Robustness of exponential dichotomies of boundary-value problems for general first-order hyperbolic systems. Ukr. Math. J. **65**, 236–251 (2013)
11. Lakshmikantham, V., Bainov, D., Simeonov, P.S.: Theory of Impulsive Differential Equations. World Scientific, Singapore (1989)
12. Myslo, Y.M., Tkachenko, V.I.: Global attractivity in almost periodic single species models. Funct. Differ. Equ. **18**, 269–278 (2011)
13. Myslo, Y.M., Tkachenko, V.I.: Almost periodic solutions of Mackey–Glass equations with pulse action. Nonlinear Oscillations **14**, 537–546 (2012)
14. Naulin, R., Pinto, M.: Splitting of linear systems with impulses. Rocky Mt. J. Math. **29**, 1067–1084 (1999)
15. Pinto, M., Robledo, G.: Existence and stability of almost periodic solutions in impulsive neural network models. Appl. Math. Comput. **217**, 4167–4177 (2010)
16. Pliss, V.A., Sell, G.R.: Robustness of exponential dichotomies in infinite-dimensional dynamical systems. J. Dyn. Diff. Equat. **11**, 471–513 (1999)
17. Rogovchenko, Y.V., Trofimchuk, S.I.: Periodic solutions of weakly nonlinear partial differential equations of parabolic type with impulse action and their stability. Akad. Nauk Ukrain. SSR Inst. Mat., vol. 65, 44 pp (1986, Preprint)
18. Samoilenko, A.M., Perestyuk, N.A.: Impulsive Differential Equations. World Scientific, Singapore (1995)
19. Samoilenko, A.M., Perestyuk, N.A., Akhmetov, M.U.: Almost periodic solutions of differential equations with impulse action. Akad. Nauk Ukrain. SSR Inst. Mat., vol. 26, 49 pp (1983, Preprint)
20. Samoilenko, A.M., Perestyuk, N.A., Trofimchuk, S.I.: Generalized solutions of impulse systems, and the beats phenomenon. Ukr. Math. J. **43**, 610–615 (1991)
21. Samoilenko, A.M., Trofimchuk,S.I.: Almost periodic impulse systems. Differ. Equ. **29**, 684–691 (1993)
22. Stamov, G.T.: Almost Periodic Solutions of Impulsive Differential Equations. Lecture Notes in Mathematics, vol. 2047. Springer, Berlin (2012)
23. Stamov, G.T., Alzabut, J.O.: Almost periodic solutions for abstract impulsive differential equations. Nonlinear Anal. **72**, 2457–2464 (2010)
24. Tkachenko, V.I.: On linear almost periodic pulse systems. Ukr. Math. J. **45**, 116–125 (1993)
25. Tkachenko, V.I.: On the exponential dichotomy of pulse evolution systems. Ukr. Math. J. **46**, 441–448 (1994)
26. Tkachenko, V.I.: On multi-frequency systems with impulses. Neliniyni Kolyv. **1**, 107–116 (1998)
27. Tkachenko, V.I.: Almost periodic solutions of parabolic type equations with impulsive action. Funct. Differ. Equ. **21**, 155–170 (2014)
28. Trofimchuk, S.I.: Almost periodic solutions of linear abstract impulse systems. Differ. Equ. **31**, 559–568 (1996)

Printed in the United States
By Bookmasters